COMPUTATIONAL IMAGING AND ANALYTICS IN BIOMEDICAL ENGINEERING

Algorithms and Applications

AAP Series on Digital Signal Processing, Computer Vision and Image Processing: Advances and Applications

COMPUTATIONAL IMAGING AND ANALYTICS IN BIOMEDICAL ENGINEERING

Algorithms and Applications

Edited by

T. R. Ganesh Babu, PhD

U. Saravanakumar, PhD

Balachandra Pattanaik, PhD

APPLE ACADEMIC PRESS

First edition published 2024

Apple Academic Press Inc.
1265 Goldenrod Circle, NE,
Palm Bay, FL 32905 USA

760 Laurentian Drive, Unit 19,
Burlington, ON L7N 0A4, CANADA

CRC Press
2385 NW Executive Center Drive,
Suite 320, Boca Raton FL 33431

4 Park Square, Milton Park,
Abingdon, Oxon, OX14 4RN UK

Library and Archives Canada Cataloguing in Publication

Title: Computational imaging and analytics in biomedical engineering : algorithms and applications / edited by T.R. Ganesh Babu, PhD, U. Saravanakumar, PhD, Balachandra Pattanaik, PhD.
Names: Babu, T. R. Ganesh, editor. | Saravanakumar, U., 1984- editor. | Pattanaik, Balachandra, 1971- editor.
Description: First edition. | Series statement: AAP series on digital signal processing, computer vision and image processing: advances and applications | Includes bibliographical references and index.
Identifiers: Canadiana (print) 20230581536 | Canadiana (ebook) 20230581587 | ISBN 9781774914717 (hardcover) | ISBN 9781774914700 (softcover) | ISBN 9781032669687 (ebook)
Subjects: LCSH: Diagnostic imaging—Mathematical models. | LCSH: Medical informatics—Mathematical models.
Classification: LCC RC78.7.D53 C66 2024 | DDC 616.07/54—dc23

Library of Congress Cataloging-in-Publication Data

Names: Babu, T. R. Ganesh, editor. | Saravanakumar, U., 1984- editor. | Pattanaik, Balachandra, 1971- editor.

Title: Computational imaging and analytics in biomedical engineering : algorithms and applications / edited by T.R. Ganesh Babu, U. Saravanakumar, Balachandra Pattanaik.

Other titles: AAP series on digital signal processing, computer vision and image processing.

Description: First edition. | Palm Bay, FL : Apple Academic Press Inc., 2024. | Series: AAP series on digital signal processing, computer vision and image processing: advances and applications | Includes bibliographical references and index. | Summary: "Computational Imaging and Analytics in Biomedical Engineering: Algorithms and Applications focuses on mathematical and numerical methods for medical images and data. The book presents the various mathematical modeling techniques, numerical analysis, computing and computational techniques, and applications of machine learning for medical images and medical informatics. It also focuses on programming concepts using MATLAB and Phython for medical image and signal analytics. The volume demonstrates the use of various computational techniques and tools such as machine learning, deep neural networks, artificial intelligence and human-computer interaction, fusion methods for CT and pet images, etc. for diagnosis of brain disorders, cervical cancer, lung disease, melanoma, atrial fibrillation and other circulatory issues, dental images, diabetes, and other medical issues. Key features: Addresses the various common challenges related to biomedical image analysis Presents a variety of mathematical models for medical images Discusses applications of algorithms on medical images for various medical issuses Describes the development of intelligent computing machines such as embedded systems Explores the programming techniques using MATLAB and Phython for biomedical applications This book presents a plethora of uses of algorithms and applications in computational imaging and analytics for the medical/health field. It will serve as a resource on recent advances and trends in the field of computational imaging, where computation is playing a dominant role in imaging systems"-- Provided by publisher.

Identifiers: LCCN 2023053511 (print) | LCCN 2023053512 (ebook) | ISBN 9781774914717 (hardcover) | ISBN 9781774914700 (paperback) | ISBN 9781032669687 (ebook)

Subjects: MESH: Diagnostic Imaging--methods | Medical Informatics Computing | Biomedical Engineering--methods

Classification: LCC RC78.7.D53 (print) | LCC RC78.7.D53 (ebook) | NLM WN 180 | DDC 616.07/54--dc23/eng/20240109

LC record available at https://lccn.loc.gov/2023053511

LC ebook record available at https://lccn.loc.gov/2023053512

ISBN: 978-1-77491-471-7 (hbk)
ISBN: 978-1-77491-470-0 (pbk)
ISBN: 978-1-03266-968-7 (ebk)

AAP SERIES ON DIGITAL SIGNAL PROCESSING, COMPUTER VISION, AND IMAGE PROCESSING: ADVANCES AND APPLICATIONS

BOOK SERIES EDITORS:

Dr. Manoj Gupta, PhD
Associate Professor, Department of Electronics and Communication Engineering, JECRC University, Jaipur (Rajasthan), India

Dr. Pradeep Kumar, PhD
Discipline of Electrical, Electronic and Computer Engineering, Howard College Campus, University of KwaZulu-Natal, Durban, South Africa

Brief description of the Book Series:

Digital signal processing, computer vision and image processing as a whole is considered to be one of the most rapidly evolving areas of research and technology today with growing applications in almost all disciplines of engineering. Medical imaging, computer vision, healthcare, medical applications, remote sensing, agriculture, robotics, communication systems and space exploration are some of the applications of digital signal processing, computer vision and image processing, to name a few. The present day curriculum covers many aspects of digital signal processing, computer vision and image processing, addressing the theoretical aspects in particular. This book series is intended to supplement the theoretical knowledge with special emphasis on the practical side. The series content has been specifically chosen to give a thorough understanding of the fundamental aspects to advanced applications of digital signal processing, computer vision and image processing.

The series is open to monograph. handbooks, authored books, edited volumes, and conference proceedings. This series aims to embrace all aspects, sub-fields and new challenges in the followings research domains (related topics) but are not limited to:

- **Image and Video Processing**: Image filtering, restoration and enhancement, image segmentation, video segmentation and tracking, morphological processing, feature extraction and analysis, interpolation and super-resolution, motion detection and estimation, computer vision, pattern recognition, content-based image retrieval, image/signal computations and services, features and models for image/signals, machine learning based image and signal processing, data mining techniques, imaging algebra, mathematical morphology, probabilistic, statistical, optimization, approximation theory, models in imaging science, video signal processing, visualization, watermarking, video surveillance, video compression and streaming, video analysis and event recognition, biometrics , medical image analysis, artificial intelligence and related areas.
- **Signal Processing:** Filters theory, spectral analysis, time-frequency and time-scale representation, EEG/ECG signals, FIR/IIR and adaptive filters, statistical signal processing, filtering, detection and estimation, nonlinear signal processing, radar, antennas, telecommunications systems, acoustics. Signal processing theory and methods, high signal processing: integrating 5G and IoT with satellite networks, hands-free speech communication and microphone arrays, wearable sensor signal processing, architecture and frameworks, audio/speech processing and coding, watermarking, data mining techniques, statistical and optical signal processing, communication signal processing, DSP system and embedded systems, multimedia processing, artificial intelligence , IoT, cyber physical systems and related areas.
- **Computer Vision:** Algorithms, feature extraction and pattern recognition, ML and deep learning in vision, CBIR, object and face recognition, AR/VR, object detection and localization, 3D object extraction, tracking and visual navigation, cognitive and biological inspired vision, artificial intelligence, machine learning , ubiquitous computing and related areas.
- **Applications Areas:** Biometric, bioinformatics and biomedical imaging, medial images and applications, healthcare applications, agriculture applications, augmented and mixed reality, mental health and cognitive stimulations, security and video-surveillance, quality control and inspection, archaeology, remote sensing, embedded systems and applications, automated vehicles, speech and music

processing, robotics, rehabilitation, occupational therapy and telemedicine, artificial intelligence and machine learning/deep learning based applications, cyber physical systems, Internet of Things (IoT), Industry 4.0, Medicine 5.0 and other related applications.

BOOKS IN THE SERIES:

Artificial Intelligence and Machine Learning Techniques in Image Processing and Computer Vision
Editors: Karm Veer Arya, PhD, Ciro Rodriguez Rodrigues, PhD, Saurabh Singh, PhD, and Abhishek Singhal, PhD

Computational Imaging and Analytics in Biomedical Engineering: Algorithms and Applications
Editors: T. R. Ganesh Babu, PhD, U. Saravanakumar, PhD, and Balachandra Pattanaik, PhD

ABOUT THE EDITORS

T. R. Ganesh Babu, PhD,
Professor, Department of Electronics and Communication Engineering at Muthayammal Engineering College (Autonomous), Rasipuram, Namakkal District, India.

T. R. Ganesh Babu, PhD, is currently working as a Professor in the Department of Electronics and Communication Engineering at Muthayammal Engineering College (Autonomous), Rasipuram, Namakkal District. He has to his credit about 25 years of teaching experience in various engineering colleges across India. He has over 118 research publications in international and national journals, conferences, books, and book chapters. He has authored six engineering books on topics in communication engineering, linear integrated circuits, digital communication, digital image processing, control systems, and FAQs in analog and digital signals. He has published three edited books as a chief editor under Dipti Press and one book from IGI Global. He has to his credit one patent granted and has filed nine more patents and publishing. He has already guided 10 research scholars, and seven more are pursuing their degrees at Anna University India. He is a reviewer for Springer, Elsevier, and Inderscience Publisher. He received his Distinguished Faculty Award from Venus International Foundation, Chennai. He received a PhD from Anna University, Chennai, India, in 2014 in the field of medical image processing.

U. Saravanakumar, PhD
Professor and Head of Electronics and Communication Engineering, Muthayammal Engineering College (Autonomous), Namakkal, India

U. Saravanakumar, PhD, is working as a Professor and Head of Electronics and Communication Engineering in Muthayammal Engineering College (Autonomous), Namakkal. His research interests are in the field of VLSI design, silicon photonics, reconfigurable computing, system on chip, and embedded systems. He worked in various institutions designated in various positions. He graduated in Electronics and Communication Engineering from Anna University, Chennai, in 2006. In the year 2008, he received his master's degree in VLSI Design from Anna University, Chennai. He received his doctoral degree in the faculty of Information and Communication Engineering, specializing in VLSI Design and on-chip communication from Anna University in 2014.

Balachandra Pattanaik, PhD

Balachandra Pattanaik, PhD, is currently working as a full-time Professor in the Department of Electrical and Computer Engineering, in the College of Engineering and Technology, Wollega University, Ethiopia, Africa, Dr. Pattanaik holds PhDs in Electrical Engineering (Embedded Systems) and in Mathematics, and has an ongoing postdoctoral fellowship on electric vehicles. He holds 20 innovation patents issued by the Indian Government. He has the following: innovation patent grants: two, international patent grants in Australia and Germany: four. He has published 13 engineering books and six book chapters as well as many national and international publications. He is an advisor to the IEEE, Power and Energy Student Chapter at Bule Hora University in Ethiopia, a senior member of IEEE, and a recipient of the Chartered Engineer Award from the Institution of Engineers of India. He holds many administrative positions, such as Head of Department of Electrical Engineering, Principal, and Anna University Chief Superintendent for university examinations, NSS coordinator, research coordinator, accreditation management representative, editorial board member, and chair/vice chair for conference sessions. He has guided many undergraduate and postgraduate research projects and has been a foreign PhD thesis evaluator for many universities.

CONTENTS

CONTRIBUTORS

M. Suresh Anand
Department of Computing Technologies, School of Computing, SRM Institute of Science & Technology, Kattankulathur, India

Sridhar P. Arjunan
Department of Electronics and Instrumentation Engineering, SRM Institute of Science and Technology, Kattankulathur, India

B. Arputhamary
Department of Computer Applications, Bishop Heber College, Tiruchirappalli, India

V. Ramesh Babu
Department of Computer Science and Engineering, Sri Venkateswara College of Engineering, Sriperumbudur, India

Yohannes Bekuma Bakare
Department of Electrical and Computer Engineering, College of Engineering and Technology, Wollega University, Ethiopia, Africa

G. Balanagireddy
Department of Electronics and Communication Engineering, Rajiv Gandhi University of Knowledge Technologies-Ongole Campus, Ongole, India

B. Balakumar
Centre for Information Technology and Engineering, Manonmaniam Sundaranar University, Tirunelveli, India

S. Mary Cynthia
Department of ECE, Jeppiaar Institute of Technology, Chennai, India

M. Ganthimathi
Department of CSE, Muthayammal Engineering College, Namakkal, Tamil Nadu, India

N. Gopinath
Department of Computer Science and Engineering, Sri Sairam Engineering College, Chennai, India

G. Gunasekaran
Department of Computer Science and Engineering, Dr. M. G. R. Educational and Research Institute, Maduravoyal, Chennai, India

M. V. Ishwarya
Department of Artificial Intelligence and Data Science, Agni College of Technology, Chennai, India

Vajiram Jayanthi
SENSE, Vellore Institute of Technology, Chennai, India

S. Jacily Jemila
Vellore Institute of Technology, Chennai, India

Mahendrakan K.
Department of Electronics and communication Engineering, Hindusthan Institute of Technology, Coimbatore, Tamil Nadu, India

A. Karunamurthy
BWDA Arts and Science College, Vilupuram, India

Syed Khasim
Department of CSE, Dr. Samuel George Institute of Engineering &Technology, Andhra Pradesh, India

T. Ganesh Kumar
Department of Computing Science and Engineering, Galgotias University, Greater Noida, Uttar Pradesh, India

K. Sampath Kumar
Department of Computing Science and Engineering, Galgotias University, Greater Noida, Uttar Pradesh, India

M. Kumarasamy
Department of Computer Science, College of Engineering and Technology, Wollege University, Ethiopia, Africa

A. Kumaresan
School of Computer Science and Engineering, Vellore Institute of Technology, Vellore, India

M. Malathi
Department of Electronics and communication Engineering, Rajalakshmi Institute of Technology, Chennai, India

M. Moorthy
Muthayammal Engineering College, Rasipuram, India

K. Sakthi Murugan
Department of ECE, PSN College of Engineering and Technology, Tirunelveli, India

R. Murugasami
Department of Electronics and Communication Engineering, Nandha Engineering College (Autonomous), Erode, India

S. Muthukumar
Department of CSE, B.S. Abdur Rahman Crescent Institute of Science and Technology, Chennai, India

N. Naveenkumar
Department of CSE, Muthayammal Engineering College, Rasipuram, India

S. Omkumar
Department of ECE, SCSVMV (Deemed University), Kanchipuram, India

Gururama Senthilvel P.
Department of Computing Science and Engineering, Galgotias University, Greater Noida, Uttar Pradesh, India

Balachandra Pattanaik
Department of Electrical and Computer Engineering, College of Engineering and Technology, Wollege University, Ethiopia, Africa

S. Pragadeeswaran
Department of CSE, Muthayammal Engineering College, Namakkal, Tamil Nadu, India

G. Lakshmi Vara Prasad
Department of IT, QIS College of Engineering and Technology, Andhra Pradesh, India

R. Praveena
Department of ECE, Muthayammal Engineering College, Namakkal, India

G. Soniya Priyatharsini
Department of ECE, DR. M. G. R. Educational and Research Institute, Maduravoyal, Chennai, India

S. Punitha
Department of ECE, Muthayammal Engineering College, Rasipuram, India

A. Purushothaman
Department of ECE, Hindhusthan Institute of Technology, Coimbatore, India

Sreenithi R.
Department of Computer Technology, Madras Institute of Technology, Chennai, India

A. Rajan
Department of ECE, Sreerama Engineering College, Thirupathi, India

J. Martin Sahayaraj
Department of Electronics and Communication Engineering, Sri Indu College of Engineering and Technology, Hyderabad, Telangana, India

K. Savima
Department of Computer Science, S.T.E.T. Women's College, Mannargudi, India

K. Sekar
Department of Electrical and Electronics Engineering, Hindusthan College of Engineering and Technology, Coimbatore, Tamil Nadu

Sivakumar Shanmugasundaram
SENSE, Vellore Institute of Technology, Chennai, India

S. Sharmila
Department of Civil Engineering, Nandha Engineering College, Erode, India

K. Shebagadevi
Department of ECE, Muthayammal Engineering College, Namakkal, India

P. Sinthia
Department of Biomedical Engineering, Saveetha Engineering College, Chennai, India

M. Sivakumar
Department of ECE, Mohamed Sathak A. J. College of Engineering, Chennai, India

S. R. Sridhar
Department of CSE, Muthayammal Engineering College, Namakkal, Tamil Nadu, India

P. Srinivasan
Department of CSE, Muthayammal Engineering College, Rasipuram, India

P. Subramanian
Department of CSE, Mohamed Sathak A. J. College of Engineering, Chennai, India

P. Sukumar
Department of Computer Science and Engineering, Nandha Engineering College (Autonomous), Erode, India

R. Sumathi
Department of Computer Science and Engineering, Kalasalingam Academy of Research and Education, Krishnankoil, India

J. Surendharan
HKBK College of Engineering, Bangalore, Karnataka, India

N. Sureshkumar
Department of ECE, Muthayammal College of Engineering, Rasipuram, India

A. Brintha Therese
Vellore Institute of Technology, Chennai, India

K. Umapathy
Department of ECE, SCSVMV (Deemed University), Kanchipuram, India

C. Vijayakumaran
Department of CSE, SRM Institute of Science and Technology, Kattankulathur, Chennai, India

ABBREVIATIONS

AD	axial diffusivity
AGI	artificial general intelligence
AI	artificial intelligence
AML	acute myeloid leukemia
ANN	artificial neural network
ARR	arrhythmia
AS	attacker scale
ASD	autism spectrum disorder
ATS	attack target scale
BBB	blood–brain barrier
BC	binary classifier
BFC	bias field corrector
BN	Bayesian network
BOLD	blood-oxygen-level-dependent
BSE	brain surface extractor
BSL	British sign language
CAD	computer-aided diagnostic
CAF	cooperation attack frequency
CAPEX	capital expenditure
CHO	channeled hotelling observer
CNN	convolution neural network
CNNPL	convolutional neural network with a prototype learning
CNS	central nervous system
CRM	customer relationship management
CS	collusion set
CSF	cerebrospinal fluid
CSL	Chinese sign language
CT	computer tomography
DCNN	deep convolutionary neural network
DL	deep learning
DT	decision tree
DTI	diffusion tensor imaging
ECG	electrocardiography
EEG	electroencephalography

EOG	electro-oculogram
ERP	enterprise resource planning
EV	exploration views
FA	fractional anisotropy
FCM	fuzzy C-means
FL	feedback limit
FLAIR	fluid attenuated inversion recovery
fMRI	functional magnetic resonance imaging
FP	false positive
FS	feedback set
FWHM	full width half maximum
GLM	general linear model
GOFAI	good old fashioned artificial intelligence
GSL	Greek sign language
HCI	human computer interaction
HE	histogram equalization
HPV	human papilloma virus
IaaS	infrastructure services
IdM	identity management service
IQA	image quality assessment
ISTFT	inverse signals using short time Fourier transform
KNN	K-nearest neighbor
LASSO	least absolute shrinkage and selection operator
LR	logistic regression
LSF	French sign language
LSTM	long short-term memory
MCT	manual contour tracing
MD	mean diffusivity
ME	maximum entropy
MEG	magnetoencephalogram
ML	machine learning
MRI	magnetic resonance imaging
NB	Naïve Bayes
NMR	nuclear appealing resonation
NN	neural network
NSCT	non-subsample contourlet transform
OFR	out-of-field recurrence
OLAP	online analytical processing
OPEX	operational expenditure

PaaS	platform as a service
PCA	principle component analysis
PET	positron emission tomography
PReDicT	predicting response to depression treatment
PSNR	peak signal noise ratio
PTSD	posttraumatic stress disorder
PVC	partial volume classifier
RBFN	radial basis function network
RF	radio frequency
RF	random forest
RMSE	root mean square error
RNN	recurrent neural network
ROI	region of interest
SaaS	software as a service
SD	standard deviation
SIM	similarity index measure
SLAs	service-level agreements
SMEQ	subjective mental effort question
sMRI	structural MRI
SNR	signal-to-noise ratio
SPECT	single photon emission computed tomography
SPM	statistical parametric mapping
STC	slice timing correction
STFT	signals using short time Fourier transform
STIs	sexually transmitted infections
SVM	support vector machine
SvReg	surface and volume registration
TAI	traumatic axonal injury
TCGA	tissue from the Cancer Genome Atlas
TD	typically developed
TDO	total dermoscopic value
TE	time to echo
TES	trust management system
TF	time-frequency
TR	repetition time

PREFACE

Medical images are useful in analyzing the internal physical anatomy of the human body and therefore, they are treated as a core component in medical diagnosis and treatment. These medical images contain information about the human anatomy such as cell and tissue arrangements, any growth in human cells, and so on. These information cannot be interpreted easily by human eyes and need some medical devices, tools, and software to analyze in-depth. On the other side, medical images are stored in databases, forming big-data for population groups, and thus, handling of medical data (or images) is also a crucial task.

To simplify these issues, various image-processing techniques are applied on medical images and datasets with suitable algorithms to automatically or semiautomatically extract useful information about the patients. For further simplification and optimization, emerging techniques such as neural networks, machine learning, deep learning (DL), and artificial intelligence are adapted for feature extraction of medical images, medical image segmentation, and image-based bio-models.

These advancements enable appropriate use of medical images in the field of healthcare including: (1) medical knowledge of biological systems through detailed analysis of structures and functions of particular biological part or organs; (2) computer-assisted frameworks for medical treatments; and (3) development of image datasets of human body (bodies) for diagnosis, guiding doctors, and research activities.

This book comprises 20 chapters, and they are organized in the following manner.

Chapter 1 discusses statistical analysis of seizure data to support clinical proceeding. The radiographic imaging is a powerful and clinically important tool within oncology. Artificial intelligence is used for the quantification of radiographic characteristics of images, using predefined algorithms. The strategies of clinical imaging data include normalization, robust models, and statistical analyses. They also help in improving quality and enhancing perfection of medical image analysis and supporting clinicians diagnose, treat, plan, and cover changes, and execute procedures more safely and effectively. Statistical models give a surgery-specific environment to the problem from a training set of cases conforming to the problem along with the result.

Chapter 2 presents a spatial preprocessing in segmentation of brain MRI using t1 and t2 images. Preprocessing is a very important process before segmentation in medical image processing. In this chapter, T1 and T2 MRI images of brain are segmented before and after preprocessing using SPM-12 neuroimaging software, and the importance of preprocessing is discussed by using five important quality metrics MSE, PSNR, SNR, EPI, and SSIM (Structural Similarity Index Metric).

Chapter 3 discusses a comparative volume analysis of pediatric brain with adult brain using t1 MRI images segmentation of pediatric brain is very useful in medical image analysis to supervise the brain growth and development in infants. Atlas-based methods are used for automatic segmentation of MRI images of brain. In this chapter, automatic segmentation of four pediatric and three adult brains are done using Brainsuite19a brain imaging software, and the volumes of different brain parts are computed.

Chapter 4 presents a comparison of region of interest and cortical area thickness of seizure and hemosiderin-affected brain images. Brain disorder and tumors are caused by the severe neurological abnormality in their functions. Hemosiderin is induced by changes in the characteristic functional magnetic field and can be detected using susceptibility weighted T1 images. Clinical syndromes and its relevance are unclear. Hemosiderin and epilepsy brain images are preprocessed with available software tool. Based on the region of interest, mean thickness area and cortical thickness area of the different parts of the brain are determined.

Chapter 5 presents a design and analysis of classifier for atrial fibrillation detection and ECG classification based on deep neural networks (DNN) atrial fibrillation (AF), which is the most common cardiac arrhythmia as well as a significant risk factor in heart failure and coronary artery disease. Deep learning is the current interest of different healthcare applications that includes the heartbeat classification based on ECG signals. AF can be detected by using a short ECG recording. This chapter describes constructing a classifier for detection of atrial fibrillation in signals of ECG using deep neural networks.

Chapter 6 discusses design and analysis of efficient short time Fourier transform-based feature extraction for removing EOG artifacts using deep learning regression. It also discusses how to eliminate electrooculogram (EOG) noise from the signals of electroencephalogram (EEG) by employing the benchmark dataset of EEGdenoiseNet. This work involves

EOG-contaminated signals of EEG signals for training a regression model based on deep learning in order to eliminate the artifacts of EOG. Initially, original input signals are used to train the model of regression and then transformed signals using short-time Fourier transform (STFT) are employed.

Chapter 7 discusses that research in medical imaging is increasingly turning to machine learning approaches. To recognize patterns in medical images, this method relies on pattern recognition. In order to enhance medical imaging applications, these algorithms are now being used to detect faults in the sickness diagnosing system, which may lead to significantly ambiguous medical therapy.

Chapter 8 discusses about the innovations in artificial intelligence (AI) and human computer interaction (HCI) in the digital era. AI and HCI have frequently been described as having contradicting perspectives on how people and PCs ought to interface. As the two advance, there is a profound differentiation that cuts across these networks, in how scientists imagine the connection between information and plan. Evolution in AI has prompted improvement in strategies and apparatuses that have influenced regions outside AI's core functions.

Chapter 9 presents a computer-aided automatic detection and diagnosis of cervical cancer by using feature markers. The proposed system constitutes reprocessing, feature succession, and nuclei partition. Internal structure processes are used to partition cell center area. The gray equal, wavelet, and GLCM structures are mined from standard plus diseased cell center. The mined structures are proficient and categorized using ANFIS categorization.

Chapter 10 presents a detailed study of sentiment analysis. It explains the basics of sentiment analysis, its types, and different approaches of sentiment analysis and more. The open-source tools and datasets are also discussed.

Magnetic resonance imaging (MRI) techniques and its advancements in biomedical research and clinical applications are presented in Chapter 11. The emergence of new techniques increases the applications of MR imaging remarkably. MRI is performed with the use of magnetization properties. To capture MRI, a strong magnetic field is applied to randomly oriented protons of human body available in water molecules to make changes in the alignment. Then this magnetization is disordered by applying external radio frequency wave. This absorbed RF energy is then emitted with the help of several relaxation processes and the protons realigned. Subsequently, the emitted signals are determined.

Chapter 12 presents a hybrid clustering approach for medical image segmentation. The predominant algorithms used in hybrid clustering techniques are K-means, adaptive K-means, spatial fuzzy c-means algorithm. By combining the K-means algorithm along with a spatial fuzzy c-means algorithm, the hybrid clustering segmentation yields higher accuracy level for tumor detection and also minimizes the processing time required for the segmentation process.

Chapter 13 presents an analysis of dental image with medical image processing. This chapter deals with such aspects considering radio graphical images. Measuring the distance between mandible inferior border and the superior border of "alveolar" for image is done manually. Thresholding via various methods such as "Haung," "Otsu" and their corresponding values are provided for analysis.

Chapter 14 discusses an investigation on diabetes using multilayer perceptron. The goal of this work is to identify, detect, and forecast the emergence of diabetes in its earliest stages by employing machine learning techniques and algorithms. When it comes to diabetes classification, an MLP is used.

Chapter 15 elaborates a dermoscopic implementation and classification on melanoma disease using gradient boost classifier. In this study, the authors proposed deep learning techniques for classification of deadly diseases. Dermoscopic images are used in this study as a dataset directory. The authors of this chapter used the classifier of gradient boost and feature extraction by NSCT to predict an accurate and specific value of melanoma disease.

Chapter 16 presents a segmented lung disease prediction using deep learning techniques such as convolutional neural network (CNN) and KNN classifier.

Chapter 17 describes a design detecting and classifying melanoma skin using CNN with K-means clustering. An automated approach with cuckoo search optimization and K-means clustering for detecting the skin cancer and for classifying the cancer as normal and abnormal by support vector machine have been proposed. For preprocessing, median filters are used to reduce noise inference in the input image. For validating the accuracy, the authors utilize the IISC-DSI dataset.

Chapter 18 describes detection of lung cancer using fusion methods for CT and PET images. In the first part of the work, segmentation of lungs, a tumor in CT image is used to spatially weigh fuzzy c-means clustering (SWFCM) techniques. In the second part of the work, segmentation of lungs, a tumor in PET image is used to spatially weigh fuzzy c-means clustering (SWFCM) techniques. In the third part, the diagnosis is strengthened for mass screening; the CT images and the PET images are fused effectively.

The four fusion methods namely wavelet transform, curvelet transform, non-subsample contourlet transform (NSCT), and multimodal image fusion are applied. The performance analysis entropy, peak signal noise ratio (PSNR), standard deviation (SD), structural similarity index measure (SIM), and root mean square error (RMSE) are computed.

Chapter 19 presents a framework promoting position trust evaluation system in the cloud environment. The design and development of Cloud Armor, a legacy trust evaluation framework that offers a set of functions to provide trust as a service, is discussed in this chapter (TaaS). It also discusses the problems in calculating trust based on input from cloud users. This technology effectively protects cloud services by detecting hostile and inappropriate behavior by the use of trust algorithms which may recognize on/off assaults and colluding attacks by using different security criteria.

Chapter 20 presents machine learning for medical images. This chapter comprises key phases of digital image processing applications of medical images, usages, and difficulties of artificial intelligence in healthcare applications.

Our deep gratitude to all the contributors for providing quality content to the readers of the book. We extend our sincere thanks to all the reviewers for their suggestions which helped us to pack this book with its merits. From the bottom of our heart, we thank our family members, employers, research scholars, and colleagues for their unrestricted support. We wish to thank the tireless team of Apple Academic Press for their wonderful support and guidance throughout the process of making this book.

— T. R. Ganesh Babu
U. Saravanakumar
Balachandra Pattanaik

CHAPTER 1

STATISTICAL ANALYSIS OF SEIZURE DATA TO SUPPORT CLINICAL PROCEEDINGS

VAJIRAM JAYANTHI[1], SIVAKUMAR SHANMUGASUNDARAM[2], and SYED KHASIM[3]

[1]*SENSE, Vellore Institute of Technology, Chennai, India*

[2]*SENSE, Vellore Institute of Technology, Chennai, India*

[3]*Department of CSE, Dr. Samuel George Institute of Engineering &Technology, Andhra Pradesh, India*

ABSTRACT

Radiographic imaging is highly valued and widely utilized in the field of oncology. It plays a crucial role in assisting clinicians with diagnostic, treatment planning, and procedural decision-making tasks. Artificial intelligence (AI) technology is employed to analyze radiographic images by quantifying various characteristics through predetermined algorithms. Strategies for managing clinical imaging data involve normalization techniques, robust modelling approaches, and statistical analyses. These methods aim to improve the quality and accuracy of medical image analysis, thereby enhancing the precision of clinical outcomes. Statistical models provide a customized framework for addressing specific surgical concerns by training on a dataset comprising cases with corresponding outcomes. Such models are particularly useful for image recovery tasks, assessing intensity, graph curves, and global shapes to guide clinical recommendations toward the most likely outcome.

Computational Imaging and Analytics in Biomedical Engineering: Algorithms and Applications.
T. R. Ganesh Babu, PhD, U. Saravanakumar, PhD & Balachandra Pattanaik, PhD (Eds.)

Visual interpretation by the medical team is currently the prevailing method for analyzing clinical images, heavily reliant on subjective interpretations. Therefore, automated analysis of these images using computer algorithms offers a more reliable option to ensure a highly precise analysis. This chapter focuses on the statistical analysis of seizure data to support clinical decision-making processes.

1.1 INTRODUCTION

Visual interpretation is not always the right choice in the medical process, so automated image analysis using computers with a suitable algorithm is only the perfect choice.[1] The radiographic analysis of disease images needs more attention. Because data are collected during routine clinical practice, undergone, pre- and post-processing work. The artificial intelligence (AI), statistical methodology, of image processing capabilities, with big data analysis of datasets, is growing exponentially.[2,3] The automated quantification of the radiographic process is more useful in the discovery, characterization, and monitoring of conditions are called radiomics,[4] it uses the predefined features of the images[5] based on the aspects of texture or features, shape, and intensity used for the image illustration.[6] The early success of radiomics giving clinical opinions of different brain-related diseases[7] has increased the rapid-fire expansion in this medical field.[8] The analysis of radiologic multitudinous challenges has been addressed in quantitative fields of biostatistics.[9] Microarray data normalization, transformation,[10] batch effects in high-throughput data,[11] Bayes methods used for the adjusting batch effects in microarray expression,[12] a statistical method in DNA hybridization,[13] the study of cancer subtypes.[14] Researchers in biostatistics have also contributed to the development of statistical methods for radiologic analysis, addressing challenges such as normalizing and transforming microarray data, mitigating batch effects in high-throughput data, and studying cancer subtypes.[15] [15] In an earlier state, data analysis in radiology faces a lot of challenges and avoidable data analysis pitfalls.[16] The more than 150 biomedical image analyses are discussed up to 2016. To quantify the robustness of a ranking, Kendall's tau of statistical methods is used such as statistical analysis with radiomic quantification, biomarker identification, and validation.[17] The different ways used in AI are especially useful when it comes to the field of "big data," which describes large volumes of complex and variable data.

1.2 MATERIALS AND METHODS

1.2.1 MEDCALC

MedCalc is a useful clinical calculator used for evidence-based medicine. This tool provides hundreds of clinical decisions, which include risk scores, algorithms, dosing calculators, diagnostic criteria, formulas, and classifications are all framed in the easy handling way. MedCalc serves for clinical decision-making. It is designed to do statistical analysis and offer the relevant clinical decision-making purpose. The use of MedCalc is recommended for a fast and accurate method for analyzing patient data. It is very useful for clinical decision-making. The tool was designed based on the calculating confidence intervals of regression,[18] correlation and residuals,[19] regression, and analysis of variance.[20] Funnel plots are used for detecting bias in meta-analysis. The tool used to find the statistical analysis such as meta-analysis, clustering, and regression.[21] The receiver-operating characteristic (ROC) method is used for the evaluation purpose.[22]

1.2.2 CLUSTER ANALYSIS

Clustering is a method of grouping the set of objects in the same cluster (group) that are similar to each other clusters. The clustering algorithms operate in noisy, deficient, and high dimensional data, their performance based on the datasets. The literature shows the two cluster algorithms, cluster ensembles method, covariance matrix, comparison of supervised classifiers, clustering data mining, simulated annealing, genetic algorithms, the problem of overfitting, data clustering, statistical computing, financial risk analysis, classification purpose, SOM neural network with fuzzy means, k means, and traditional clustering algorithms.[23] Researchers in biostatistics have also contributed to the development of statistical methods for radiologic analysis, addressing challenges such as normalizing and transforming microarray data, mitigating batch effects in high-throughput data, and studying cancer subtypes.[15] A modified interpretation[29] was used to produce 400 distinct datasets, used to quantify the clustering algorithms. The data generation approaches are also used.[30] The clustering algorithm relies on performance and parameter values,[31] and overfitting datasets.[32] The new dataset gives good to bad performance sometimes. Clustering algorithms are used in several R programming languages.[33] The algorithms were estimated based on the number of features, classes, objects for each class, and the average distance between classes. The data generation and performance measures are used to compare the algorithms are shown in this paper.

1.2.3 META-ANALYSIS

Meta-analysis solves the genetic studies issues by identifying inter-study heterogeneity and two-stage process. The first measure of treatment effect is 95% confidence intervals (CI) used to measure odds ratios (OR), relative risks (RR), and risk differences. Meta-analysis can be qualitative or quantitative (meta-analysis). A meta-analysis uses statistical methods to perform: data acquisition, problem formulation, quality, statistical analysis of quantitative data, and clinical interpretation. Meta-analyses of the diagnostic test measure the accuracy, specificity, and sensitivity of the uncertainty.[34] These measures are used in radiology and biomedical studies. These studies provide synthesized, quantitative information on certain topics to get meta-analyses of diagnostic accuracy.[35] To evaluate the quality, statistical tools are used[36] to access the impact of evidence[37] and the statistical results.[38] The study of 100 top-cited meta-analyses of diagnostic accuracy, published in radiology journals.[39] The meta-analysis literature is based on PubMed.[40] The bivariate analysis of sensitivity and specificity was published in 2005 also used for this chapter analysis.[41] The meta-analyses to perform a quantitative imaging modality of X-ray, ultrasound computed tomography, magnetic resonance imaging, used for detection of a specific condition or disease, concerned with diagnostic accuracy measures.

1.2.4 GENETIC META ANALYSIS

Meta-analyses of genotyping studies provide advanced techniques of genotyping at untyped HapMap loci, enabling the combination of genotype data from datasets with millions of variants. HapMap loci refer to specific locations on the human genome that were examined and genotyped as part of the HapMap project. The loci (plural of locus) in the HapMap were selected based on their representation of common genetic variations across different human populations. These loci were genotyped using various genetic techniques to determine the specific genetic variation present at each location. The data from HapMap loci has been widely used by researchers to study the genetic basis of diseases, drug responses, and other complex traits. These analyses can be carried out in a sequential, cumulative manner to enhance statistical power. The establishment of consortia has greatly facilitated advancements in various fields. Numerous non-overlapping consortia exist in certain genetic diseases, each contributing to the epidemiological studies of specific aspects of their designs, data, and sample collections[42]. The genotype statistics can currently be combined throughout datasets with millions of variants, can be conducted in a sequential and cumulative manner. The development of consortia[43] was advanced in many fields.[44] In some genetic sicknesses, numerous non-overlapping consortia exist, GWA datasets used for epidemiological studies of vital elements of their layout, statistics, and sample series cannot be altered. E-book bias, shows the traditional meta-analysis literature,[45] the advantage of GWA research in genome studies.[46] Public repositories have GWA datasets, especially those funded by the public.[47] But public statistics availability poses numerous challenges. The "genome-wide statistical significance" used for phenotype analyses.[48] A couple of analytical alternatives for the identical records are extraordinary genetic fashions, used for various adjustments. Population stratification was used for the structured mapping.[49] The unavoidable overlap and covariance accounted in the meta-analysis. The dataset variance increases the correlation which is protected within the calculations.

The restrained overlap between specific genotyping platforms and evaluation of imputed versions is done in all experiments separately and managed for population stratification of genotype assignments. The heterogeneity is Cochran's Q, which tries to answer statistically, the threshold of heterogeneity based on Q is $p < 0$. The statistic is used to quantify the percentage of variation across studies, indicating inconsistency. It is calculated as (Q-degree of freedom)/Q ratio.[51] The variance, $\tau 2$, heterogeneity, is one kind of meta-analyses, besides heterogeneity biases in a different way affecting the outcomes of various datasets, genetic results, and distinct biological placing of virtually informative heterogeneity. The meta-analysis aims to determine the association at a genome-wide level. The information may be odds ratios, standardized impact sizes, or other metrics together with variance or 95% self-belief C program language period of p-values for different datasets.[52]

1.2.5 META ANALYSIS RISK AND DIFFERENCE IN IMAGE PROCESSING

The risk difference between two proportions is a measure of effect size in studies where the variable is a dichotomous response. In the meta-evaluation of binomial proportions, the impact is the odds ratio (OR), usually analyzed as log (OR), risk ratio (RR), and its logarithm interpretation. The log (OR) and log (RR) are each unbounded, observe-stage occasion in the C programming language (0, 1). The binomial generalized linear combined fashions (GLMMs) and beta-binomial (BB) distribution, with the logit hyperlink feature, was used for the meta-analysis. The Cochrane meta-analyses of RR bias effects from the conventional inverse-variance–weighted technique. The epidemiologic research is executed with meta-regression. The linear relation is between publicity and the algorithm, a nonlinear relation among an exposure and relative chance and statistical solution also effortlessly carried out.[53] The relative dangers are equal to referent class and the meta-regression contributions of meta-evaluation. Greenland and Longenecker[54] developed a technique of dose–response data with meta-analysis.

1.2.6 META ANALYSIS OF RISK RATIO DIFFERENCE

The risk ratio (RR, or relative risk) of two event groups, whereas the odds ratio (OR) of the odds of an event, measures a value of 1 that estimates effects for both interventions. The measures regarding patient characteristics, comorbidities, symptoms crucial signs, and symptoms have been extracted from a meta-analysis. The risk metrics of meta-analysis became carried out to assess the two consequences: excessive and mortality. A meta-analysis was conducted using a random-effects model to pool the regression data on abnormal ratios (ORs). The ORs for mortality studies were calculated for any severe illness (patient case history, ICU-Incentive care unit admission) and the same scientific variable estimation, whether multivariate or univariate. The data were analyzed using R statistical software along with the meta-analysis and plots generated using the R package meta values.[55,56]

1.2.7 META ANALYSIS OF ROC CURVE

The ROC curve is used for the estimation of the individual curve. The parameters are then pooled with bivariate random effects. Receiver-running feature (ROC) is used for evaluating the overall diagnostic performance tests and for comparing the accuracy of statistical version of logistic regression, linear discriminant analysis that classifies subjects into 1 of 2 categories, diseased or not diseased. Predictive modeling to estimate anticipated outcomes consisting of mortality on patient risk traits in research with ROC analysis. The measures of accuracy, sensitivity, specificity, and location underneath the curve (AUC) that use the ROC curve. To estimate accuracy with ROC strategies, the ailment reputation of the affected person is measured without blunders is called the gold trendy. The diagnostic accuracies are specificity (i.e., true bad charge) and sensitivity (i.e., true fine charge). The final results of a threshold of the diagnostic test used to classify subjects. The ROC curve is discussed in the coronary restenosis and peak oxygen consumption analysis articles (57–58).

1.2.8 NON-LINEAR REGRESSION

Nonlinear regression is a mathematical modeling technique that relates two variables, X and Y, with a nonlinear curved relationship rather than a straight line ($y = mx + b$). This method is commonly used in image processing, particularly for analyzing images that contain a combination of shading and noise. When applied to images, nonlinear regression can be used to measure both qualitative and quantitative aspects, with a particular focus on addressing shading problems. In this context, an image can be seen as a combination of a noise image, which represents the characteristics of the pixels in the scene, and a shadowing image, which describes the illumination and shape of the objects in the scene. To model a specific scene image within a certain range, nonlinear regression is applied to decompose the image into its natural factors, allowing for the calculation of the inherent characteristics of the scene[59–64]. This process includes the utilization of a weighting function to adjust the original estimate, leading to a more precise estimate that reduces errors. Nonlinear regression can be applied to image patches without encountering any issues with dimensionality. In such cases, the estimation of the shadowing image involves modifying the image constraints in the most effective manner. Moreover, the Retinex algorithm, which employs Poisson's equation and secondary estimates, is commonly used to estimate the image derivatives and enhance the accuracy of the original estimates.

1.3 RESULTS AND DISCUSSION

MedCalc is a fast and reliable statistical software that includes all features and with more than 200 statistical tests, procedures, and graphs. Medical calculators help physicians' memory work and calculation skills to the clinical test.

Method 1: Seizure patient treatment based on cluster analysis

The cluster analysis is based on the treatment of seizure-affected persons and their clinical proceedings are measured as shown in Table 1.1 and Figures 1.1 and 1.2 are given below

TABLE 1.1 Input Data of Seizure Patient Treatment Based on Cluster Analysis.

Gender	Treatment	Measurment 1	Measurement 2
Female	A	21	25
Male	B	22	26
To be continued			

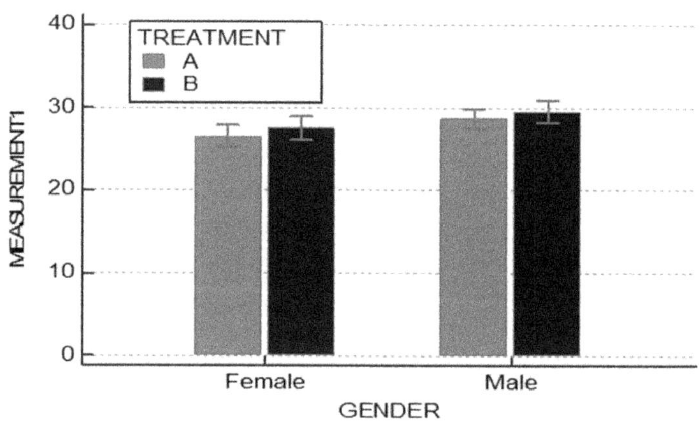

FIGURE 1.1 Cluster comparison graph.

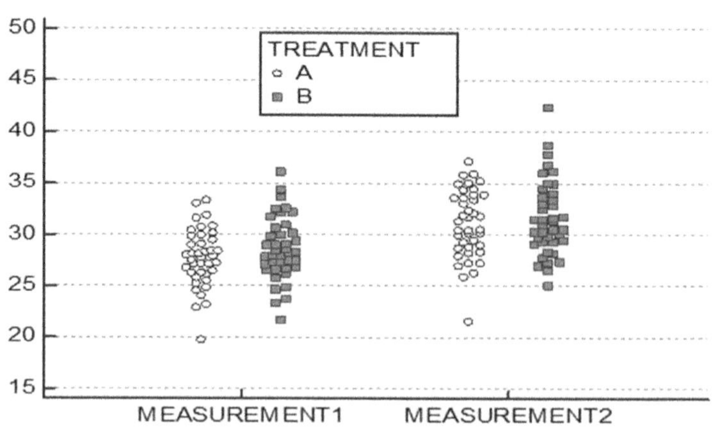

FIGURE 2.2 Cluster multiple variable graph.

Method 2: Meta-analysis risk ratio and difference

Average assumption of seizure-treated positive and the control positive patient's details are discussed in Table 1.2 and Figure 1.3.

TABLE 1.2 Meta-Analysis Risk Ratio and Difference.

Seizure study	Treated_positive	Treated_Total	Controls_positive	Controls_total
Study 1	95	103	47	104
Study 2	119	127	34	129
Study 3	51	223	12	76
Study 4	122	139	61	142
Study 5	47	53	10	51
Study 6	121	135	29	68
Study 7	337	378	170	376

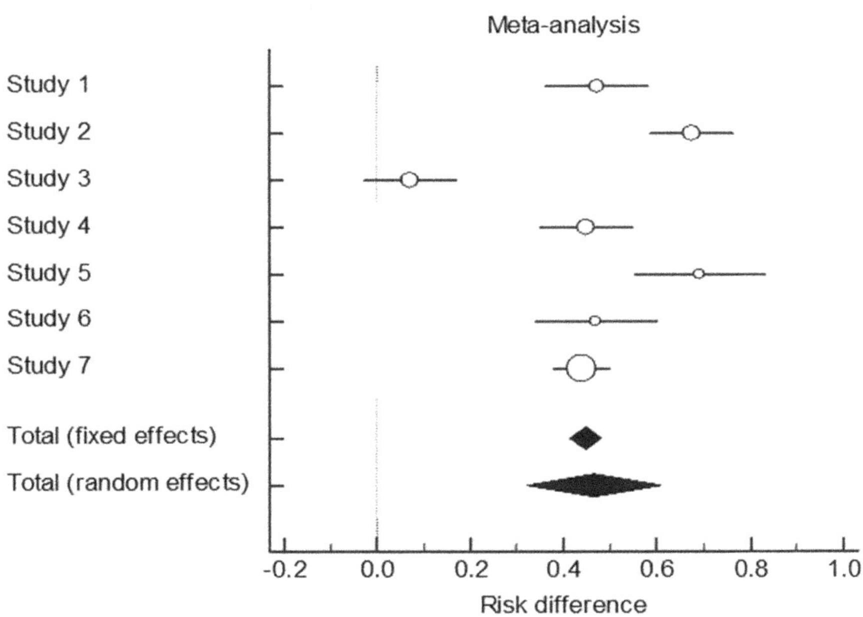

FIGURE 1.3 Meta-analysis risk ratio and difference.

Method 3: Generic meta analysis

The variable for estimate and variable for standard error based on heterogeneity calculation, and the graph was shown in Figure 1.3.

TABLE 1.3 Generic Meta-Analysis.

Seizure	Hazard_Ratio__Log_	SE_of_LOG_HR	(Generic meta-analysis)
Study A	1	− 0.077	0.212
Study B	2	0.012	0.221
Study C	3	0.323	0.426
Study D	4	0.154	0.23
Study E	5	0.051	0.348
Study F	6	− 0.661	0.232
Study G	7	− 0.199	0.337

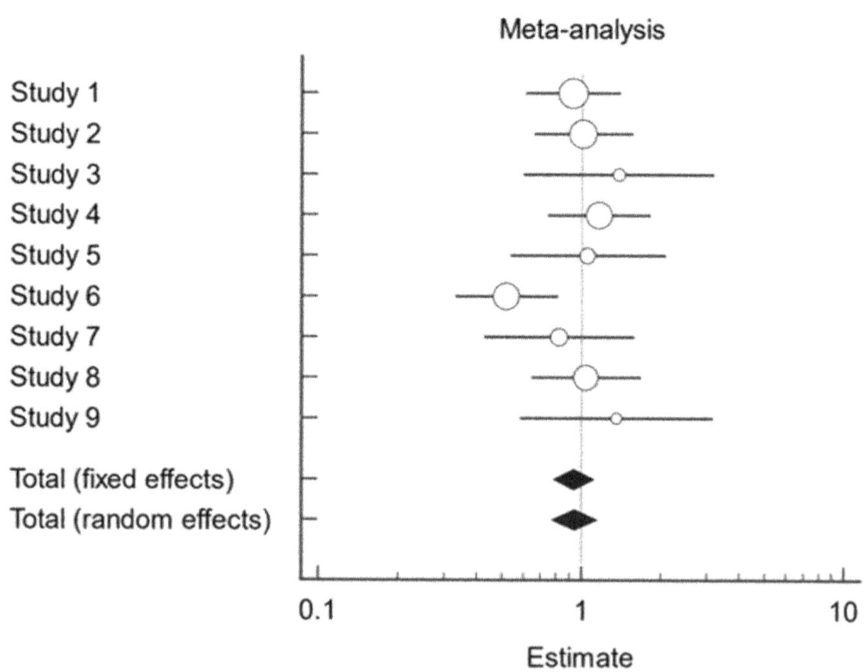

FIGURE 1.4 Meta-analysis generic invariance

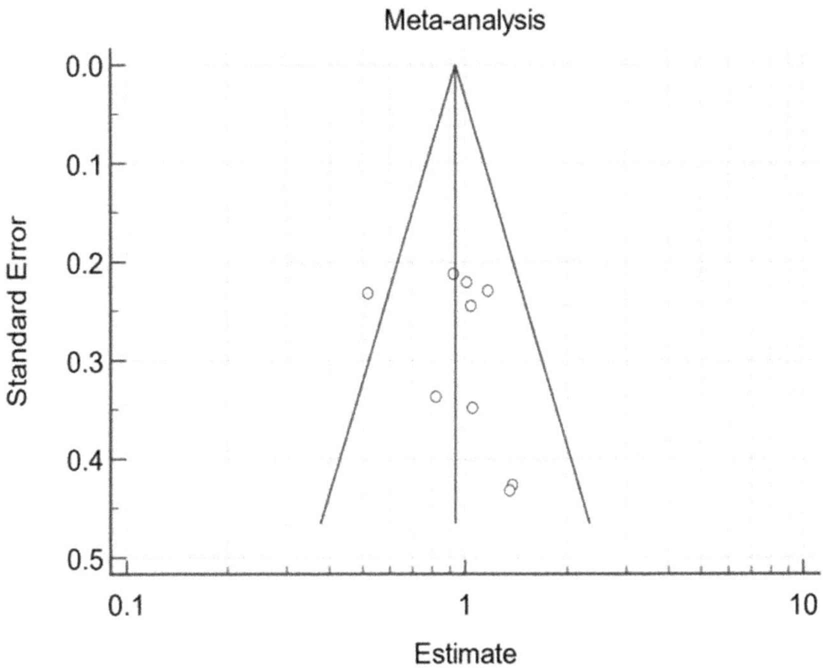

FIGURE 1.5 Meta-analysis generic invariance.

Method 4: Meta-analysis continuous measure

The result of continuous measure of the meta-analysis is shown in Table 1.4 and the output in Table 1.5 and Figure 1.6.

TABLE 1.4 Input Data to Meta-Analysis Continuous Measure.

Seizure type_%	FLE	PLE	OLE	MTLE	ME	TOTAL
Subclinical	18	3	2	85	20.12	135
SPS	109	41	10	60	23.5	258
CPS	35	22	21	213	19.7	541
SGS	65	61	5	62	23.21	317
TOTAL	227	127	28	420	24.9	1251

FLE, frontal lobe epilepsy; *PLE*, peritial lobe epilepsy; *OLE*, occipital lobe epilepsy; *MTLE*, mesial temporal lobe epilepsy; *ME*, mesial epilepsy.

TABLE 1.5 Meta-Analysis: Continuous Measure-OUTPUT.

Variable for studies	**SeizureType_%**
1. Intervention groups	
Variable for number of cases	FLE
Variable for mean	PLE
Variable for SD	OLE
2. Control groups	
Variable for number of cases	MTLE
Variable for mean	ME
Variable for SD	TOTAL
Test for heterogeneity	
Q	1.2429
DF	4
Significance level	P = 0.8710
I2 (inconsistency)	0.00%
95% CI for I2	0.00 to 37.00
Publication bias	
Egger's test	
Intercept	− 0.663
95% CI	− 2.6639 to 1.3379
Significance level	P = 0.3691
Begg's test	
Kendall's Tau	− 0.6
Significance level	P = 0.1416

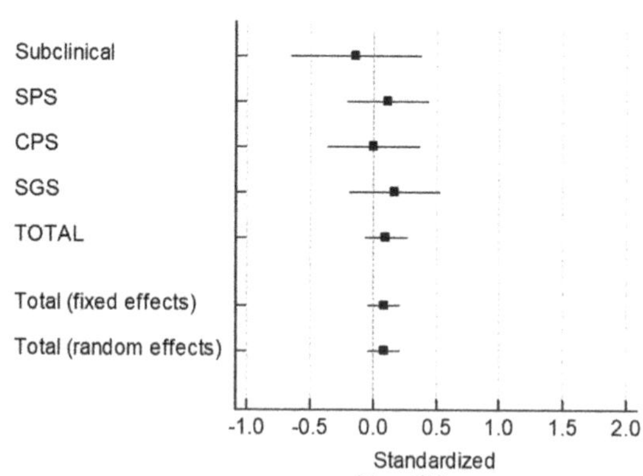

FIGURE 1.6 Meta-analysis continuous measure.
Notes: *SPS*, simple partial seizure; *CPS*, complex partial seizure; *SGS*, secondarily generalized seizure.

Method 5: Meta-analysis based on the correlation

The result of meta-analysis based on the correlation is shown in Table 1.7 as referred from the literature survey and Figure 1.7.

TABLE 1.7 Meta-Analysis Based on the Correlation.

Seizure_study	Correlation_coefficient	Number of samples
Alickovic et al.	0.56	133
Fasil and Rajesh	0.43	149
Manzouri et al.	0.53	131
Parvez and Paul	0.51	120
Siddiqui et al.	0.66	111
Hussein et al.	0.46	152
Mursalin et al.	0.33	60

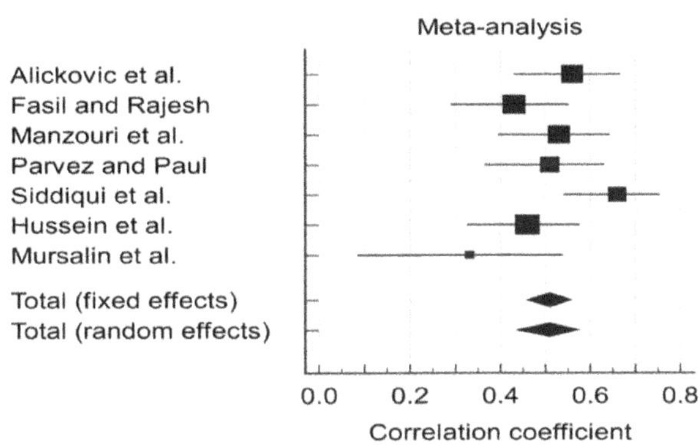

FIGURE 1.7 Meta-analysis based on correlation co-efficient.

Method 6: Meta-analysis risk ratio and difference

The result of meta-analysis based on risk ratio and difference are shown in Table 1.8 and Figure 1.8.

TABLE 1.8 Meta-Analysis Based on Risk Ratio and Difference.

Study	Treated_positive	Treated_Total	Controls_positive	Controls_total
Study 1	95	103	47	104
Study 2	119	127	34	129
Study 3	51	223	12	76

TABLE 1.8 *(Continued)*

Study	Treated_positive	Treated_Total	Controls_positive	Controls_total
Study 4	122	139	61	142
Study 5	47	53	10	51
Study 6	121	135	29	68
Study 7	337	378	170	376

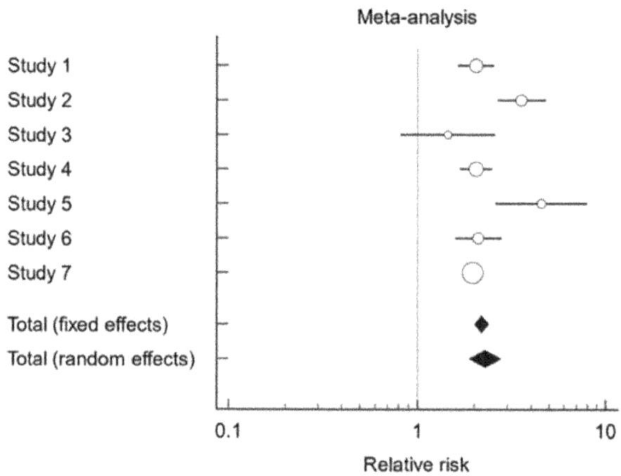

FIGURE 1.8 Meta-analysis relative risk.

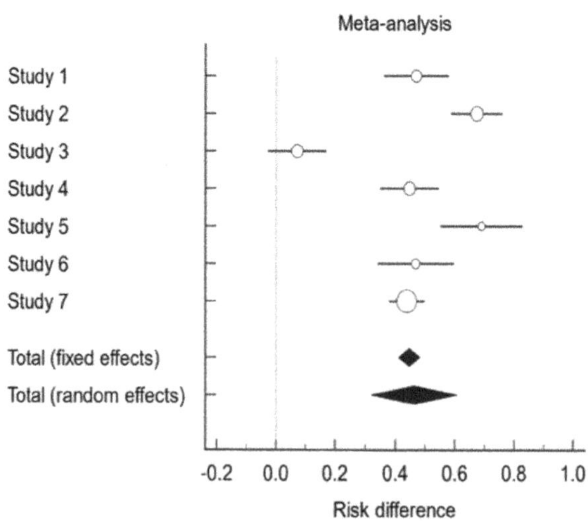

FIGURE 1.9 Meta-analysis relative difference.

Method 7: Non-linear regression method

The non-linear regression method is captured based on the dose given to the patients and their responses.

TABLE 1.9 Meta-Analysis Based on Non-linear Regression Method.

Seizure dose	Response
0	0.1
1.3	0.5
2.8	0.9
5	2.6
10.2	7.1
16.5	12.3
21.3	15.3
31.8	20.4
52.2	24.2

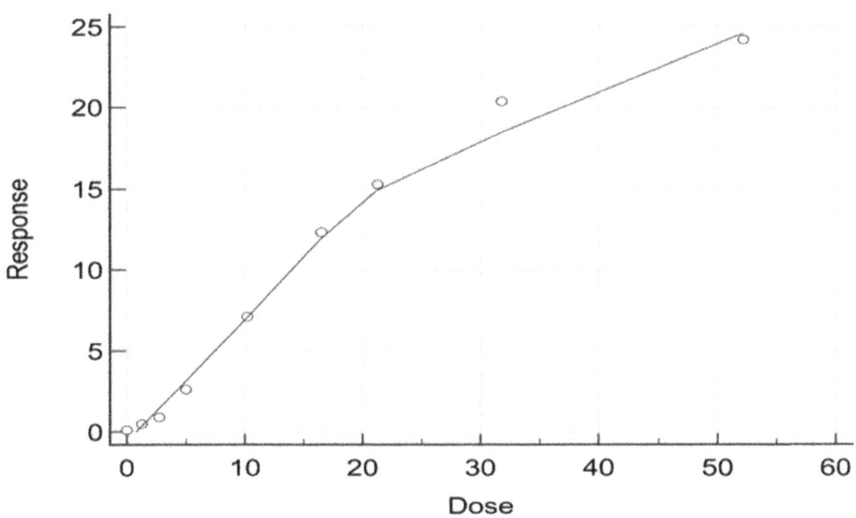

FIGURE 1.10 Scatter diagram of dose–response.

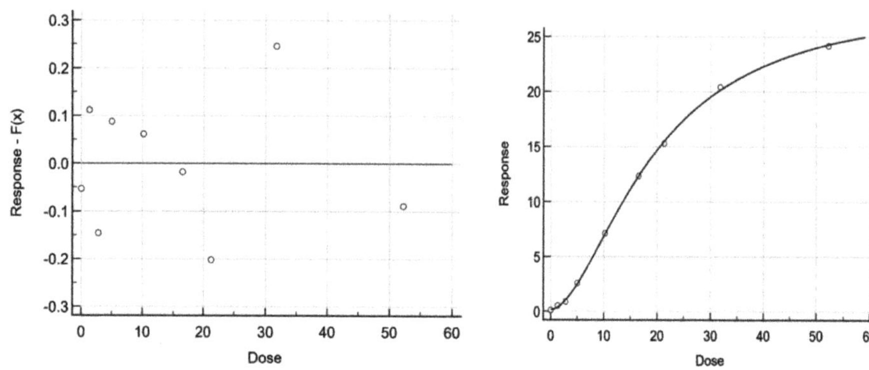

FIGURE 1.11 Non-linear regression dose–response.

1.4 CONCLUSION

Statistical analysis is very essential in the field of healthcare applications core areas. However, meaningful and accurate data analyses are challenging and with errors and incorrect conclusions. This article presents the radiographic medical image statistical analysis. The completely error-free unbiased experimentation and handling of research-compromising pitfalls. The statistical analysis will enhance the quality of clinical data for better integration of imaging data with patient-specific.

KEYWORDS

- **statistical analysis**
- **meta-analysis**
- **cluster analysis**
- **radiology**

REFERENCES

1. Renukalatha, S.; Suresh, K. V.; A Review on Biomedical Image Analysis. *Biomed. Eng. App. Basis Commun.* **2018**.
2. Wang, W.; Krishnan, E. A Review on the State of the Science. *Big Data Clin. JMIR Med. Inform.* **2014,** *2*, e1.

3. Luo, J.; Wu, M.; Gopukumar, D.; Zhao, Y. A Literature Review: Big Data Application in Biomedical Research and Health Care. *Biomed. Inform. Insights* **2016,** *8*, 1–10.
4. Lambin, P.; Rios-Velazquez, E.; Leijenaar, R.; Carvalho, S.; van Stiphout, RGPM.; Granton, P. Extracting More Information from Medical Images Using Advanced Feature Analysis Radiomics. *Eur. J. Cancer* **2012,** *48*, 441–446.
5. van Griethuysen, J. J. M.; Fedorov, A.; Parmar, C.; Hosny, A.; Aucoin, N.; Narayan, V. Radiomics System to Decode the Radiographic Phenotype Computational. *Cancer Res.* **2017,** *77*, e104–e107.
6. Rusk, N. Deep Learning. *Nat Methods* **2015,** *13*, 35.
7. Huynh, E.; Coroller, T. P.; Narayan, V.; Agrawal, V.; Romano, J.; Franco, I. et al. Associations of Radiomic Data Extracted from Static and Respiratory-Gated CT Scans with Disease Recurrence in Lung Cancer Patients Treated with SBRT. *PLoS One* **2017,** *12*, e169–e172.
8. Kolossváry, M.; Kellermayer, M.; Merkely, B.; Maurovich-Horvat, P. A Comprehensive Review on Radiomic Techniques: Cardiac Computed Tomography Radiomics. *J. Thorac. Imaging* **2018,** *33*, 26–34.
9. O'Connor, J. P. B.; Aboagye, E. O.; Adams, J. E.; Aerts, H. J. W. L.; Barrington, S. F.; Beer, A. et al. Imaging Biomarker Roadmap for Cancer Studies. *Nat. Rev. Clin. Oncol.* **2017,** *14*, 169–186.
10. Quackenbush, J. Microarray Data Normalization and Transformation. *Nat. Genet.* **2002,** *32*, 496–501.
11. Leek, J. T.; Scharpf, R. B.; Bravo, H. C.; Simcha, D.; Langmead, B.; Johnson, W. E. et al Tackling the Widespread and Critical Impact of Batch Effects in High-Throughput Data. *Nat. Rev. Genet.* **2010,** *11*, 733–739.
12. Johnson, W. E.; Li, C.; Rabinovic, A. Adjusting Batch Effects in Microarray Expression Data Using Empirical Bayes Methods. *Biostatistics* **2007,** *8*, 118–127.
13. Lee, M. L.; Kuo, F. C.; Whitmore, G. A.; Sklar, J. Importance of Replication in Microarray Gene Expression Studies, Statistical Methods and Evidence from Repetitive cDNA Hybridizations. *Proc. Natl. Acad. Sci. USA* **2000,** *97*, 9834–9839.
14. Neve, R. M.; Chin, K.; Fridlyand, J.; Yeh, J.; Baehner, F. L.; Fevr, T. et al. A Collection of Breast Cancer Cell Lines for the Study of Functionally Distinct Cancer Subtypes. *Cancer Cell* **2006,** *10*, 515–527.
15. Allison, D. B.; Cui, X.; Page, G. P.; Sabripour, M. From Disarray to Consolidation and Consensus: Microarray Data Analysis. *Nat. Rev. Genet.* **2006,** *7*, 55–65.
16. Aerts, H. J. Data Science in Radiology—A Path Forward. *Clin. Cancer Res.* **2018,** *24*, 532–534.
17. Parmar, C.; Barry, J.; Hosny, D.; Ahmed, Q.; John, A.; Hugo, J. W. L. Data Analysis Strategies in Medical Imaging. *Clin. Cancer Res.* **2018**. clincanres.0385.2018.
18. Altman, D. G.; Gardner, M. J. Calculating Confidence Intervals for Regression and Correlation. *Br. Med. J.* **1988,** *296*, 1238–1242.
19. Altman, D. G. Construction of Age-Related Reference Centiles Using Absolute Residuals. *Statistics Med.* **1983,** *12*, 917–924.
20. Glantz, S. A.; Slinker, B. K.; Neilands, T. B. *Primer of Applied Regression & Analysis of Variance*, 2nd ed.; McGraw- Hill.
21. Walter, S. D. The Partial Area Under the Summary ROC Curve. *Stat. Med. 24*, 2025–2040.

22. Zweig, M. H.; Campbell, G. Receiver-Operating Characteristic (ROC) Plots: A Fundamental Evaluation Tool in Clinical Medicine. *Clin. Chem.* 39:561–577.
23. Fraley, C.; Raftery, A. E. How Many Clusters? Which Clustering Method? Answers via Model-Based Cluster Analysis. *Comput. J.* **1998,**41 (8), 578–588.
24. Halkidi, M.; Batistakis, Y.; Vazirgiannis, M. On Clustering Validation Techniques. *J. Intell. Inf. Syst.* **2001,**17 (2–3), 107–145.
25. Strehl, A.; Ghosh, J.; Cardie, C.; Cluster Ensembles—A Knowledge Reuse Framework for Combining Multiple Partitions. *J. Mach. Learn. Res.* **2002,** *3*, 583–617.
26. Fowlkes, E. B.; Mallows, C. L. A Method for Comparing Two Hierarchical Clusterings. *J. Am. Stat. Assoc.* **1983,** *78* (383), 553–569.
27. Lawrence, H.; Arabie, P. Comparing Partitions. *J. Classification* **1985,** *2* (1), 193–218.
28. Jaccard, P. Nouvelles recherches sur la distribution florale. *Bulletin de la Sociète Vaudense des Sciences Naturelles.* **1908,** *44*, 223–270.
29. Hirschberger, M.; Qi, Y.; Steuer, R. E. Randomly Generating Portfolio-Selection Covariance Matrices with Specified Distributional Characteristics. *Eur. J. Operat. Res.* **2007,** *177* (3), 1610–1625.
30. Amancio, D. R.; Comin, C. H.; Casanova, D.; Travieso, G.; Bruno, O. M.; Rodrigues, F. A.; et al. A Systematic Comparison of Supervised Classifiers. *Plos One* **2014,** *9* (4), e94137. pmid:24763312.
31. Berkhin, P.; Kogan, J.; Nicholas, C.; Teboulle, M. *A Survey of Clustering Data Mining Techniques*; Springer: Berlin Heidelberg, 2006; pp 25–71.
32. Hawkins, D. M. The Problem of Overfitting. *J. Chem. Inf. Comput. Sci.* **2004,** *44* (1), 1–12. pmid:14741005.
33. R Development Core Team. *R: A Language and Environment for Statistical Computing*; 2006.
34. Mallett, S.; Halligan, S.; Thompson, M.; Collins, G. S.; Altman, D. G. Interpreting diagnostic Accuracy Studies for Patient Care. *BMJ* **2012.**
35. McGrath, T. A.; McInnes, M. D.; Korevaar, D. A.; Bossuyt, P. M. Meta-Analyses of Diagnostic Accuracy in Imaging Journals: Analysis of Pooling Techniques and Their Effect on Summary Estimates of Diagnostic Accuracy. *Radiology* **2016,** *281* (1), 78–85.
36. Choudhri, A. F.; Siddiqui, A.; Khan, N. R.; Cohen, H. L. Understanding Bibliometric Parameters and Analysis. *Radiographics* **2015,** *35* (3), 736–746.
37. Frank, R. A; Sharifabadi, A. D.; Salameh, J. P.; McGrath, T. A.; Kraaijpoel, N.; Dang, W.; et al. Citation Bias in Imaging Research: Are Studies with Higher Diagnostic Accuracy Estimates Cited More Often? *Eur. Radiol.* **2019,** *29* (4), 1657–1664.
38. Letchford, A.; Moat, H. S.; Preis, T. The Advantage of Short Paper Titles. *R. Soc. Open Sci.* **2019,**2 (8), 150266.
39. Analytics C. Journal Impact Factor 2018, Journal Citation Reports 2019.
40. Reitsma, J. B.; Glas, A. S.; Rutjes, A. W.; Scholten, R. J.; Bossuyt, P. M.; Zwinderman, A. H. Bivariate Analysis of Sensitivity and Specificity Produces Informative Summary Measures in Diagnostic Reviews. *J. Clin. Epidemiol.* **2015,** *58* (10), 982–990.
41. Hillner, B. E.; Tosteson, A. N.; Song, Y.;Tosteson, T. D.; Onega, T.; Goodman, D. C. et al Growth in the Use of PET for Six Cancer Types After Coverage by Medicare: Additive or Replacement? *J. Am. Coll. Radiol.* **2012,** *9* (1), 33–41.
42. Anderson, C. A.; Pettersson, F. H.; Barrett, J. C. et al. Evaluating the Effects of Imputation on the Power, Coverage, and Cost Efficiency of Genome-Wide SNP Platforms. *Am. J. Hum. Genet.* **2008,** *83* (1), 112–119.

43. Seminara, D.; Khoury, M. J.; O'Brien, T. R. et al. The Emergence of Networks in Human Genome Epidemiology: Challenges and Opportunities. *Epidemiology* **2007,** *18*, 1–8.
44. Ioannidis, J. P.; Bernstein, J.; Boffetta, P. et al. A Network of Investigator Networks in Human Genome Epidemiology. *Am. J. Epidemiol.* **2005,** *162*, 302–304.
45. Rothstein, H. R.; Sutton, A. J.; Borestein, M., Eds. *Publication Bias in Meta-Analysis–Prevention, Assessment and Adjustments*; Wiley: Chichester, 2005.
46. Barrett, J. C.; Cardon, L. R. Evaluating Coverage of Genome-Wide Association Studies. *Nat. Genet.* **2006,** *38* (6), 659–662.
47. Manolio, T. A.; Rodriguez, L. L.; Brooks, L. et al. GAIN Collaborative Research Group. New Models of Collaboration in Genome-Wide Association Studies: The Genetic Association Information Network. *Nat. Genet.* **2007,** *39* (9), 1045–1051.
48. Hoggart, C. J.; Clark, T. G.; De Iorio, M.; Whittaker, J. C.; Balding, D. J. Genome-Wide Significance for Dense SNP and Resequencing Data. *Genet. Epidemiol.* **2008,** *32* (2), 179–185.
49. Ioannidis, J. P. Why Most Discovered True Associations Are Nflated. *Epidemiology* **2008,** *19* (5), 640–648.
50. Pritchard, J. K.; Stephens, M.; Rosenberg, N. A.; Donnelly, P. Association Mapping in Structured Populations. *Am. J. Hum. Genet.* **2000,** *67*, 170–181.
51. Ioannidis, J. P.; Patsopoulos, N. A.; Evangelou, E. Uncertainty in Heterogeneity Estimates in Meta-Analyses. *BMJ.* **2007,** *335*, 914–916.
52. Moonesinghe, R.; Khoury, M. J.; Liu, T.; Ioannidis, J. P. Required Sample Size and Non-Replicability Thresholds for Heterogeneous Genetic Associations. *Proc. Natl. Acad. Sci. U.SA* **2008,** *105* (2), 617–622.
53. Il'yasova, D.; Hertz-Picciotto, I.; Peters, U. et al. Choice of Exposure Scores for Categorical Regression in Meta-Analysis: A Case Study of a Common Problem. *Cancer Causes Control* **2005,** *16*, 383–388.
54. Greenland, S.; Longnecker, M. P. Methods for Trend Estimation from Summarized Dose-Response Data, with Applications to Meta-Analysis. *Am. J. Epidemiol.* **1992,** *135*, 1301–1309.
55. Metlay, J. P.; Waterer, G. W.; Long, A. C.; Anzueto, A.; Brozek, J.; Crothers, K. et al. Diagnosis and Treatment of Adults with Community-acquired Pneumonia. An Official Clinical Practice Guideline of the American Thoracic Society and Infectious Diseases Society of America. *Am. J. Respir. Crit. Care Med.* **2019,** *200*, e45–e67. PMID:31573350.
56. R Core Team. *R: A Language and Environment for Statistical Computing*; R Foundation for Statistical Computing: Vienna, Austria, 2019. http://www.R-project.org/
57. Mauri, L.; Orav, J.; O'Malley, A. J.; Moses, J. W.; Leon, M. Z. B.; Holmes, D. R.; Teirstein, P. S.; Schofer, J.; Breithardt, G.; Cutlip, D. E.; Kereiakes, D. J.; Shi, C.; Firth, B. G.; Donohoe, D. J.; Kuntz, R. Relationship of Late Loss in Lumen Diameter to Coronary Restenosis in Sirolimus-Eluting Stents. *Circulation.* **2005,** *111*, 321–327.
58. O'Neill, J.; Young, J. B.; Pothier, C. E.; Lauer, M. S. Peak Oxygen Consumption as a Predictor of Death in Patient with Heart Failure Receiving β-Blockers. *Circulation.* **2005,** *111*, 2313–2318.
59. Tappen, M. F.; Freeman, W. T.; Adelson, E. H. Recovering Intrinsic Images from a Single Image. *IEEE Trans. Pattern Analy. Mach. Intell.* **2005**.
60. Kimmel, R.; Elad, M.; Shaked, D.; Keshet, R.; Sobel, I. A Variational Framework for Retinex. *Int. J. Comput. Vision* **2003,** *52* (1), 7–23.

61. Bell, W.; Freeman, W. T. Learning Local Evidence for Shading and Reflectance. In *Proc. Int. Conf. Comput. Vision* **2001**.
62. Freeman, E. T.; Pasztor, E. C.; Carmichael, O. T.; Learning Low-Level Vision. *Int. J. Comput. Vision* **2000,** *40* (1), 25–47.
63. Land, E. H.; McCann, J. J. Lightness and Retinex Theory. *J. Optic. Soc. Am.* **1971,** *61*, 1–11.
64. Barrow, H. G.; Tenenbaum, H. T. Recovering Intrinsic Scene Characteristics from Images. In *Computer Vision Systems*; Hanson, A., Riseman, E., Eds.; Academic Press, 1978; pp 3–26.
65. https://www.researchgate.net/figure/Seizure-outcomes-by-epilepsy-etiology_tbl2_239066212[DATA]

CHAPTER 2

SPATIAL PREPROCESSING IN SEGMENTATION OF BRAIN MRI USING T1 AND T2 IMAGES

S. JACILY JEMILA and A. BRINTHA THERESE

Vellore Institute of Technology, Chennai, India

ABSTRACT

Preprocessing is a very important process before segmentation in medical image processing. To upgrade the characteristics of the image by diminishing the noise, preprocessing is employed. In this chapter, T1 and T2 MRI images of the brain are segmented before and after preprocessing using SPM-12 neuroimaging software, and the importance of preprocessing is discussed using five important quality metrics: peak signal to noise ratio (PSNR), mean squared error (MSE), edge preservation index (EPI), signal to noise ratio (SNR), and structural similarity index metric (SSIM). We can conclude that if we preprocess the MRI brain T1 and T2 images before segmentation, we can reduce the MSE and increase the PSNR, SNR, and SSIM, which are very important in the evaluation of medical images. EPI value is also retained.

2.1 INTRODUCTION

Preprocessing in medical images is of considerable importance for the following motives:

Computational Imaging and Analytics in Biomedical Engineering: Algorithms and Applications.
T. R. Ganesh Babu, PhD, U. Saravanakumar, PhD & Balachandra Pattanaik, PhD (Eds.)

a. The image feature ought to be strengthened.
b. Noise in the image does require elimination.
c. Images are to be higher fit for further computation.
d. Marks or labels existing in the images can intrude in the post-processing of these images.

Nowadays, medical imaging plays a very important role in human lives. In medical imaging, MRI is the most preferable technique for diagnosing most medical issues because it has the following advantages:

1. MRI functions in the RF range.
2. No ionizing radiation present.
3. Can produce wonderful soft tissue contrast.
4. Can generate 3-D images.
5. Capability of producing images at any orientation (axial, sagittal, and coronal).

MRI is a risk-free diagnostic and uncomplicated procedure to scan the internal organs and structure of the human body through strong magnets and radiowaves. An MRI scan is often used to examine the brain, spinal cord, bones, joints, breast, heart, and blood vessels. For the past 40 decades, it has been used as a significant tool in scientific research and medical diagnosis. A commercial MRI imaging machine is displayed in Figure 2.1.

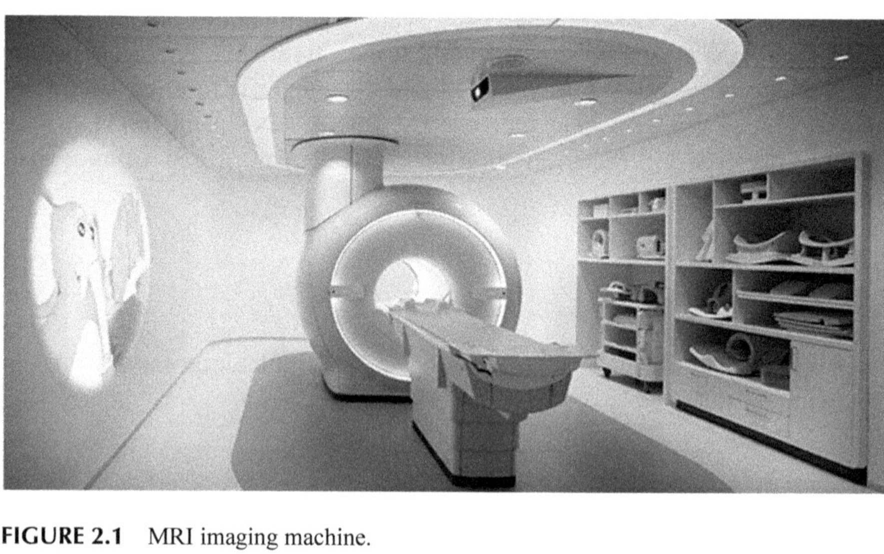

FIGURE 2.1 MRI imaging machine.

Segmentation of MRI is a significant step in medical image analysis using MRI. It impacts the conclusion of the overall analysis. Manual segmentation is a time-consuming method, and there is also an option for error. Preprocessing is a very important step before segmentation. If we do preprocessing,[1,2] then we could improve the quality of the segmented parts. To make the images more suitable for further analysis and to upgrade the characteristics of the image by reducing the noise, preprocessing is very helpful.

2.2 NOISE IN MAGNETIC RESONANCE IMAGES

Many image attainment processes (MRI, PET, SPECT, etc.) sustain image degradation by noise. For MRI, the foremost seedbed of random noise is thermal noise, which frames a statistically-independent random source piercing the MR data in the time domain. Thermal noise is white and can be depicted by a Gaussian random field with a zero mean and constant variance. Hence, the noise is not associated with the signal or with itself. Aside from thermal noise, structured noise ordinarily demotes the image standard as well as MR system attributes, physiological pulses, and object movement. The features of noise depend on their source. Noise, inhomogeneous pixel intensity allocation, and abrupt boundaries in the medical MR images invoked by the MR data acquisition procedure are the capital problems that influence the grade of MRI segmentation. One leading origin of noise is the ambient electromagnetic field groomed up by the radiofrequency (RF) detectors acquiring the MR signal, and the other is the object or body being imaged.

In MRIs, raw data are constitutionally complex, valued, and spoiled with zero-mean Gaussian-distributed noise with equal variance. Next to the inverse Fourier transformation, the real and imaginary images are quite Gaussian-distributed, habituated to the orthogonality and linearity of the Fourier transform. MR magnitude images are framed by exclusively fetching the square root of the sum of the square of the two self-reliant Gaussian random variables (real and imaginary images) pixel by pixel. After this nonlinear transformation, MR magnitude data can be proven to be Rician-distributed.

There is a correlation between SNR, acquisition time, and spatial resolution in MR images. The SNR is fairly dominant in huge MRI usages, and this is cultivated implicitly and clearly by averaging. The MRI data obtainment procedure can be performed by two averaging methods: (1) Spatial volume averaging is mandatory because of the discrete manner of the obtainment procedure, and (2) In the case of few usages, the signal for the same k-space location is obtained several times and averaged in order to minimize noise.

The two averaging methods are related. When a larger sampling rate of the frequency domain is applied, improved resolution images are acquired. However, in order to get the desired SNR at dominant spatial resolution, a lengthy acquisition time is needed, as is the extra time required for averaging. Contrarily, the acquisition time, with the subsequent SNR and the imaging resolution, is practically bounded by patient comfort and system throughput. Consequently, high SNR MRI images can be obtained at the cost of constrained temporal for spatial resolution. Also, improved resolution MRI imaging is obtainable at a cost of lower SNR for long acquisition times.

A further considerable cause of noise in MRI imaging is thermal noise in the human body. Common MRI imaging concerns sampling in the frequency domain (also called "k-space") and employing an inverse discrete Fourier transform. Signal measures have features in both real and imaginary channels, and each channel is strained by additive white Gaussian noise. Thus, the complex reconstructed signal contains complex white additive Gaussian noise. Because of phase errors, the magnitude of the MRI signal is ordinarily preferred for the MRI image reconstruction. The magnitude of MRI signal is real-valued and is applied for the image processing tasks as well as for visual inspection.

The method by which the magnitude MRI image is rebuilt results in a Rician distribution of noise. Since the Rician noise is signal-based, splitting the signal away from the noise is the most challenging process. In large-intensity areas of the magnitude image, the Rician distribution can be approached as a Gaussian distribution, and in low-intensity regions, it can be represented as a Rayleigh distribution. A practical effect is a lowered contrast of the MRI image; as the noise increases, the mean intensity values of the pixels in low-intensity regions also increase. As established, it is a reality that Rician noise demotes the MRI images in both qualitative and quantitative senses, making image processing, interpretation, and segmentation more complex.

2.3 MAJOR CHALLENGES IN BIOMEDICAL IMAGE PROCESSING

It is challenging to integrate a medical image straightaway into the automatic computation of image processing. This is named the semantic gap, which implies the disparity between the cognitive interpretation of a diagnostic image by the physician and the uncomplicated structure of discrete pixels, which is exploited in computer programs to characterize an image. In the medical domain, there are three main facets that hinder bridging this gap.

2.3.1 *HETEROGENEITY OF IMAGES*

Medical images demonstrate organs, living tissues, or body parts. Although acquired with the same modality and pursuing a standardized shape, acquisition protocol, size, and internal structures, these objects might differ notably not simply from patient to patient but also among various views of the same patient and similar views of the same patient at different times. Thus, the universal formulation of prior knowledge is impossible.

2.3.2 *UNKNOWN DELINEATION OF OBJECTS*

Biological structures cannot be extracted from the background because the diagnostically related object is pictured in the overall image. Though determinable objects are noticed in biomedical images, their segmentation is difficult. Hence, they are segmented at the texture level.

2.3.3 *ROBUSTNESS OF ALGORITHMS*

Automatic analysis of images in medicine should be accurate. This is followed as a rule in medical image processing. The images that cannot be treated properly must be inevitably discarded and withdrawn from further processing.

2.4 CONCEPT OF MRI

Our body comprises beyond 70% of water. The water molecule is in the form of H_2O, where H is the proton. This proton is an important component in MRI. Based on the resonance property of the proton, only MRI images will be made. All protons in our body normally rotate in random directions with a Larmor frequency of 63.855 MHZ, as shown in Figure 2.2, and they are in an out-of-phase condition.

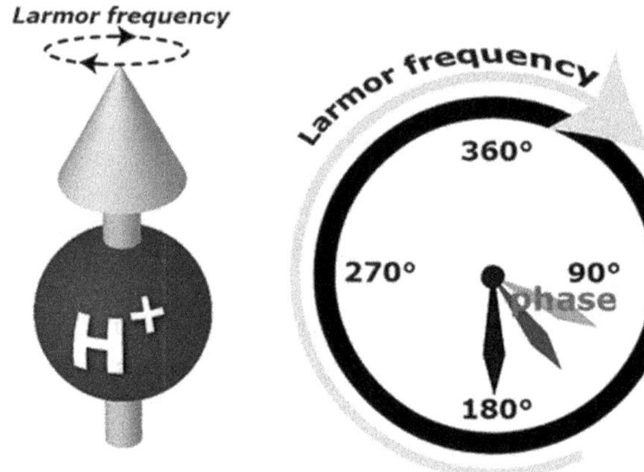

FIGURE 2.2 Proton spin at Larmor frequency.

Source: Reprinted from Radiology Expert. https://www.radiology.expert/en/modules/mri-technique-introduction/mri-technique/

When the patient is positioned in a magnetic field, that is, in an MRI scanner, then all the protons will align in the z direction as in Figure 2.3, which is at the side of the magnetic field. Even though all the protons are in the z direction, they are in an out-of-phase condition.

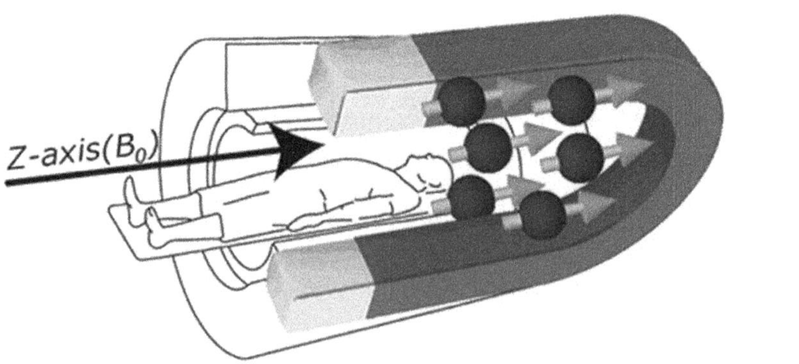

FIGURE 2.3 Position of protons when the patient is positioned in an MRI scanner.

Now, if we apply some RF pulse with a frequency equal to Larmor's frequency, then excitation occurs, and all the protons are aligned in an in-phase condition. The protons will change their position from the z plane (longitudinal plane) to the xy plane (transversal plane). Now the longitudinal magnetization will reduce and the trasversal magnetization will increase. If we stop the RF excitation, then relaxation will take place. In relaxation, the protons revert back to the longitudinal plane from the transversal plane, which is given in Figure 2.4.

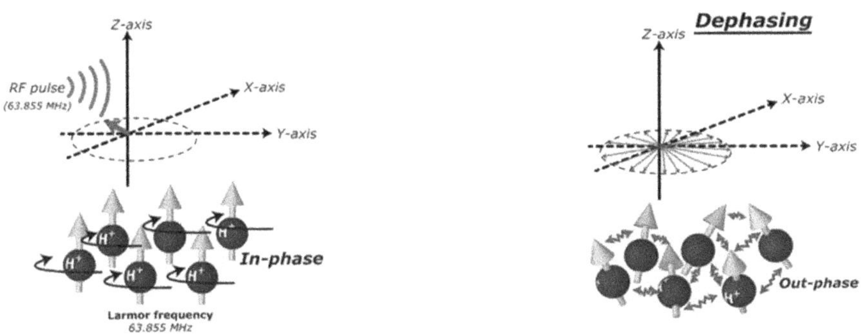

FIGURE 2.4 Alignment of protons in in-phase and out-phase conditions.

Two independent processes take place during relaxation.

1. Longitudinal relaxation (T1 relaxation)
2. Transversal relaxation (T2 relaxation)

2.4.1 *T1 RELAXATION*

When the RF pulse is removed, protons try to come to the longitudinal plane from the transversal plane by giving some energy to the surroundings, so now longitudinal magnetization increases. T1 is the time required to achieve 63% of the original longitudinal magnetization. This is the time required for T1 imaging.

2.4.2 *T2 RELAXATION*

When we stop the RF pulse, not only do the protons try to come to the longitudinal plane, but they also change to an out-of-phase condition. T2 is the time required to dephase up to 37% of the original value. This is the time required for T2 imaging.

For different tissues, there are different T1 and T2 timing and relaxation curves.

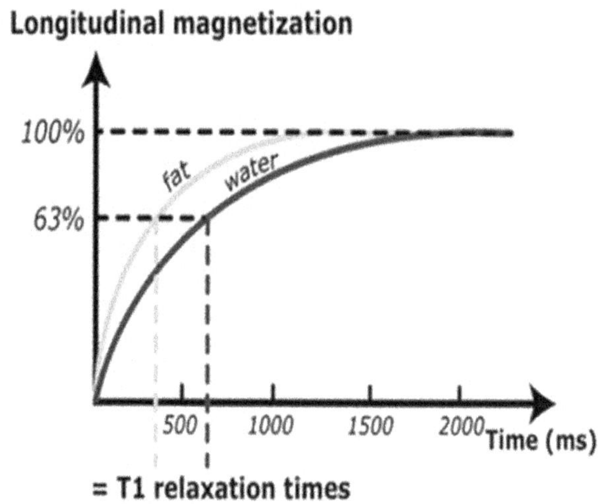

FIGURE 2.5 T1 relaxation curve.

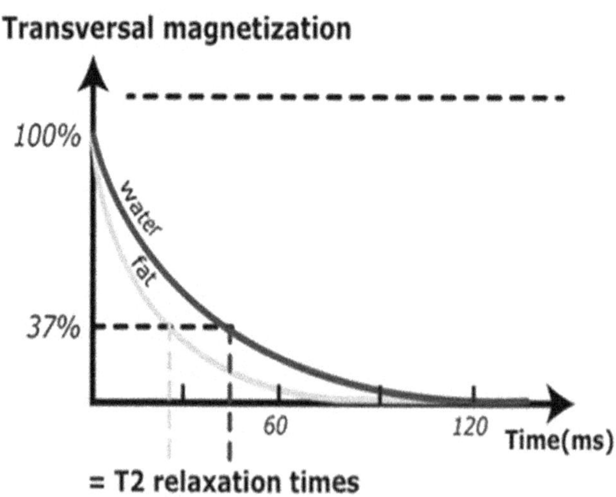

FIGURE 2.6 T2 relaxation curve.

Figures 2.5 and 2.6 show T1 and T2 relaxation curves for fat and water, respectively. From the figures, we can observe that fat and water contain individual relaxation curves. The T1 and T2 relaxation times of water are high when compared to those of fat.

2.5 T1-WEIGHTED VERSUS T2-WEIGHTED IMAGES

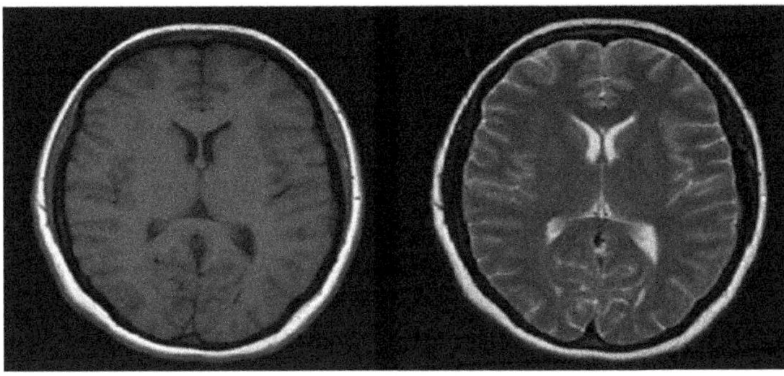

FIGURE 2.7 T1 versus T2 MRI.

Even though many kinds of MRIs are available, T1-weighted and T2-weighted are the two fundamental types of MRI images, often called T1 and T2 images, which are given in Figure 2.7. In T1 images, fat is white, and in T2 images, fat and water are white. In T1-weighted images, cerebrospinal fluid (CSF) is black, gray matter is gray, white matter is white, and bones are black. But in the case of T2-weighted images, CSF is white, gray matter is gray, white matter is dark gray, and bones are black. Table 2.1 shows the different tissue colors in T1-weighted and T2-weighted images.

TABLE 2.1 Different Tissue Colors in T1-Weighted and T2-Weighted Images.

Tissue	T1-weighted	T2-weighted
CSF	Dark	Bright
White matter	Light	Dark gray
Cortex	Gray	Light gray
Fat	Bright	Light
Inflammation	Dark	Bright

2.6 SPATIAL PREPROCESSING

SPM-12 follows spatial preprocessing. The main goals of spatial preprocessing are:

1. To match all scans of a particular subject.
2. To match scans of all subjects into standard space.

The most significant steps are:

1. Realign
2. Coregister
3. Normalization
4. Smoothing.

2.6.1 REALIGN

Realignment is most significant in preprocessing steps. Variations due to movement and shape differences between series of scans are rectified in this step. In the realignment step, it will compare our input with the mean image or first slice of the input. To correct these errors, rigid-body transformation[8] is used. It will allow movement and rotation only in the x, y, and z directions. It minimizes the sum of squared differences by trial and error method. It is used only within modalities.

2.6.2 COREGISTER

Coregister is used to match scans of different modalities, that is, T1 and T2 images can be matched by using this step. In this step, T1 and T2 images are matched using mutual information.[6] If we try to match the sum of squared differences, it will not give an output because the variation is high between T1 and T2 images. The matching is obtained by comparing the sharpness of the lines in the 2-D joint histogram. The joint histogram is formed by placing the gray level of the first scan in the x-axis and the gray level of the second scan in the y-axis, so two similar images would thus result in a diagonal line from bottom left to upper right. Similar to realigning, this step also allows only rigid-body transformation.

2.6.3 NORMALIZATION

This step is used to manipulate the scans into a standard space. SPM-12 uses a unified segmentation procedure. It combines three steps.

1. Segmentation
2. Bias correction
3. Spatial normalization[8]

Segmentation is the process of separating different tissue classes. Bias correction is the process of smoothly removing varying intensity differences among images. Spatial normalization is achieved by using deformation fields. Deformation fields are nothing but images used to quantify the displacement at each location in 3-D space. The bright color indicates the position needs to be shifted right, and the dark color indicates the location needs to be shifted left.

2.6.4 SMOOTHING

Smoothing is obtained by taking the average of every voxel with a weighted sum of its neighbors. The weighted sum is defined by the Gaussian kernel. The size of the Gaussian kernel is given by the full width half maximum (FWHM); the larger the FWHM, the more smoothing we can get. For MRI, we use Gaussian with a FWHM that is twice the voxel size. The reason to do smoothing is to correct slight functional anatomical differences between subjects. The amount of smoothing should be determined by our application. If our ROI is a very small area, then you cannot apply much smoothing because it will affect resolution.

2.7 SEGMENTATION

Segmentation[5] is a very important process in medical image analysis. Even though many segmentation algorithms are available, segmentation using SPM-12[7] software is very suitable for brain MRI segmentation. Segmentation based on tissue classes is used here. It will give proper segmentation of gray matter, white matter, CSF, the skull, and soft tissues separately.

2.8 RESULTS AND DISCUSSIONS

If 49 slices of brain MRI in DICOM format of a 6-day-old baby are given to the SPM-12 software as input, it will convert the images into two images in Nifti format. One corresponds to T1 images, and the second one corresponds to T2 images.

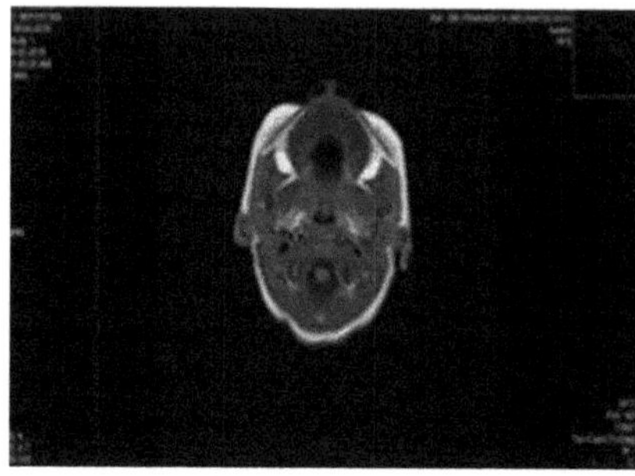

FIGURE 2.8 T1 image in DICOM format.

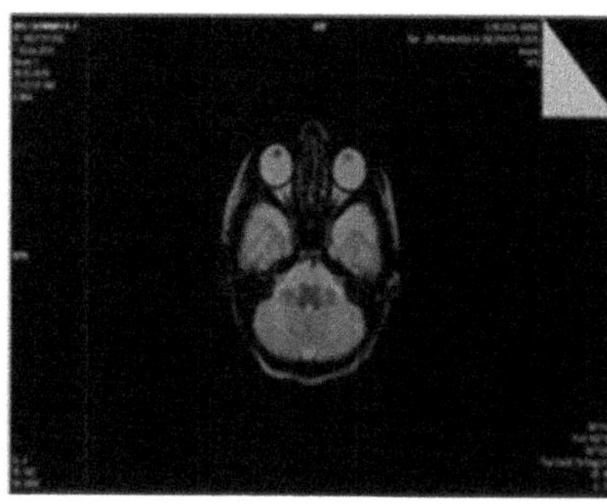

FIGURE 2.9 T2 image in DICOM format.

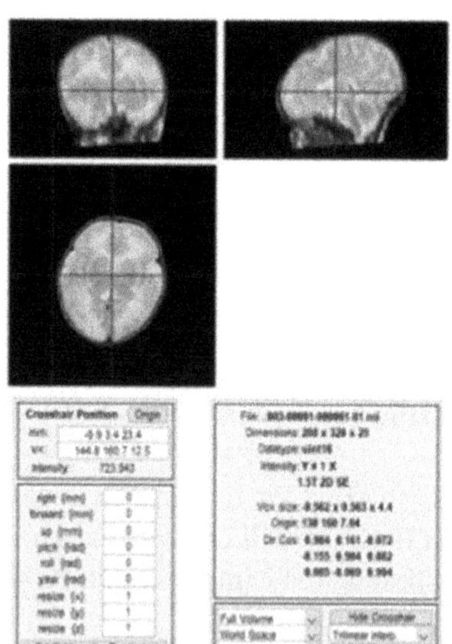

FIGURE 2.10 Nifti image from T1 images.

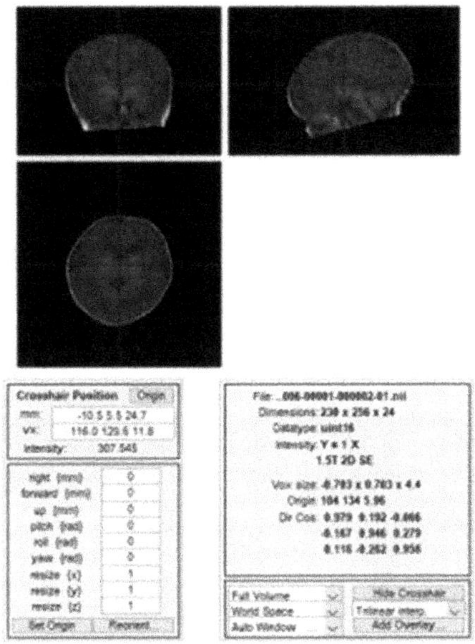

FIGURE 2.11 Nifti image from T2 images.

The input T1 and T2 images in DICOM format are given in Figures 2.8 and 2.9, and the input images in Nifti format are given in Figures 2.10 and 2.11.

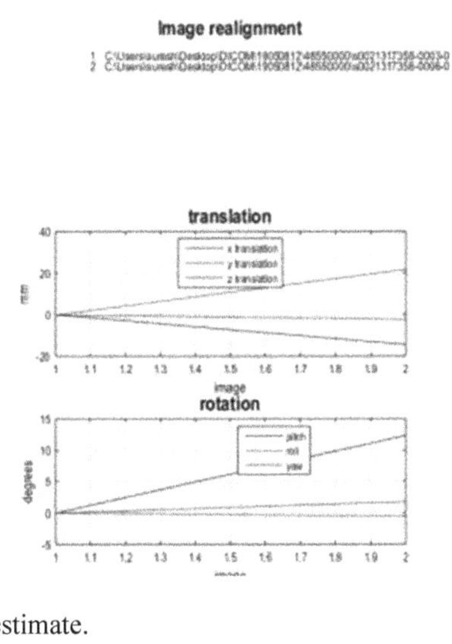

FIGURE 2.12 Realign estimate.

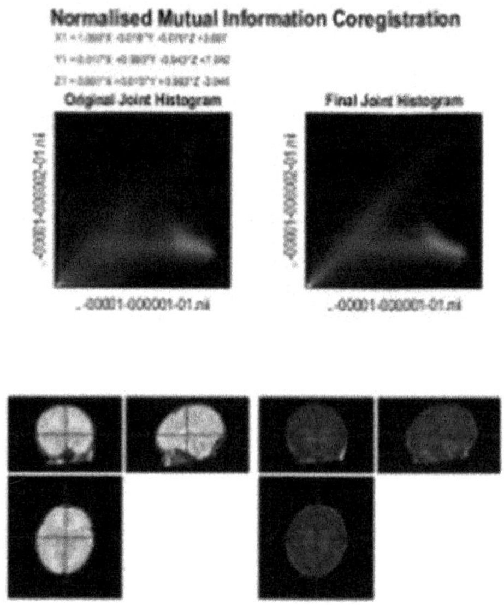

FIGURE 2.13 Joint histogram.

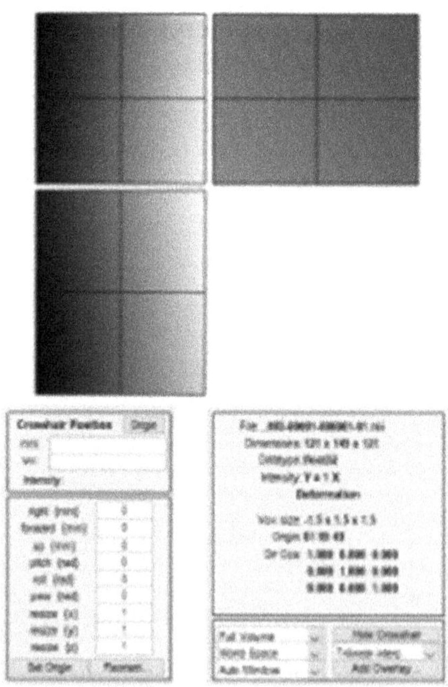

FIGURE 2.14 Deformation field from T1 image.

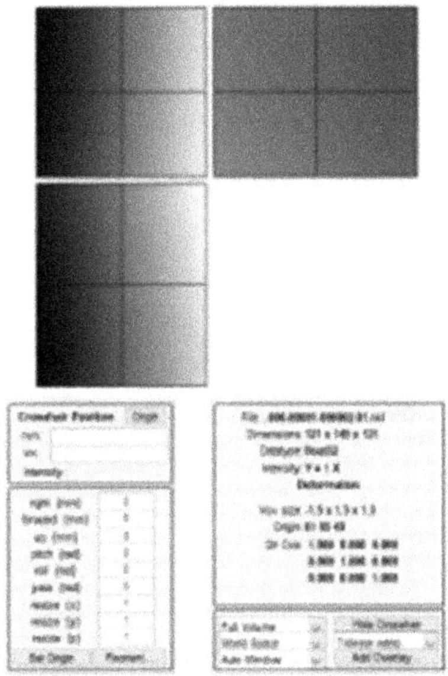

FIGURE 2.15 Deformation field.

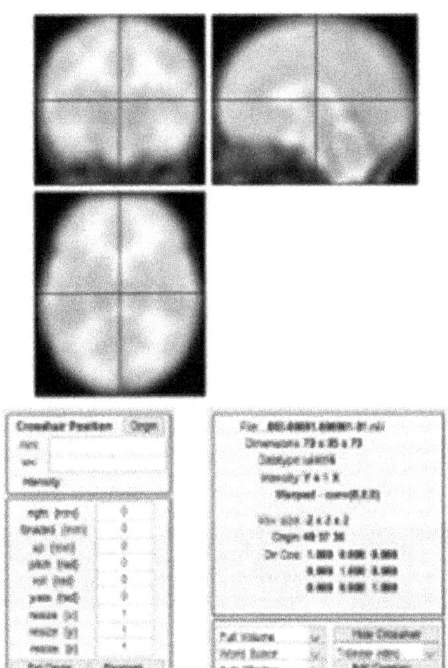

FIGURE 2.16 Smoothed Nifti image from T1 image.

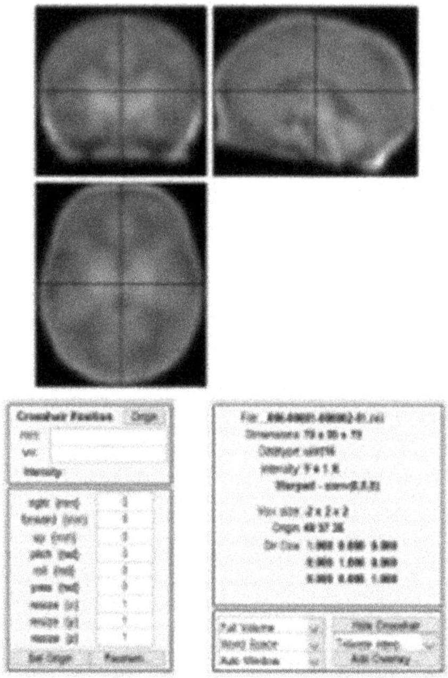

FIGURE 2.17 Smoothed Nifti image from T1 and T2 images.

The realign estimate output is as given in Figure 2.12. The joint histogram obtained during coregistration of T1 and T2 images is given in Figure 2.13. Deformation fields generated during the normalization process for T1 and T2 images are given in Figures 2.14 and 2.15. The smoothed images are given in Figures 2.16 and 2.17.

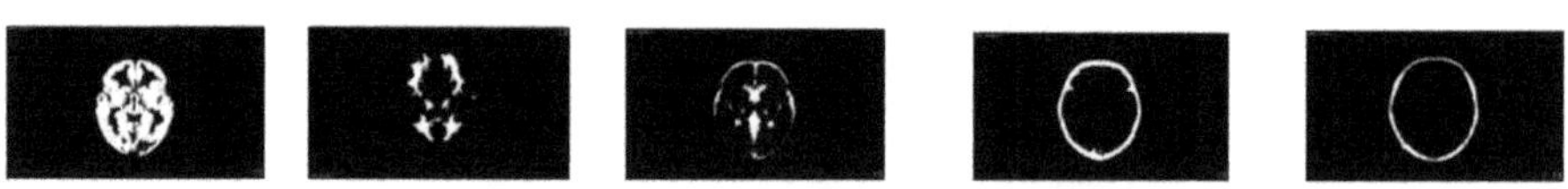

(a) gray matter (b) white matter (c) CSF (d) skull (e) soft tissue

FIGURE 2.18 (a–e) Segmented brain parts from T1 image without preprocessing.

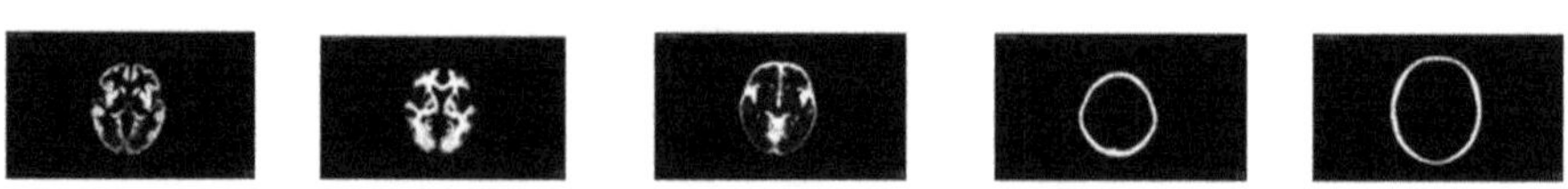

(a) gray matter (b) white matter (c) CSF (d) skull (e) soft tissue

FIGURE 2.19 (a–e) Segmented brain parts from T2 image without preprocessing.

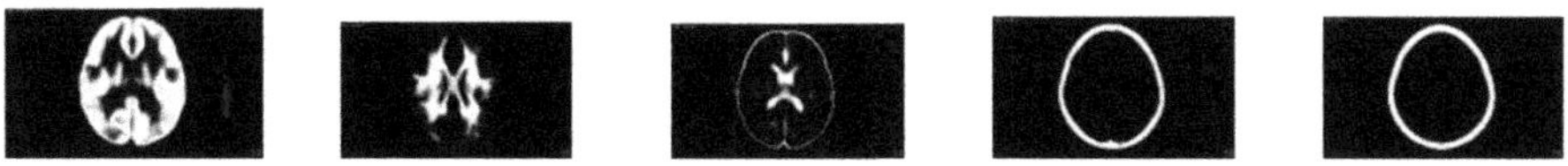

(a) gray matter (b) white matter (c) CSF (d) skull (e) soft tissue

FIGURE 2.20 (a–e) Segmented brain parts from T1 image with preprocessing.

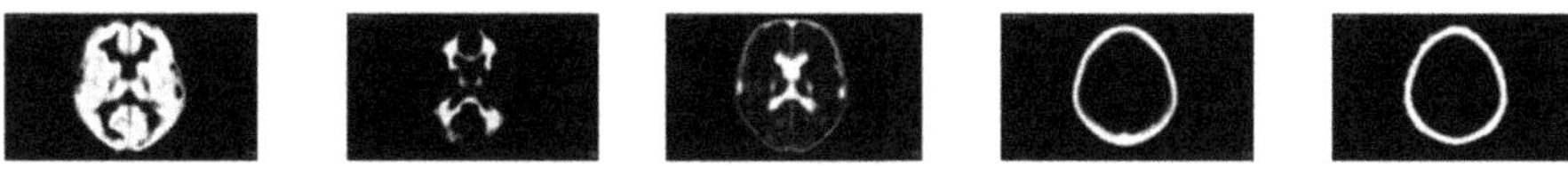

(a) gray matter (b) white matter (c) CSF (d) skull (e) soft tissue

FIGURE 2.21 (a–e) Segmented brain parts from T2 image with preprocessing.

The segmented brain parts for T1 and T2 images without preprocessing are as shown in Figures 2.18 (a–e) and 2.19 (a–e), and the segmented parts for T1 and T2 images with preprocessing are given in Figures 2.20 (a–e) and 2.21 (a–e). The outputs are in Nifti format only; for easy display, they are converted into DICOM using the Nifti to DICOM converter.

2.9 QUALITY METRICS

Image quality metrics[3,4] are used to analyze the quality of the image. Quality metrics are calculated for the segmented images, which are segmented before and after preprocessing. If quality metrics are maintained well, then only they are suitable for further analysis. MSE, PSNR, SNR, EPI, and SSIM are calculated for the brain parts segmented with and without preprocessing for T1 and T2 images.

The results are plotted as a graph as shown in Figure 2.22 (a–j), from which we can understand the effect of preprocessing. From the above images, we know that if we preprocess the MRI brain T1 and T2 images before segmentation, we can reduce the MSE and increase the PSNR, SNR, and SSIM, which are very important in the analysis of medical images. The EPI value is also maintained well. For example, the MSE for white matter in T1 is 653.2879 without preprocessing and 609.7175 with preprocessing. PSNR for gray matter in T1 is 20.1991 without preprocessing and 20.3984 with preprocessing. SNR for white matter in T1 is 4.3804 without preprocessing and 8.5689 with preprocessing. EPI for gray matter in T1 is 0.7253 without preprocessing and 0.6864 with preprocessing. The SSIM for white matter in T1 is 0.9784 without preprocessing and 0.9913 with preprocessing.

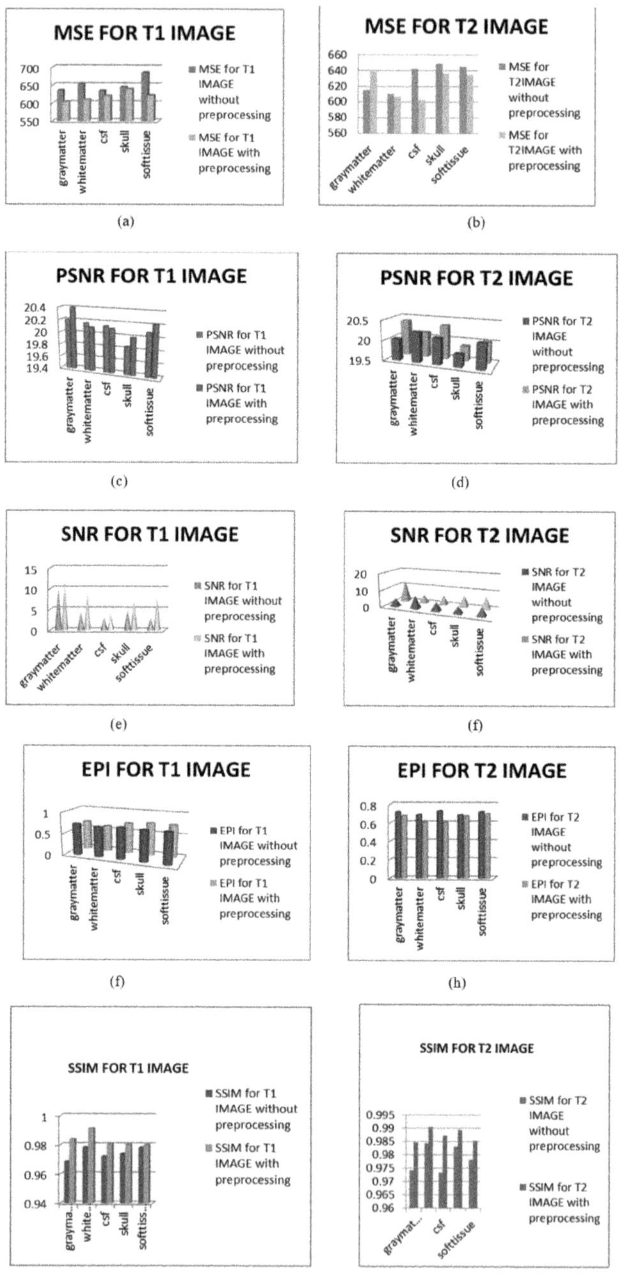

FIGURE 2.22 (a–j) Quality metrics comparison for T1 and T2 images with and without preprocessing.

2.10 CONCLUSIONS

In this chapter, T1 and T2 MRI brain images are segmented before and after preprocessing. Realignment, coregistration, normalization, and smoothing are the spatial preprocessing tools applied. With and without preprocessing, MSE, PSNR, SNR, EPI, and SSIM values are calculated for the segmented brain parts, and the results are analyzed. From the results, we can conclude that preprocessing is very important before the segmentation of the brain MRI.

KEYWORDS

- **preprocessing**
- **segmentation**
- **MSE**
- **SNR**
- **PSNR**
- **EPI**
- **SSIM**

REFERENCES

1. Gonzalez, R. G.; Woods, R. E. *Third Edition Digital Image Processing*; Pearson: South Asia, 2014.
2. Sridhar, S. *Digital Image Processing*; Oxford: India, 2011.
3. Najarian, K.; Splinter, R. *Biomedical Signal and Image Processing*; Taylor & Francis: New York, 2006.
4. Demirkaya, O.; Asyali, M. H.; Sahoo, P. K. *Image Processing with MATLAB® Applications in Medicine and Biology*; CRC Press: New York, 2009.
5. Semmlow, J. L. *Biosignal and Medical Image Processing*; CRC Press: New York, 2009.
6. Josien, P. W. P.; Antoine Maintz, J. B.; Viergever, M. A. Mutual-Information-Based Registration of Medical Images: A Survey. *IEEE Trans. Med. Imaging* **2003,** *22* (8), 986–1004.
7. Eklund, A.; Nichols, T.; Andersson, M.; Knutsson, H. Empirically Investigating the Statistical Validity of SPM, FSL and AFNI for Single Subject FMRI Analysis. *IEEE* **2015**, 1376–1380.
8. Muro, A.; Zapirain, B. G.; Méndez, A.; Ruiz, I.; FMRI Processing Tool for the Analysis, Parametrisation and Comparison of Preprocessed SPM Images. In *18th European Signal Processing Conference* 2010; pp 1335–1339.

CHAPTER 3

COMPARATIVE VOLUME ANALYSIS OF PEDIATRIC BRAIN WITH ADULT BRAIN USING T1 MRI IMAGES

S. JACILY JEMILA and A. BRINTHA THERESE

VIT University, Chennai, India

ABSTRACT

Segmentation of the pediatric brain is very useful in medical image analysis to supervise the brain's growth and development in infants. Atlas-based methods are used for automatic segmentation of a brain MRI. In this chapter, automatic segmentation of four pediatric and three adult brains is done using Brainsuite 19a brain imaging software, and the volumes of different brain parts are computed.

3.1 INTRODUCTION

Automatic segmentation[8] is a very tough task in pediatric brains due to the low SNR, partial volume effects, and large changes in the shape and appearance of the brain. Manual segmentation requires more time and cost, and the probability of error is also higher due to the variation of results from person to person. So there is a need for an accurate automatic segmentation tool. Brainsuite[13] 19a is such a type of automatic segmentation software that can segment the brain into 170 parts.

Computational Imaging and Analytics in Biomedical Engineering: Algorithms and Applications.
T. R. Ganesh Babu, PhD, U. Saravanakumar, PhD & Balachandra Pattanaik, PhD (Eds.)

3.2 MATERIALS AND METHODS

3.2.1 DATASETS

Real-time datasets of seven different patients are considered for this analysis. Among the seven, four are babies and three are adults. Real-time data that are in DICOM format are converted to Nifty format by using SPM-12, a neuroimaging software. Real-time datasets at the following ages are considered for volume analysis:

1. 6-day-old baby
2. 2-month-old baby 1
3. 2-month-old baby 2
4. 2 months, 20 days old baby
5. 42-year old adult
6. 46 years, 7 months 13 days old adult
7. 47-year-old adult

3.2.2 HUMAN BRAIN'S IMPORTANT PARTS AND FUNCTIONS

The surface area of the brain is around 233–465 square inches (1500–2000 cm^2). To fit this surface area to the skull, the cortex is folded, and sulci (grooves) and gyri (folds) are formed. The cerebral cortex, or cortical surface, is divided into four major lobes. They are the frontal lobe, parietal lobe, occipital lobe, and temporal lobe. Each lobe has a different function. The brain is divided into left and right halves using the interhemispheric fissure, which is a large groove. The corpus callosum is used to facilitate communication between these two halves. Also, the right and left temporal lobes communicate through the anterior commissure, which is a tract of fiber. The cortical area above the corpus callosum is divided by a groove called the cingulate sulcus. The area between the groove and the corpus callosum is called the cingulate gyrus.

Some important areas of the brain and their functions:

- Parietal lobe—it will receive and respond to somatosensory input (pain and touch).
- Frontal lobe—it is involved in motor skills (including speech) and cognitive functions.
- Occipital lobe—it receives and processes visual information directly from the eyes.

- Temporal lobe—it processes auditory information from the ears.
- Basal ganglia—located within the temporal lobe, coordinate fine motions, such as fingertip movements.
- Limbic system—important in emotional behavior and controlling movements of muscles of the digestive tract and body cavities.
- Hippocampus—it is important for short-term memory.

Amygdala—controls social and sexual behavior and other emotions. Insula—influences automatic functions of the brainstem, processes taste information, and separates the temporal and frontal lobes.

3.2.3 CORTICAL SURFACE EXTRACTION

Cortical surface extraction[13] is the first step in volume calculation using Brainsuite 19a. The cortical surface is the exterior surface of the brain. Some important steps in cortical surface extraction are skull stripping, nonuniformity correction, tissue classification, cerebrum labeling, topology correction, and wisp removal.

3.2.3.1 SKULL STRIPPING

The first step in many of the MRI analysis sequences is the skull stripping.[9] In T1 images, the edges between the skull and brain regions are well-defined. The brain surface extractor (BSE) is used to extract the brain region in Brainsuite 19a. It performs the following steps: anisotropic diffusion, edge detection, and morphological operations for brain surface extraction. Anisotropic diffusion applies smoothing to lower contrast edges while retaining the higher contrast edges. For edge detection, it uses the Marr–Hildreth edge detection operator, which applies Gaussian blur followed by a Laplacian operator. In morphological operations, first erosion is applied and then dilation is applied; these are used to eliminate the noise-related connections between the scalp and brain.

3.2.3.2 NONUNIFORMITY CORRECTION

The intensity of nonuniformity is always present in most of the MRI images due to variation in the magnetic field. To correct this problem, bias field correction (BFC) is applied. BFC estimates the correction field based on the tissue gain variation.

3.2.3.3 TISSUE CLASSIFICATION

This is done by a partial volume classifier (PVC). PVC assigns integer values for every voxel based on the tissue type. It considers the following tissue types: background voxels, cerebrospinal fluid, gray matter, and white matter. PVC models the brain as a set of piecewise continuous regions of single tissue types bounded by partial volume contributions. In this method, the neighborhood of voxels should be similar.

3.2.3.4 CEREBRUM LABELING

Once the brain region is extracted and classified, then it is separated into cerebellum, cerebrum, and other structures to enable identification of the cerebral cortex. To achieve this, volumetric registration is needed for an MRI image.

3.2.3.5 TOPOLOGY CORRECTION

The boundary of the cortex should be topologically a sphere. It should not have any holes or handles. But because of some segmentation errors, there may be some holes or handles. A graph-based algorithm is applied to force the segmented voxels to have a spherical topology.

3.2.3.6 WISP REMOVAL

The wisp removal step is used to remove the segmentation errors in the brain extraction process by removing thin, wispy structures. It decomposes the binary mask into a graph, separating weakly connected components.

3.2.4 SURFACE/VOLUME REGISTRATION

Surface and volume registration[14] (SvReg) is used to coregister our brain image with a standard atlas. For coregistration, it uses anatomical information present in the image. SvReg is very helpful for regional analysis. SvReg uses a series of multistep registration and refinement processes. We use Brainsuit atlas1 for SvReg. It is based on the Colin 27 atlas. It is an average of 27 scans of an individual.

3.2.5 VOLUME ESTIMATION

SvReg of Brainsuite 19a gives a cortical surface with segmented regions. Each segmented region is differentiated by a different color. We can check the volume for every segmented region by clicking on the ROI details.

3.3 RESULTS AND DISCUSSION

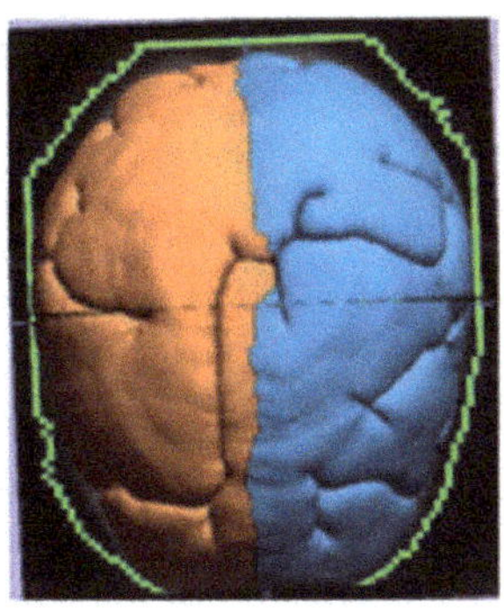

FIGURE 3.1 Cortical surface extraction output for an adult.

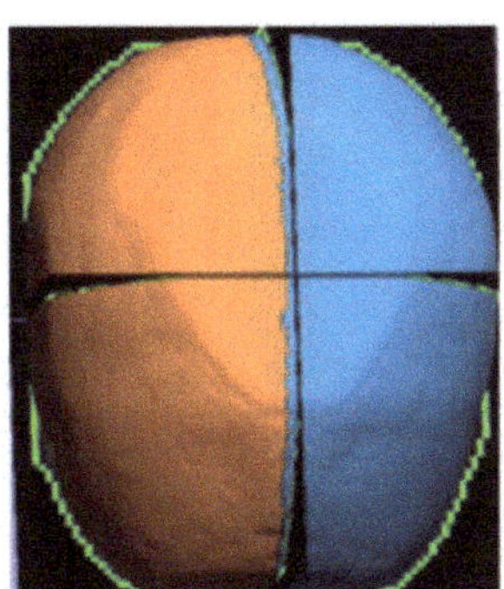

FIGURE 3.2 Cortical surface extraction output for a baby.

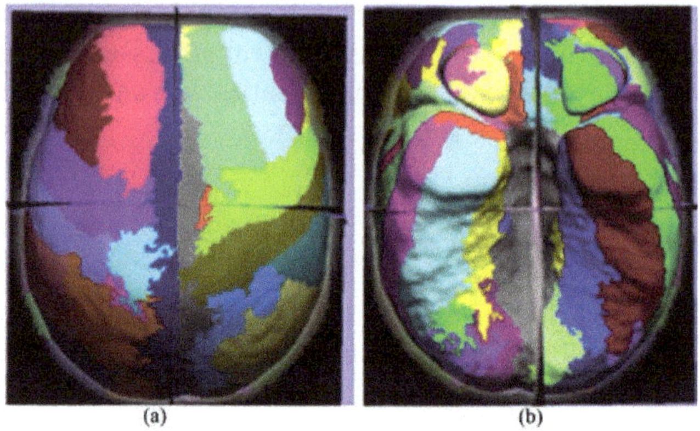

FIGURE 3.3 SvReg output of a 6-day-old baby: (a) front view and (b) back view.

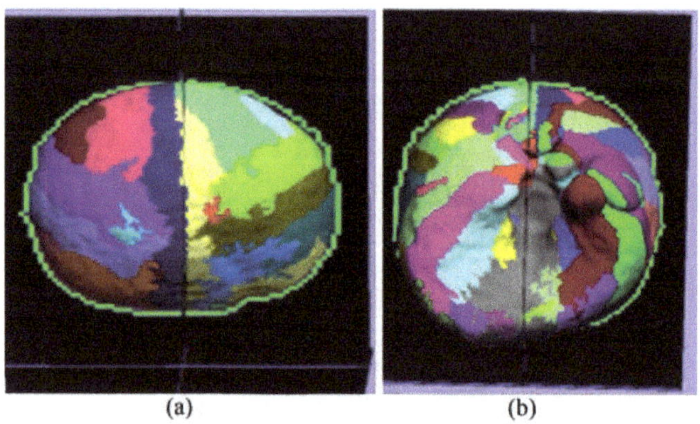

FIGURE 3.4 SvReg output of a 2-month-old baby 1: (a) front view and (b) back view.

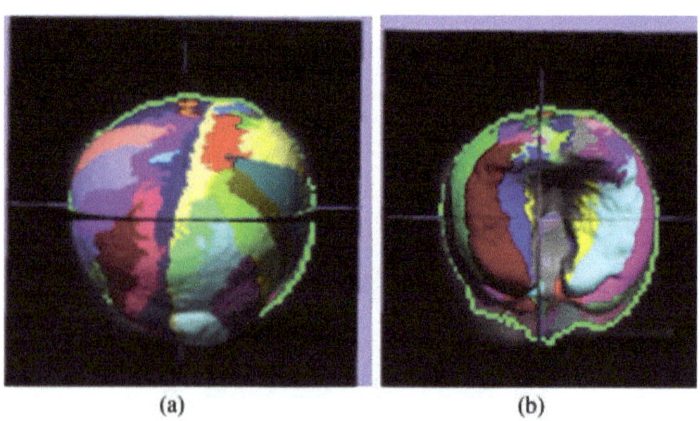

FIGURE 3.5 SvReg output of a 2-month-old baby 2: (a) front view and (b) back view.

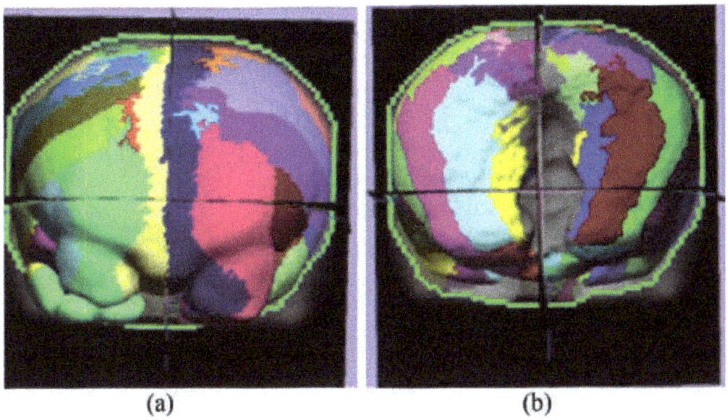

FIGURE 3.6 SvReg output of a 2 months 20 days old baby: (a) front view and (b) back view.

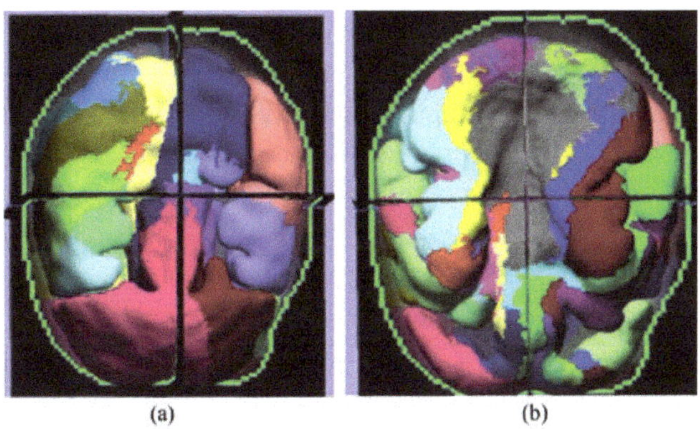

FIGURE 3.7 SvReg output of a 42 years adult: (a) front view and (b) back view.

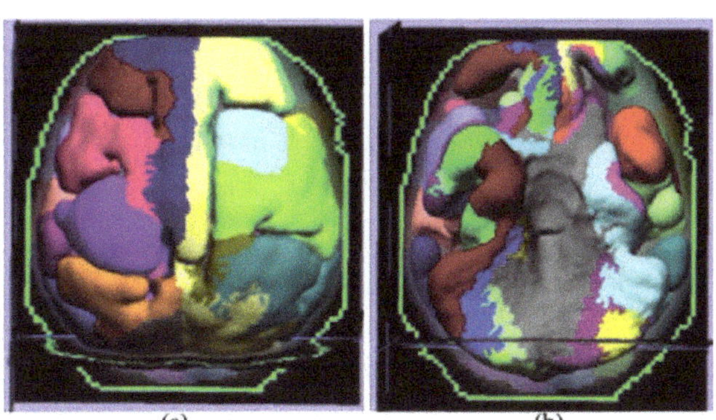

FIGURE 3.8 SvReg output of a 46 years, 7 months 13 days adult: (a) front view and (b) back view.

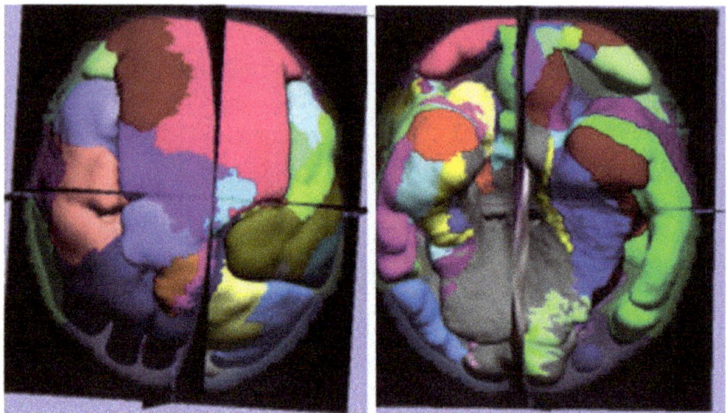

FIGURE 3.9 SvReg output of a 47 years adult: (a) front view and (b) back view.

The outputs of cortical surface extraction for an adult and a baby are given in Figures 3.1 and 3.2. When compared to a baby, there are some curves on the surface. The outputs for SvReg are given in Figures 3.3–3.9 for different datasets. Different colors in the image indicate different regions of the brain. Volumes corresponding to these different regions are given in Table 3.1.

TABLE 3.1 Volumes Corresponding to Different Brain Regions for Different Datasets.

Sl. no	Brain parts	6 days baby	2 months baby1	2 months baby 2	2 month 20 days baby	46 years, 7 months 13 days adult	42 years adult	47 years adult
1	Background	2099.84814	3589.632	3745.7	2766.4321	2622.3	2784.9	2601.41
2	R. caudate nucleus	7.79200029	1.568	0.648	10.72	3.176	7.416	2.304
3	L. caudate nucleus	13.3520002	1.6	1.352	10.68	1.088	1.528	5.808
4	R. putamen	7.46400023	4.648	2.992	6.1040001	2.896	3.688	4.736
5	L. putamen	5.83200026	3.528	2.616	4.48	2.904	2.416	4.96
6	R. globus pallidus	9.54400063	3.2	4.12	6.8480005	1.192	2.464	3.056
7	L. globus pallidus	8.03200054	2.192	4.16	5	1.2	1.528	3.384
8	R. nucleus accumbens	0.712000012	0.248	0.272	0.288	0.224	0.112	0.16
9	L. nucleus accumbens	0.624000013	0.28	0.192	0.184	0.128	0.192	0.344

TABLE 3.1 *(Continued)*

Sl. no	Brain parts	6 days baby	2 months baby1	2 months baby 2	2 month 20 days baby	46 years, 7 months 13 days adult	42 years adult	47 years adult
10	R. claustrum	0.880000055	0.456	0.32	0.7920001	1.256	0.512	0.688
11	L. claustrum	1.39200008	0.624	0.472	1.0400001	0.608	0.888	1.016
12	R. thalamus	76.776001	37.464	28.296	64.632004	18.048	14.944	14.088
13	L. thalamus	85.7360077	36.048	27.4	57.816002	19.656	23.888	19.856
14	R. basal forebrain	4.71200037	1.848	1.584	1.7600001	1.656	1.192	1.872
15	L. basal forebrain	4.98400021	3.024	1.592	1.7360001	1.616	1.464	2.016
16	R. lateral geniculate nucleus	0.296000004	0.064	0.12	0.168	0.328	0.104	0.2
17	L. lateral geniculate nucleus	0.168000013	0.152	0.12	0.128	0.224	0.208	0.072
18	R. medial geniculate nucleus	0.248000011	0.136	0.144	0.176	0.208	0.04	0.232
19	L. medial geniculate nucleus	0.280000001	0.064	0.12	0.064	0.04	0.152	0.104
20	R. superior colliculus	0.440000027	0.048	0.048	0.032	0.408	0.216	0.12
21	L. superior colliculus	0.192000002	0.04	0.088	0.144	0.28	0.328	0.08
22	R. inferior colliculus	0.320000023	0.016	0.032	0.024	0.264	0.232	0.08
23	L. inferior colliculus	0.104000002	0.024	0.032	0.008	0.408	0.376	0.208
24	R. mamillary body	0.256000012	0.048	0	0.016	0.016	0.408	0.056
25	L. mamillary body	0	0	0	0	0.024	0.136	0
26	L. ventricular system	4.11199999	0	0	3.6240001	0	0	0
27	R. lateral ventricle	1.39200008	0.04	0	2.296	2.128	5.776	3.4
28	Third ventricle	3.57600021	2.168	0.872	1.9840001	3.264	4.424	6.312

TABLE 3.1 *(Continued)*

Sl. no	Brain parts	6 days baby	2 months baby1	2 months baby 2	2 month 20 days baby	46 years, 7 months 13 days adult	42 years adult	47 years adult
29	Brainstem	49.4880028	23.776	39.648	31.504002	49.424	49.376	56.256
30	Cerebellum	171.544006	83.464	207.01	83.952004	171.64	156.66	268.704
31	R. superior frontal gyrus (gm)	22.2480011	14.048	6.144	47.392002	34.928	34.24	11.592
32	L. superior frontal gyrus (gm)	24.3520012	13.328	6.184	45.016003	59.192	61.568	89.08
33	R. middle frontal gyrus (gm)	19.8800011	6.344	6.36	21.912001	23.688	12.072	5.568
34	L. middle frontal gyrus (gm)	24.2640018	9.200001	1.968	15.376	41.744	37.96	32.088
35	R. pars opercularis (gm)	8.27200031	3.896	2.28	12.664001	15.024	4	4.624
36	L. pars opercularis (gm)	5.50400019	4.4	1.056	6.3360004	3.704	4.904	10.856
37	R. pars triangularis (gm)	15.9040003	5.992	6.208	17.584002	9.704	0.856	2.328
38	L. pars triangularis (gm)	10.2160006	3.288	4.008	20.968	6.968	24.456	21.584
39	R. pars orbitalis (gm)	2.08800006	0	2.36	0	0.288	0	0.016
40	L. pars orbitalis (gm)	2.00800014	0	0.912	0.784	0.744	1.496	1.424
41	R. precentral gyrus (gm)	16.2880001	12.768	5.24	20.576	81.44	21.224	13.16
42	L. precentral gyrus (gm)	16.9280014	13.544	9.736	20.808001	78.952	22.672	31.552
43	R. transverse frontal gyrus (gm)	0.30400002	0	0.344	1.6	0.976	0.064	0.008

TABLE 3.1 *(Continued)*

Sl. no	Brain parts	6 days baby	2 months baby1	2 months baby 2	2 month 20 days baby	46 years, 7 months 13 days adult	42 years adult	47 years adult
44	L. transverse frontal gyrus (gm)	0.552000046	0	0.136	0.424	0.392	0.928	1.136
45	R. gyrus rectus (gm)	5.74400043	0.112	2.768	2.9200001	1.776	1.056	1.712
46	L. gyrus rectus (gm)	13.8160009	0.4	2.088	2.9360001	2.368	2.08	7.016
47	R. middle orbitofrontal gyrus (gm)	2.00800014	0	0.112	0.136	0.064	0.056	0
48	L. middle orbitofrontal gyrus (gm)	2.72800016	0.056	1.032	0.536	0.152	0.688	3.224
49	R. anterior orbitofrontal gyrus (gm)	0.856000066	0	0.08	0.312	0.608	0.016	0.248
50	L. anterior orbitofrontal gyrus (gm)	1.1680001	0	0.224	0.096	0.088	0.296	1.656
51	R. posterior orbitofrontal gyrus (gm)	1.46400011	0	2.152	1.192	0.472	0.264	1.616
52	L. posterior orbitofrontal gyrus (gm)	2.28000021	0.152	3.344	2.4160001	0.176	2.008	4.768
53	R. lateral orbitofrontal gyrus (gm)	0	0	0	0	0.344	0	0
54	L. lateral orbitofrontal gyrus (gm)	3.16800022	0.008	1.784	0.016	0.088	0.168	0.824
55	R. paracentral lobule (gm)	0	0.504	0.52	0.7040001	0.816	1.448	5.032
56	L. paracentral lobule (gm)	2.53600001	0.816	6.856	0	0.184	1.24	1.272
57	R. cingulate gyrus (gm)	16.3439999	9.072001	1.76	34.816002	60.432	31.136	39.888
58	L. cingulate gyrus (gm)	34.8160019	16.592	4.384	35.400002	21.936	99.088	70.808

TABLE 3.1 *(Continued)*

Sl. no	Brain parts	6 days baby	2 months baby1	2 months baby 2	2 month 20 days baby	46 years, 7 months 13 days adult	42 years adult	47 years adult
59	R. subcal-losal gyrus (gm)	2.57600021	0.552	1.064	1.9680001	0.768	0	0
60	L. subcal-losal gyrus (gm)	2.83200002	0.072	0.184	0.368	0.056	0.552	0
61	R. postcentral gyrus (gm)	18.6880016	11.048	3.928	17.856001	8.056	20.696	31.544
62	L. postcentral gyrus (gm)	20.2000008	12.728	8.264	28.672001	41.288	70.304	52.088
63	R. supramar-ginal gyrus (gm)	13.5360003	12	6.32	16.296001	87.256	32.552	19.304
64	L. supra-marginal gyrus (gm)	20.4160004	9.712001	8.52	11.816001	23.104	78.728	57.832
65	R. angular gyrus (gm)	28.5040016	12.784	5.808	18.240002	20.248	23.864	3.112
66	L. angular gyrus (gm)	10.9360008	4.576	4.76	7.5360003	47.304	8.6	40.72
67	R. superior parietal gyrus (gm)	14.3840008	5.08	3.192	8.1280003	9.88	32.848	51.376
68	L. superior parietal gyrus (gm)	20.0400009	8.496	3.472	9.0880003	41.528	5.272	9.896
69	R. precuneus (gm)	0.024	0.816	0.952	6.9440002	6.728	1.128	1.072
70	L. precuneus (gm)	0.448000014	0.368	0.04	1.184	2.624	1.408	2.328
71	R. temporal pole (gm)	6.19200039	0	1.072	1.368	28.688	2.16	8.76
72	L. temporal pole (gm)	2.84800005	0	1.04	1.2160001	0.392	0.096	1.416
73	R. superior temporal gyrus (gm)	12.6560001	5.056	8.24	11.568001	38.688	18.944	13.832

TABLE 3.1 *(Continued)*

Sl. no	Brain parts	6 days baby	2 months baby1	2 months baby 2	2 month 20 days baby	46 years, 7 months 13 days adult	42 years adult	47 years adult
74	L. superior temporal gyrus (gm)	12.368001	3.92	3	4.5360003	13.432	8.84	23.16
75	R. transverse temporal gyrus (gm)	2	0.568	0.384	1.016	2.096	5.88	1.584
76	L. transverse temporal gyrus (gm)	1.50400007	0.968	0.512	0.776	1.352	1.368	10.096
77	R. middle temporal gyrus (gm)	30.1680012	3.888	11.584	9.1440001	30.488	12.872	15.2
78	L. middle temporal gyrus (gm)	33.0320015	5.44	8.248	10.024	34.848	31.712	59.792
79	R. inferior temporal gyrus (gm)	22.0480003	0.44	6.32	0.488	26.632	26.184	8.096
80	L. inferior temporal gyrus (gm)	23.6400013	0.28	7.896	1.0320001	36.888	27.368	25.792
81	R. fusiforme gyrus (gm)	2.15200019	0	0	0.032	0.688	5.256	6.448
82	L. fusiforme gyrus (gm)	3.0480001	0	0.032	0.024	0.824	2.384	8.112
83	R. parahip-pocampal gyrus (gm)	0	0	0	0	0.032	0.192	0.368
84	L. parahip-pocampal gyrus (gm)	0	0	0	0	0.408	0.968	0.592
85	R. hippo-campus (gm)	0.472000033	2.232	0.392	0.168	1.464	1.28	1.648
86	L. hippo-campus (gm)	0.248000011	1.248	0	0.44	1.56	1.576	0.752
87	R. amygdala (gm)	0.344000012	0.712	0.352	0.328	0.176	0.664	0.072
88	L. amygdala (gm)	0.527999997	1.12	0.048	0.272	0.344	0.488	0.912

TABLE 3.1 *(Continued)*

Sl. no	Brain parts	6 days baby	2 months baby1	2 months baby 2	2 month 20 days baby	46 years, 7 months 13 days adult	42 years adult	47 years adult
89	R. superior occipital gyrus (gm)	2.28000021	0.688	0.568	3.3200002	0.336	0.64	0.92
90	L. superior occipital gyrus (gm)	0.104000002	0.352	0.096	0.6160001	0.672	0.864	0.928
91	R. middle occipital gyrus (gm)	5.18400002	3.016	0.392	6.5120001	2.592	3.184	0.744
92	L. middle occipital gyrus (gm)	3.88800025	1.68	0.464	4.1680002	11.888	3.752	6.824
93	R. inferior occipital gyrus (gm)	3.08000016	0.808	0.336	1.6160001	21.608	10.32	1.336
94	L. inferior occipital gyrus (gm)	0.088000007	0.016	0.08	0.288	3.776	3.968	1.888
95	R. lingual gyrus (gm)	2.08800006	0.624	0.4	1.5520001	6.576	1.008	12.064
96	L. lingual gyrus (gm)	0.056000002	0.128	0.312	0.152	3.76	2.088	6.848
97	R. cuneus (gm)	2.60000014	3.008	0.728	1.192	2.992	11.656	7.744
98	L. cuneus (gm)	0.232000008	0.744	1.208	0.44	0.904	7.528	5.072
99	R. insula (gm)	9.00800037	1.488	6.416	10.512	5.632	9.696	7.128
100	L. insula (gm)	9.84000015	1.992	4.152	10.8	10.144	9.352	6.568
101	White matter (cerebrum)	1082.71204	590.944	384.35	753.28803	300.67	378.38	363.76
102	R. superior frontal gyrus (wm)	8.40799999	5.192	1.984	37.224003	12.176	13.624	9.24
103	L. superior frontal gyrus (wm)	8.3760004	4.12	2.472	48.224003	23.376	39.424	53.192
104	R. middle frontal gyrus (wm)	5.96000051	2.704	2.952	8.1920004	10.864	7.016	2.816

TABLE 3.1 *(Continued)*

Sl. no	Brain parts	6 days baby	2 months baby1	2 months baby 2	2 month 20 days baby	46 years, 7 months 13 days adult	42 years adult	47 years adult
105	L. middle frontal gyrus (wm)	5.67200041	3.24	1.648	8.5520001	18.184	20.752	15.744
106	R. pars opercularis (wm)	2.28800011	1.328	1.432	3.1120002	7.928	0.8	0.752
107	L. pars opercularis (wm)	2.3360002	1.44	2.336	2.664	3.496	1.344	4.696
108	R. pars triangularis (wm)	5.44000006	1.648	3.704	6.2960005	0.936	1.144	1.056
109	L. pars triangularis (wm)	3.03200006	0.768	1.304	5.0640001	3.912	14.76	8.816
110	R. pars orbitalis (wm)	0.088000007	0	0.624	0	0	0	0
111	L. pars orbitalis (wm)	0.256000012	0	0.048	0.112	0	0.264	0.272
112	R. precentral gyrus (wm)	9.20800018	6.208	2.48	13.696001	41.288	8.464	7.648
113	L. precentral gyrus (wm)	12.1760006	6.72	2.528	21.224001	54.16	6.008	23.184
114	R. transverse frontal gyrus (wm)	0.048	0.04	0.272	0.48	0.32	0	0
115	L. transverse frontal gyrus (wm)	0.184000015	0	0.16	0.064	0.096	0.112	0.296
116	R. gyrus rectus (wm)	3.72000027	0.04	0.408	0.448	0.184	0.008	0.048
117	L. gyrus rectus (wm)	5.61600018	0.216	0.16	0.536	0.248	0.768	1.192
118	R. middle orbitofrontal gyrus (wm)	0.576000035	0	0	0	0	0	0

TABLE 3.1 *(Continued)*

Sl. no	Brain parts	6 days baby	2 months baby1	2 months baby 2	2 month 20 days baby	46 years, 7 months 13 days adult	42 years adult	47 years adult
119	L. middle orbitofrontal gyrus (wm)	1.44800007	0.016	0.312	0.032	0.016	0.112	0.616
120	R. anterior orbitofrontal gyrus (wm)	0.064000003	0.016	0.32	0	0.176	0	0
121	L. anterior orbitofrontal gyrus (wm)	0.368000031	0	0.032	0	0	0.016	0.144
122	R. posterior orbitofrontal gyrus (wm)	0.160000011	0	0.144	0.04	0	0	0
123	L. posterior orbitofrontal gyrus (wm)	0.45600003	0	0.072	0.072	0	0.016	0.352
124	R. lateral orbitofrontal gyrus (wm)	0.30400002	0.008	0.16	0	0.336	0	0
125	L. lateral orbitofrontal gyrus (wm)	0.496000022	0.008	0.096	0	0.024	0.096	0.216
126	R. paracentral lobule (wm)	2.18400002	0.296	0.976	2.3760002	1.064	0.48	3.176
127	L. paracentral lobule (wm)	10.9520006	0.92	0.224	0.936	0.216	0.952	1.888
128	R. cingulate gyrus (wm)	6.64800024	3.968	0	37.512001	28.912	13.512	19
129	L. cingulate gyrus (wm)	93.3280029	8.848001	2.112	62.952004	15.808	62.272	43.568
130	R. subcal-losal gyrus (wm)	0.84800005	0.336	0.496	0.8800001	0.112	0	0
131	L. subcal-losal gyrus (wm)	1.24800003	0.056	0.072	0.592	0.008	0.12	0
132	R. postcentral gyrus (wm)	12.2720003	5.688	1.432	13.384001	6.136	11.08	20.376
133	L. postcentral gyrus (wm)	17.8400002	8.280001	6.776	21.192001	26.128	46.936	37.944

TABLE 3.1 *(Continued)*

Sl. no	Brain parts	6 days baby	2 months baby1	2 months baby 2	2 month 20 days baby	46 years, 7 months 13 days adult	42 years adult	47 years adult
134	R. supramarginal gyrus (wm)	9.14400005	6.976	3.592	10.264001	42.072	13.144	11.456
135	L. supramarginal gyrus (wm)	25.6240005	6.528	5.624	8.3440008	10.76	52.856	42.232
136	R. angular gyrus (wm)	20.8480015	4.192	1.32	2.7840002	14.216	9.648	1.368
137	L. angular gyrus (wm)	18.960001	1.608	0.536	1.8160001	17.928	4.312	25.992
138	R. superior parietal gyrus (wm)	20.6640015	4.36	2.456	11.152	5.088	17.888	23.824
139	L. superior parietal gyrus (wm)	47.5600014	4.984	1.816	12.504001	24.728	1.528	5.424
140	R. precuneus (wm)	3.17600012	1.36	0.464	7.3520002	1.064	0.856	1.088
141	L. precuneus (wm)	1.37600005	0.656	0.04	1.4080001	3.024	1.04	1.128
142	R. temporal pole (wm)	0.224000007	0	0.072	0.5840001	9.048	0.32	0.384
143	L. temporal pole (wm)	0.184000015	0	0.456	0.48	0.016	0	0.264
144	R. superior temporal gyrus (wm)	6.54400015	1.656	2.848	2.6960001	17.456	8.168	7.224
145	L. superior temporal gyrus (wm)	6.6880002	1.584	1.16	1.0400001	5.336	4.984	16.664
146	R. transverse temporal gyrus (wm)	0.968000054	0.256	0.576	0.24	1.152	1.304	0.688
147	L. transverse temporal gyrus (wm)	1.2240001	0.528	0.328	0.216	0.152	0.984	7.656

TABLE 3.1 *(Continued)*

Sl. no	Brain parts	6 days baby	2 months baby1	2 months baby 2	2 month 20 days baby	46 years, 7 months 13 days adult	42 years adult	47 years adult
148	R. middle temporal gyrus (wm)	5.87200022	2.104	3.688	4.5520001	11.144	1.712	4.824
149	L. middle temporal gyrus (wm)	7.14400053	1.76	2.68	3.0640001	6.624	16.08	28.4
150	R. inferior temporal gyrus (wm)	8.68000031	1.256	2.816	3.3760002	10.048	7.856	3.328
151	L. inferior temporal gyrus (wm)	4.60000038	0.688	3.352	2.296	10.928	8.928	7.664
152	R. fusiforme gyrus (wm)	7.96800041	0.456	2.8	2.1920002	2.048	5.328	2.568
153	L. fusiforme gyrus (wm)	4.4000001	0.592	2.304	1.8480001	3.328	4.272	4.912
154	R. parahip-pocampal gyrus (wm)	0.504000008	0	0.072	0.016	0.248	0	2.456
155	L. parahip-pocampal gyrus (wm)	0.496000022	0.136	0.456	0.064	1.768	0.312	1.144
156	R. hippo-campus (wm)	1.12800002	0.464	1.16	1.6	0.512	1.2	4.24
157	L. hippo-campus (wm)	1.91200006	2.152	0.4	0.896	0.616	0.752	1.136
158	R. amygdala (wm)	0.600000024	0.256	0.232	0.7600001	0	0.488	0.568
159	L. amygdala (wm)	0.520000041	0.128	0.48	0.92	0	0.008	0.248
160	R. superior occipital gyrus (wm)	4.96000004	0.944	1.008	3.0240002	0.088	0.472	2.16
161	L. superior occipital gyrus (wm)	1.48800004	0.336	0.24	0.32	1.424	0.608	1.096
162	R. middle occipital gyrus (wm)	10.0400009	1.656	1.192	3.0480001	6.048	3.432	0.808

TABLE 3.1 *(Continued)*

Sl. no	Brain parts	6 days baby	2 months baby1	2 months baby 2	2 month 20 days baby	46 years, 7 months 13 days adult	42 years adult	47 years adult
163	L. middle occipital gyrus (wm)	3.84000015	1.512	0.376	3.2640002	12.752	1.84	5.4
164	R. inferior occipital gyrus (wm)	7.04800034	0.856	1.024	2.776	6.048	4.056	0.288
165	L. inferior occipital gyrus (wm)	1.82400012	1.144	0.56	0.8800001	1.288	1.984	1.496
166	R. lingual gyrus (wm)	16.6160011	2.184	4.656	4.3840003	12.088	4.208	11.84
167	L. lingual gyrus (wm)	11.000001	1.752	3.024	2.424	6.544	3.272	5.632
168	R. cuneus (wm)	6.46400023	3.504	2.256	1.784	2.448	4.176	3.64
169	L. cuneus (wm)	1.77600014	0.576	2.648	0.536	1.424	2.168	2.848
170	R. insula (wm)	1.1680001	0.632	0.832	0.744	1.96	2.08	1
171	L. insula (wm)	1.64800012	0.768	0.808	0.648	2.408	2.632	1.512

In Table 3.1, "wm" represents white matter regions, and "gm" represents gray matter regions. From the above table, we noticed that the left mamillary body, right parahippocampal gyrus (gm), and left parahippocampal gyrus (gm) are not present in babies. Left ventricular system, right lateral orbitofrontal gyrus, right middle orbitofrontal gyrus (wm), left anterior orbitofrontal gyrus (wm), right pars orbitalis (wm), and left pars orbitalis (wm) are not present in most of the babies and adults. Right mamillary body, right lateral ventricle, right pars orbitalis (gm), left pars orbitalis (gm), right transverse frontal gyrus (gm), left transverse frontal gyrus (gm), right middle orbitofrontal gyrus (gm), right anterior orbitofrontal gyrus (gm), left anterior orbitofrontal gyrus (gm), right posterior orbitofrontal gyrus (gm), right paracentral lobule (gm), right temporal pole (gm), left temporal pole (gm), right fusiforme gyrus (gm), left fusiforme gyrus (gm), left transverse frontal gyrus (wm), left lateral orbitofrontal gyrus (wm), right cingulate gyrus (wm), right temporal pole (wm), left temporal pole (wm), and right parahippocampal gyrus (wm) are missing in some babies.

3.4 CONCLUSION

In this chapter, automatic segmentation of four pediatric and three adult brains is done using Brainsuite 19a brain imaging software, and the volumes of different brain parts (170) are computed.

KEYWORDS

- **segmentation**
- **pediatric brain**
- **atlas-based method**
- **volume**
- **MRI**

REFERENCES

1. Gonzalez, R. G.; Woods, R. E. *Third Edition Digital Image Processing*; Pearson: South Asia, 2014.
2. Sridhar, S. *Digital Image Processing*; Oxford, India, 2011.
3. Najarian, K.; Splinter, R. *Biomedical Signal and Image Processing*; Taylor & Francis: New York, 2006.
4. Demirkaya, O.; Asyali, M. H.; Sahoo, P. K. *Image Processing with MATLAB® Applications in Medicine and Biology*; CRC Press: New York, 2009.
5. Semmlow, J. L. *Biosignal and Medical Image Processing*; CRC Press: New York, 2009.
6. Gousiasa,I. S.; Hammersa, A.; Heckemanna, R. A.; Counsella, S. J.; Dyeta, L. E.; Boardmana, J. P.; David Edwardsa, A. Atlas Selection Strategy for Automatic Segmentation of Pediatric Brain MRIs into 83 ROI. IEEE, 2010.
7. Klimont, M. B.; Flieger, M.; Rzeszutek, J.; Stachera, J.; Zakrzewska, A.; JoNczyk-Potoczna, K. Automated Ventricular System Segmentation in Paediatric Patients Treated for Hydrocephalus Using Deep Learning Methods. *BioMed Res. Int.* **2019,** 1–9.
8. Makropoulos, A.; Gousias, I. S.; Ledig, C.; Aljabar, P.; Serag, A.; Hajnal, J. V.; David Edwards, A.; Counsell, S.J.; Rueckert, D. Automatic Whole Brain MRI Segmentation of the Developing Neonatal Brain. *IEEE Trans. Med. Imaging* **2014,** *33* (9), 1818–1831.
9. Devi, C. N.; Sundararaman, V. K.; Chandrasekharan, A.; Alex, Z. A. Automatic Brain Segmentation in Pediatric MRI Without the Use of Atlas Priors. *IEEE WiSPNET 2016 Conference*, 2016; pp 1481–1484.
10. Gousias, I. S.; Hammers, A.; Counsell, S. J.; David Edwards, A.; Rueckert, D. Automatic Segmentation of Pediatric Brain MRIs Using a Maximum Probability Pediatric Atlas. *IEEE* **2012.**

11. Gao, Y.; Zhang, M.; Grewen, K.; Thomas Fletcher, P.; Gerig, G. Image Registration and Segmentation in Longitudinal MRI Using Temporal Appearance Modeling. *IEEE* **2016**, 629–632.
12. Levman, J.; Vasung, L.; MacDonald, P.; Rowley, S.; Stewart, N.; Lim, A.; Ewenson, B.; Albert Galaburdad, E. Regional Volumetric Abnormalities in Pediatric Autism Revealed by Structural Magnetic Resonance Imaging. *Int. J. Dev. Neurosci.* **2018**, 34–45.
13. Kasiri, K.; Dehghani, M. J.; Kazemi, K.; Helfroush, M. S.; Kafshgari, S. Comparison Evaluation of Three Brain MRI Segmentation Methods in Software Tools. *17th Iran. Conf. Biomed. Eng.* **2010**.
14. Gibson, E.; Young, M.; Sarunic, M. V.; Beg, M. F. Optic Nerve Head Registration Via Hemispherical Surface and Volume Registration. *IEEE Trans. Biomed. Eng.* **2010**, *57* (10), 2592–2595.

CHAPTER 4

COMPARISON OF REGION OF INTEREST AND CORTICAL AREA THICKNESS OF SEIZURE AND HEMOSIDERIN-AFFECTED BRAIN IMAGES

VAJIRAM JAYANTHI[1], SIVAKUMAR SHANMUGASUNDARAM[1], and C. VIJAYAKUMARAN[2]

[1]*SENSE, Vellore Institute of Technology, Chennai, India*

[2]*Department of CSE, SRM Institute of Science and Technology, Kattankulathur, Chennai, India*

ABSTRACT

Brain disorders and tumors are caused by severe neurological abnormalities in their functions. Hemosiderin induces changes in the characteristic functional magnetic field and can be detected using susceptibility-weighted T1 images. Clinical syndromes and their relevance are unclear. Hemosiderin and epilepsy brain images are preprocessed with available software tools, based on the region of interest, mean thickness area, and cortical thickness area of the different parts of the brain. The Brainsuite tool analyzes and processes the fMRI images that produce a spherical topology with cortical surface representations. The functional magnetic resonance image will calculate the lesion or scar-affected region based on the variation of the neural activity. The single package is used for segmentation and analyzing accurate brain images for diagnostic and clinical purposes.

Computational Imaging and Analytics in Biomedical Engineering: Algorithms and Applications.
T. R. Ganesh Babu, PhD, U. Saravanakumar, PhD & Balachandra Pattanaik, PhD (Eds.)

4.1 INTRODUCTION

Hemosiderin is a spinal tumor and is associated with ependymomas of the spinal cord.[1] It may also be visible in paragangliomas, hemangioblastomas, and larger cavernous spine malformations; it may develop a hemosiderin cap and appear and produce a lack of a complete hemosiderin ring where they reach the pial surface. In brain tumors, hemorrhage is common. The seizure initiation mechanism of hemorrhage causes changes in the blood metabolism, like hemosiderin and focal cerebral irritation, in subarachnoid hemorrhage and into the basal cisterns, which directly contact the temporal and frontal lobes. Patients with an intraparenchymal component to the hemorrhage also have a subarachnoid hemorrhage.[2–6] The natural course of tumors is caused by the new onset of seizures, and this has to be addressed in statistical studies of larger patients. The neurological disorder of seizure leads to fits. Due to the unavailability of patient data documenting effectiveness, most of the patients suffer from subarachnoid hemorrhage or intracerebral disease, which leads to short-term antiepileptic prophylaxis.

4.2 PREPROCESSING OF IMAGE

Image preprocessing is the primary step in the processing of the images. Before model training and inference of images, they should be formatted based on the processing techniques. This includes resizing, orienting, and color corrections. Image restoration involves the correction of degradation, noise, and atmosphere deflections,[7,8] which occur during the processing of images. The output produces a corrected image that is similar to the original image's characteristics. The purpose of arbitrarily manipulating an image is to support a preferred reality and achieve an aesthetic standard. The processing of image translation takes place between the digital imaging devices and the human visual system,[9] and the image restoration involves taking a noisy image and estimating the clean image from the original image. Image enhancement techniques like deblurring or contrast stretching by a nearest neighbor procedure, provided by imaging packages,[10] and the image preprocessing methods are discussed in Figure 4.1.

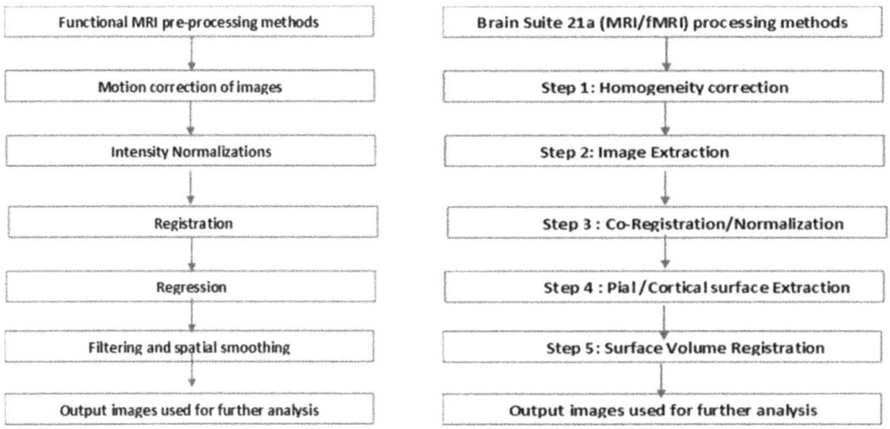

FIGURE 4.1 fMRI and brain image preprocessing methods.

4.2.1 MOTION AND SLICE TIME CORRECTION

Motion correction has a number of correction methods that modify the MR k-space or image data during reconstruction or after acquisition, independent of the potential motion performed by acquisition.[11] The nature of fMRI slices and acquisition protocols is that they are temporally misaligned from each other and cannot be acquired simultaneously.[12] This misalignment is to be corrected by slice timing correction (STC) and preprocessing pipelines. Slice acquisition parameters mainly depend on the slice timing correction, interact in a nonlinear fashion with head movement, and change the other preprocessing steps of fMRI acquisition[13] parameters and design.

4.2.2 INTENSITY NORMALIZATION

Normalization is a subject to ensure comparability across images; full brain normalization may work better than white strip interpretation, subsequent lesion analysis, and segmentation algorithms. This method is used for intensities[14] that are comparable across all tissues.

4.2.3 REGISTRATION

The registration process overlays two or more images from various sensors or imaging equipment taken at different angles and times, or from the geometrically aligned same scene of images for analysis (Zitová and Flusser, 2003), transforming the different sets of data into one coordinate system. The military's automatic target recognition, medical imaging, and analysis of image data are obtained from satellites. The MRI, CT, SPECT, or PET enables the combination of data from multiple modalities to get complete information about the patient. It helps to facilitate treatment verification, improve interventions, monitor tumor growth,[15] or compare patient data to anatomical atlases.

4.2.4 REGRESSION

The regression of the locality-context trade-off is highly task-dependent. Several deep convolutional neural networks (DCNNs) are used for image-to-image regression tasks such as denoising, relighting, colorization, etc. These approaches typically involve specific architectures coupled with fine-tuned ancillary postprocessing methods that perform more tasks than classification tasks. The locality-context is coupled with the habitual trend of incorporating VGG and ResNet architectures for nonclassification[16] tasks.

4.2.5 SMOOTHING SPATIAL FILTER

Noise reduction and blurring of images are done by smoothing the spatial filter. Blurring is a preprocess of removing noise, and smoothing filters are used to reduce image noise, sharpen edges of image structures, and enhance an image with the help of spatial domain operations or filtering of images.

The software tool analysis is covered by the Brainsuite and its step-by-step procedure of inhomogeneity correction, brain extraction-registration, tissue segmentation, intensity normalization, and cortical surface extraction. After that, surface volume registration, region of interest (ROI), mean thickness area, and cortical surface area thickness were calculated as shown in Figure 4.2.

Brain Suite 21 a Functions	Processing	Interface
Cortical surface extraction	Statistics	Masking
Surface/Volume registration (MATLAB analyses)	Functional Pipeline	Labelling
Diffusion Pipeline (Visualization)	NiPype	Sulcal Curve Delineation

FIGURE 4.2 Brainsuite 21a functions of processing and interfacing.

4.3 MATERIALS AND METHODS

The parameters of the fMRI WRE CON images of T1AXPOST and Hemosiderin are preprocessed by the Brainsuite tool. The tool includes skull and scalp removal, image nonuniformity compensation, voxel-based tissue classification, topological correction, rendering, and editing functions. The purpose of preprocessing image acquisition is to eliminate unwanted information or noise from ultrasound images without evading vital information. These variations follow further image analysis steps. The analysis of tissue types, pathological regions, and anatomical structures is carried out by the segmentation process. The first task is to find a ROI, eliminating unwanted regions from processing, and the second task is to segment the disease. The boundary estimation, classification, and categorization of diseases are done by segmentation to maintain the accuracy and sensitivity of detecting lesions (Guo and Şengür, 2013). The features technique is used to eliminate the false detection rate and improve diagnosis after the segmentation of the disease. The suspicious areas can be categorized as benign or malignant on the basis of selected features using different classification techniques. The distinct histopathological components and malignant modules with vague boundaries are often fused with adjoining tissues to produce the delineation of tissue.[17] To improve the accuracy, decrease the misdiagnosis from earlier detection, and improve the correct diagnosis, all required a fully computerized system (Nguyen et al., 2019).

The tool is designed to produce cortical representations, and each stage of the cortical surface identification process was captured, and ROI, mean area thickness, and cortical area thickness were extracted. The present results are compared with normal brain image values for further diagnosis. The average thickness of the full brain is around 2.5–3 mm, and the individual brains vary from about 2 mm at their thinnest in the calcarine cortex up to 4 mm and over in the thicker regions of the precentral gyrus, superior temporal lobes, and superior frontal lobes (Zilles, 1990). The cerebral cortex area thickness lies between 1 and 4.5 mm, and it is a highly folded neuron sheet, with an average of approximately 2.5 mm, which is compared with the measured values to detect the abnormalities.

4.4 RESULTS AND DISCUSSION

The tool is used for surface and volume registration, data diffusion, tensor fitting, and structural coregistration. The brain image analysis is used to extract label, volume, and cortical surface anatomy and diffusion coregister to structural images to allow modeling of labeled brain anatomy. Functional magnetic resonance imaging (fMRI) measures the small changes in brain activity occurring due to blood flow. It can detect abnormalities in the brain that cannot be found with other imaging processes. Hemosiderin is an iron-storage complex that is composed of lysosomes and partially digested ferritin. The breakdown of the hemisphere gives rise to biliverdin and iron. Then the body traps and releases the iron and stores it in hemosiderin tissues. Hemosiderin (golden-brown granules, iron-positive) is usually found in macrophages of red pulp and also in the marginal zone to a lesser degree. The preprocessing steps of hemosiderin are shown in Figures 4.3 and 4.4, and the preprocessing methods and extraction of surface area and volume are shown in Figure 4.5.

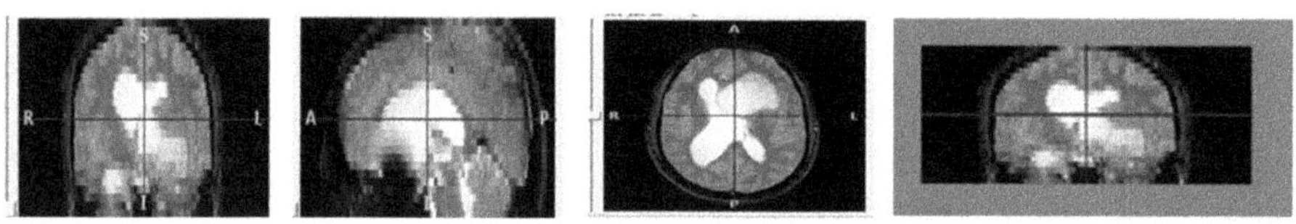

FIGURE 4.3 Stripping of skull.

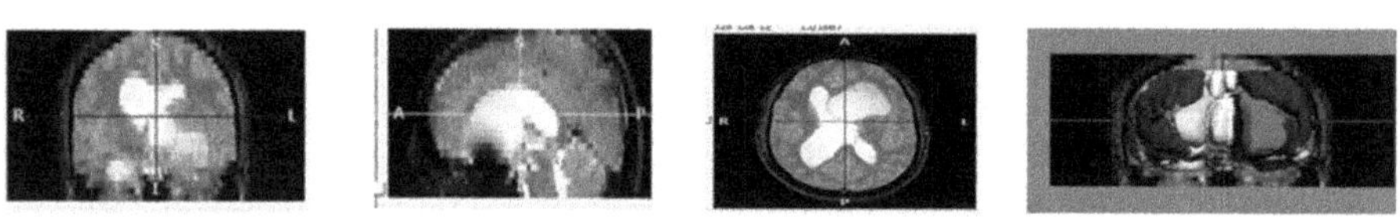

FIGURE 4.4 Surface extraction of cortical area.

Figures 4.3 and 4.4 fMRI brain Wrecon images A to B [hemosiderin] preprocessing result obtained from the Brainsuite tool.

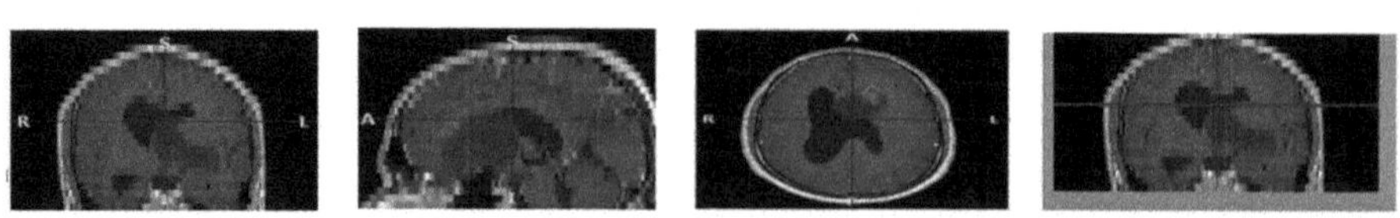

FIGURE A Stripping of skull.

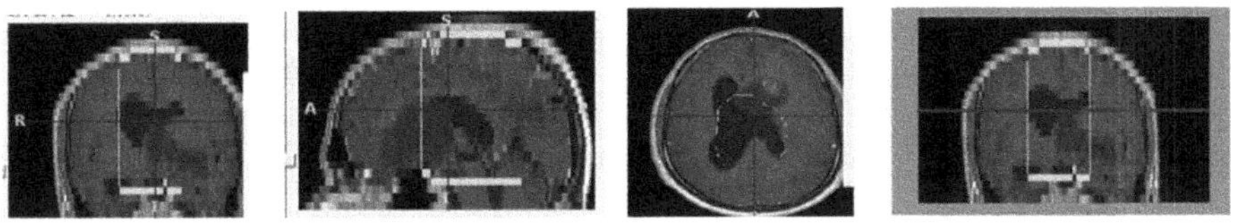

FIGURE B Scalp and skull processing.

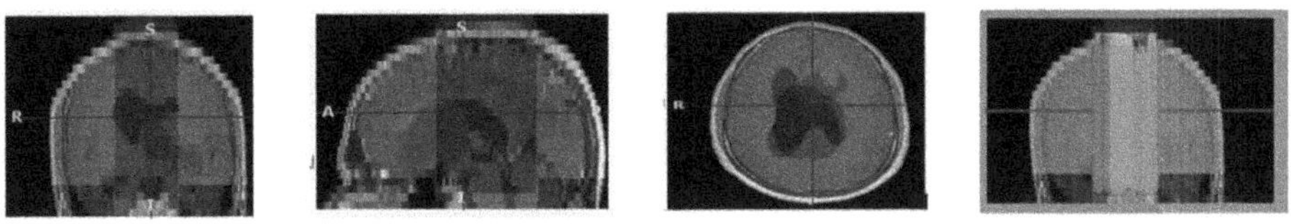

FIGURE C Classifications of tissue.

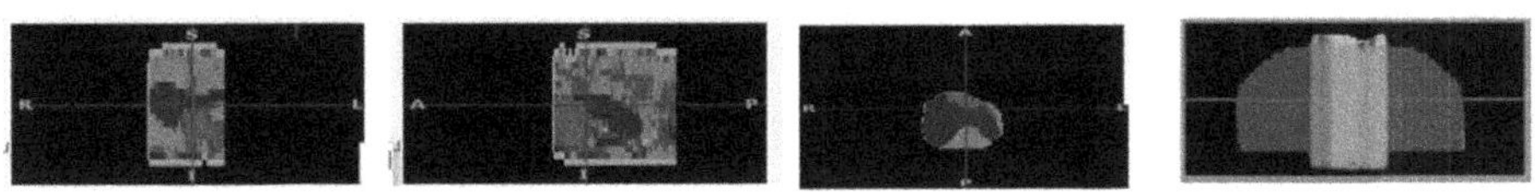

FIGURE D Cerebrum labeling.

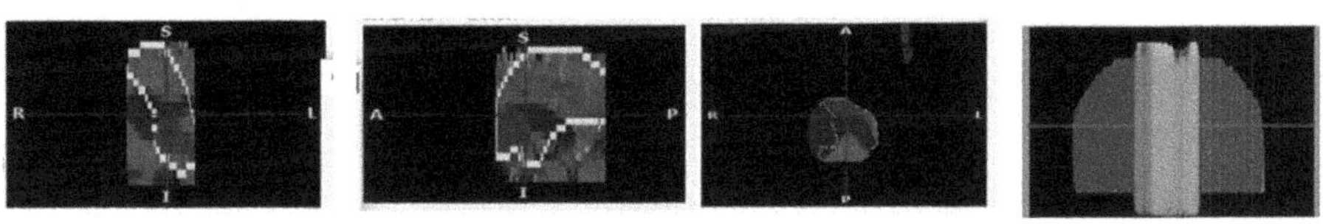

IGURE E Identification of cortex.

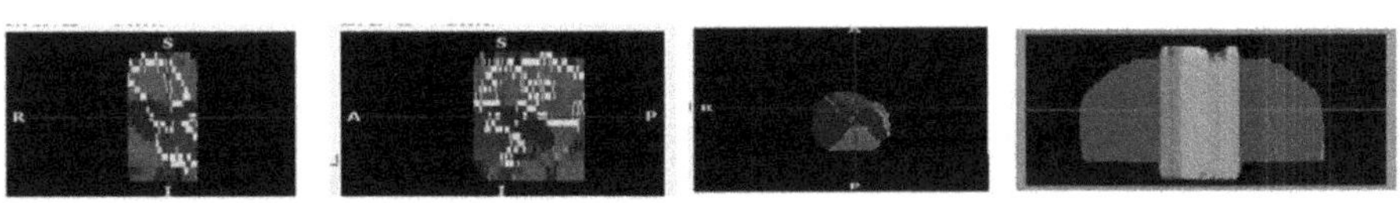

FIGURE F Scrub mask.

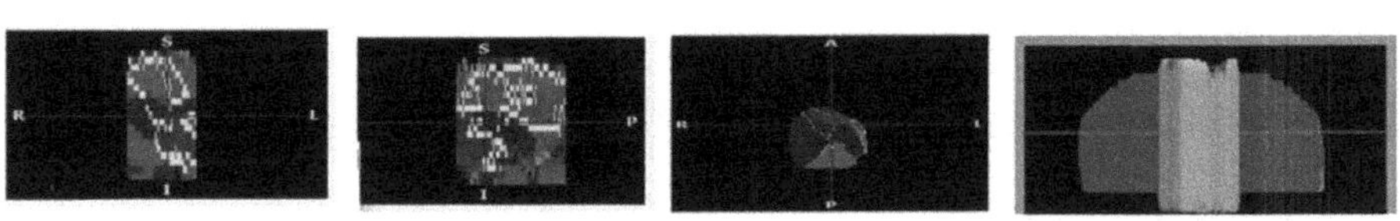

FIGURE G Correction of topology.

FIGURE H Wisp and Dewisp process.

FIGURE I Surface correction of Pial area.

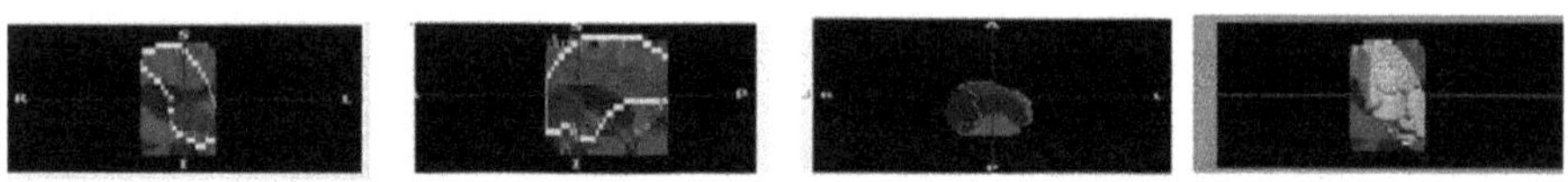

FIGURE J Splitting of hemisphere.

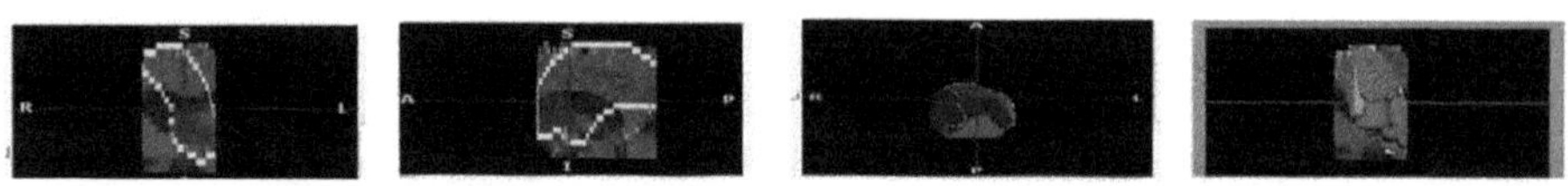

FIGURE K Cortical thickness estimation.

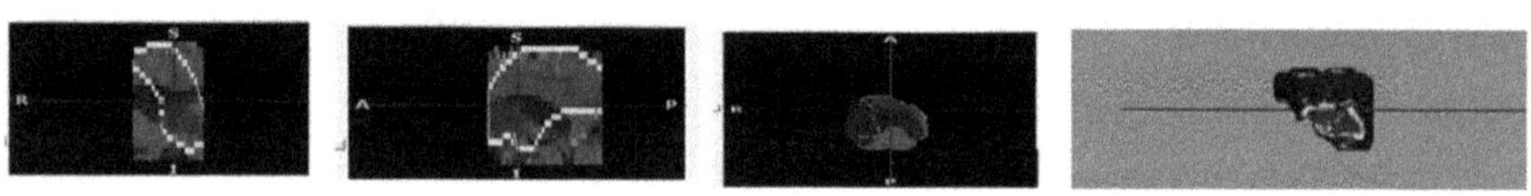

FIGURE L Surface thickness of cortical area.

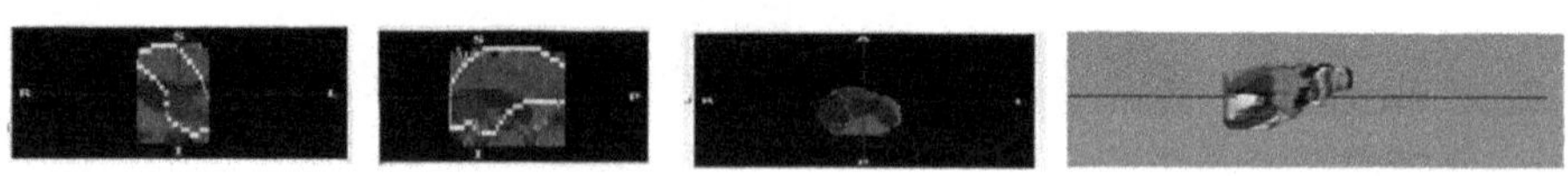

FIGURE M Surface volume registration 1.

FIGURE N Surface volume registration 2.

FIGURE 4.5 fMRI brain Wrecon image [Image of Seizure Brain—T1 AX POST], from Figure A to N are the preprocessing results of seizure-affected MRI image of T1 axial post from Brainsuite tool.

TABLE 4.1 Region of Interest, Mean Thickness, and Cortical Area Calculated Based on Various Brain Parts of fMRI Brain Seizure Images (Brain T1 AX POST) and fMRI Brain Images (hemosiderin).

ROI_ ID	Brain parts	Mean thickness (mm) fMRI brain seizure images (brain T1 AX post)	Cortical area (mm^2) fMRI brain seizure images (brain T1 AX post)	Cortical area (mm^2) fMRI brain images (hemosiderin)
120	R. superior frontal gyrus	0.610544	657.278228	3510.586298
121	L. superior frontal gyrus	1.304945	1655.557214	6183.092200
130	R. middle frontal gyrus	1.153275	337.825409	4008.851409
131	L. middle frontal gyrus	0.869990	718.740569	12,975.890310
142	R. pars opercularis	1.088048	15.596868	1286.285113
143	L. pars opercularis	0.585431	443.580352	1149.394662
144	R. pars triangularis	1.355878	142.183448	11,051.325879
145	L. pars triangularis	1.600054	927.154561	1144.524778
146	R. pars orbitalis	2.367167	1.241634	126.537966
147	L. pars orbitalis	6.727943	77.504858	184.713848
150	R. precentral gyrus	0.153721	66.350107	5353.633353
151	L. precentral gyrus	0.355621	1141.159916	5750.413625
162	R. transverse frontal gyrus	3.707373	0.000000	84.850144
163	L. transverse frontal gyrus	4.049633	136.511933	139.778236
164	R. gyrus rectus	0.000074	3.663462	557.282092
165	L. gyrus rectus	3.778476	346.149674	430.304697
166	R. middle orbitofrontal gyrus	6.180444	2.469826	216.428357
167	L. middle orbitofrontal gyrus	0.997979	63.335274	156.230785
168	R. anterior orbitofrontal gyrus	4.456243	4.495258	153.023251
169	L. anterior orbitofrontal gyrus	0.989992	39.046137	169.868460
170	R. posterior orbitofrontal gyrus	5.490447	0.764592	559.311023
171	L. posterior orbitofrontal gyrus	1.493429	108.787511	436.082553
172	R. lateral orbitofrontal gyrus	6.009783	3.741775	148.448843

TABLE 4.1 *(Continued)*

ROI_ ID	Brain parts	Mean thickness (mm) fMRI brain seizure images (brain T1 AX post)	Cortical area (mm^2) fMRI brain seizure images (brain T1 AX post)	Cortical area (mm^2) fMRI brain images (hemosiderin)
173	L. lateral orbitofrontal gyrus	1.895485	57.798568	163.160877
182	R. paracentral lobule	0.001593	21.024277	1002.323518
183	L. paracentral lobule	0.519313	79.186887	802.989708
184	R. cingulate gyrus	0.299899	338.531295	10,883.008641
185	L. cingulate gyrus	1.069448	1765.784771	7928.537898
186	R. subcallosal gyrus	0.001817	5.363604	159.993860
187	L. subcallosal gyrus	2.490743	65.046220	107.299517
222	R. postcentral gyrus	0.064345	159.239118	7476.025936
223	L. postcentral gyrus	0.447987	1271.812278	8852.655389
224	R. supramarginal gyrus	0.069290	87.657322	11,284.485650
225	L. supramarginal gyrus	0.283713	719.854933	1828.480007
226	R. angular gyrus	0.000081	3.426407	1638.347003
227	L. angular gyrus	0.929038	1175.327788	839.668871
228	R. superior parietal gyrus	0.005154	49.395522	4067.667996
229	L. superior parietal gyrus	0.754984	1064.805629	3984.797039
242	R. precuneus	0.655383	128.855513	770.375863
243	L. precuneus	0.884500	86.900164	873.517733
310	R. temporal pole	5.024895	0.903588	122.678009
311	L. temporal pole	1.080730	881.510760	90.360559
322	R. superior temporal gyrus	0.033416	11.524551	216.700062
323	L. superior temporal gyrus	1.243263	1002.493071	494.096565
324	R. transverse temporal gyrus	0.000071	2.074070	27.874815
325	L. transverse temporal gyrus	0.698706	42.794973	22.813240
326	R. middle temporal gyrus	0.925798	4.868301	2129.113825
327	L. middle temporal gyrus	0.394457	902.492128	2133.666365
328	R. inferior temporal gyrus	0.035113	15.863507	1325.005013
329	L. inferior temporal gyrus	0.531635	2871.989764	1535.948755

TABLE 4.1 *(Continued)*

ROI_ ID	Brain parts	Mean thickness (mm) fMRI brain seizure images (brain T1 AX post)	Cortical area (mm^2) fMRI brain seizure images (brain T1 AX post)	Cortical area (mm^2) fMRI brain images (hemosiderin)
330	R. fusiforme gyrus	0.092988	21.762781	1332.834595
331	L. fusiforme gyrus	0.728324	1062.793859	921.583444
342	R. parahippocampal gyrus	0.112606	0.472803	76.263671
343	L. parahippocampal gyrus	0.253237	343.688600	23.527293
422	R. superior occipital gyrus	0.000094	1.409307	98.363716
423	L. superior occipital gyrus	0.164753	330.183458	149.840778
424	R. middle occipital gyrus	0.000089	0.819907	723.966451
425	L. middle occipital gyrus	0.184693	3468.862603	704.513176
442	R. inferior occipital gyrus	0.027703	33.959151	1224.301781
443	L. inferior occipital gyrus	1.139114	515.485582	65.896897
444	R. lingual gyrus	0.019824	141.372045	990.610529
445	L. lingual gyrus	0.734774	322.399934	452.486295
446	R. cuneus	0.000096	1.359830	920.563079
447	L. cuneus	0.308793	534.099470	764.495928
500	R. insula	0.363296	47.627683	520.052231
501	L. insula	2.614143	735.265075	365.875768

Table 4.1 shows the various regions of the brain and their mean area thickness and cortical area thickness in both the seizure-affected brain and the hemosiderin brain. The values are compared with normal, healthy people's brain area values, and the severity of disease is calculated for further diagnosis and clinical analysis with the purpose of treatment.

The medical care system follows various methods to detect and analyze epileptic seizures. To identify the seizure type and etiology, diagnostic tools like electroencephalogram (EEG), magnetic resonance imaging (MRI), magnetoencephalogram (MEG), single photon emission computed tomography (SPECT), neuropsychiatric testing, and positron emission tomography (PET) are used. For identifying specific seizure types, the EEG is critical. For newly diagnosed patients, a CT scan will be used, but always an MRI is preferred for the brain analysis. It can help determine the proper seizure type and syndrome. MRI may locate brain lesions like scars or anatomic defects that are not detected by CT scans or conventional radiographs. All data are fetched into any one of the analysis techniques of statistical parametric mapping (SPM), MEG, or Curry analysis software to find and detect the stages of seizures and where they start, and the requirements of either medication or surgery are advised by medical practitioners. Machine learning algorithms are used to detect and analyze datasets by using different learning, classifier, and statistical measurement methods. Recent work is focused on fMRI to find the correlation between epileptic seizures and cerebral hemodynamic changes.[18] Seizures can be detected through the observation of the brain, heart rate, oxygen level, muscle activities, and artificial sounds or visual signatures like MRI, motion, and audio or video recording of the head and body of the person.[19] Measuring asymmetric interactions in the resting state of brain networks is discussed in Ref. [20]; correcting inhomogeneity-induced distortion in fMRI using nonrigid registration in Ref. [21]; the linear spherical deconvolution and model-free linear transform methods for diffusion MRI in Ref. [22]; and the identification of partial correlation-based networks compared with cortical thickness data in Ref. [23]. The Gaussian models are used to compare the nonrigid image registration.[24] The automatic cortical surface registration and labeling are analyzed in Ref. [25]. The geodesic curvature flow on surfaces for automatic sulcal delineation is compared with the cortical surface area.[26] An invariant shape representation using the anisotropic Helmholtz equation for the human brain is shown in Ref. [27]. The Fourier 2-sphere linear transforms of diffusion MRI are compared with different regions of the cortical surface.[28] Correcting the susceptibility-induced distortion in diffusion-weighted MRI using constrained nonrigid registration is also discussed in Ref. [29].

4.5 CONCLUSIONS

The present research is based on neutrosophy-based methods to solve various image-processing problems based on their handling capability of indeterminate information. This chapter covers the neuroanatomical abnormalities in two fMRI images, in which the cortical thickness, volume, and area were extracted from the software tool and compared between patients and healthy controls. The cortical area thickness is used to diagnose cortical abnormalities in different regions of the brain. More studies are required to analyze the brain and its abnormalities.

KEYWORDS

- **seizure**
- **hemosiderin**
- **fMRI (functional magnetic resonance image)**
- **software tool**
- **brain**

REFERENCES

1. Kondziolka, D.; Bernstein, M.; Resch, L.; Tator, C. H.; Fleming, J. F.; Vanderlinden, R. G.; Schutz, H. Significance of Hemorrhage into Brain Tumors: Clinicopathological Study. *J. Neurosurg.* **1987,** *67*, 852–857.
2. Rosen, A. D.; Frumin, N. V. Focal Epileptogenesis After Intracortical Hemoglobin Injection. *Exp. Neurol.* **1979,** *66*, 277–284.
3. Ueda, Y.; Willmore, L. J.; Triggs, W. Amygdalar Injection of FeCl3 Causes Spontaneous Recurrent Seizures. *Exp. Neurol.* **1998,** *153*, 123–127.
4. Moran, N. F.; Fish, D. R.; Kitchen, N.; Shorvon, S.; Kendall, B. E.; Stevens, J. M. Supratentorial Cavernous Haemangiomas and Epilepsy: A Review of the Literature and Case Series. *J. Neurol. Neurosurg. Psychiatry* **1999,** *66*, 561–568.
5. van Breemen, M. S.; Wilms, E. B.; Vecht, C. J. Epilepsy in Patients with Brain Tumours: Epidemiology, Mechanisms, and Management. *Lancet Neurol.* **2007,** *6*, 421–430.
6. Kucukkaya, B.; Aker, R.; Yuksel, M.; Onat, F.; Yalcin, A. S. Low Dose MK-801 Protects Against Iron-Induced Oxidative Changes in a Rat Model of Focal Epilepsy. *Brain Res.* **1998,** *788*, 133–136.
7. Robinson, R. J.; Bhuta, S. Susceptibility-Weighted Imaging of the Brain: Current Utility and Potential Applications. *J. Neuroimaging.* **2011,** *21* (4), e189–e204.
8. Park, M. J.; Kim, H. S.; Jahng, G. H.; Ryu, C. W.; Park, S. M.; Kim, S. Y. Semi Quantitative Assessment of Intra-Tumoral Susceptibility Signals Using Non-Contrast-Enhanced High-Field High-Resolution Susceptibility-Weighted Imaging in Patients with Gliomas: Comparison with MR Perfusion Imaging. *AJNR Am. J. Neuroradiol.* **2009,** *30*, 1402–1408.
9. Wen, P. Y.; Macdonald, D. R.; Reardon, D. A. Updated Response Assessment Criteria for High-Grade Gliomas: Response Assessment in Neuro-Oncology Working Group. *J. Clin. Oncol.* **2010,** *28*, 1963–1972.
10. Sato, J. R.; Hoexter, M. Q.; Fujita, A.; Rohde, L. A. Evaluation of Pattern Recognition and Feature Extraction Methods in ADHD Prediction. *Front. Syst. Neurosci.* **2012**.
11. Shinohar, R., T. et al. Statistical Normalization Techniques for Magnetic Resonance Imaging. *NeuroImage: Clin.* **2014,** *6*, 9–19.
12. Godenschweger, F.; Kägebein, U.; Stucht, D.; Yarach, U.; Sciarra, A.; Yakupov, R.; Lüsebrink, F. et al. Motion Correction in MRI of the Brain. *Phys. Med. Biol.* **2016,** *61* (5), R32–R56.

13. Parker, D. B. et al. The Benefit of Slice Timing Correction in Common fMRI Preprocessing Pipelines. *Front. Neurosci.* **2019,** *13*, 821–830.
14. Fortin, J-P. et al. Removing Inter-Subject Technical Variability in Magnetic Resonance Imaging Studies. *Neuro Image* **2016,** *132*, 198–212.
15. Zitova, B. In *Encyclopedia of Biomedical Engineering, Acquisition Methods, Methods and Modeling* 2019.
16. Santhanam, V. et al. *Generalized Deep Image to Image Regression*, **2017.**
17. Koundal, D.; Sharma, B. In *Neutrosophic Set in Medical Image Analysis*, **2019.**
18. Fergus, P.; Hussain, A.; Hignett, David; Al-Jumeily, D.; Khaled, A-A.; Hani, H. A Machine Learning System for Automated Whole- Brain Seizure Detection. *Appl. Comput. Inf.* **2015.**
19. Fergus, P.; Hignett, D.; Hussain, A.; Al-Jumeily, D.; Khaled, A-A. Automatic Epileptic Seizure Detection Using Scalp EEG and Advanced Artificial Intelligence Techniques. *BioMed Res. Int.* **2015,** 1–17.
20. Joshi, A. A.; Salloum, R.; Bhushan, C.; Leahy, R. M. Measuring Asymmetric Interactions in Resting State Brain Networks. *Inf. Process Med. Imaging* **2015,** *24*, 399–410.
21. Chambers, M. C.; Bhushan, C.; Haldar, J. P.; Leahy, R. M.; Shattuck, D. W. Correcting Inhomogeneity-Induced Distortion in FMRI Using Non-Rigid Registration. *Proc. IEEE Int. Symp. Biomed. Imaging* **2015,** 1364–1367.
22. Haldar, J. P.; Leahy, R. M. The Equivalence of Linear Spherical Deconvolution and Model-Free Linear Transform Methods for Diffusion MRI. *IEEE 10th Int. Symp. Biomed. Imaging (ISBI)*, **2013,** 508–511.
23. Wheland, D.; Joshi, A.; McMahon, K.; Hansell, N.; Martin, N.; Wright, M.; Thompson, P.; Shattuck, D.; Leahy, R. R. Identification of Partial-Correlation Based Networks with Applications to Cortical Thickness Data. In *9th IEEE International Symposium on Biomedical Imaging (ISBI)*, 2012; pp 2–5.
24. Somayajula, S.; Joshi, A. A.; Leahy, R. M. Non-Rigid Image Registration Using Gaussian Mixture Mode. In *Biomedical Image Registration.* Springer: Berlin Heidelberg, 2012; pp 286–295.
25. Joshi, A. A.; Shattuck, D. W.; Leahy, R. M. A Method for Automated Cortical Surface Registration and Labeling. *Biomed. Image Regist. Proc.* **2012,** *7359*, 180–189.
26. Joshi, A. A.; Shattuck, D. W.; Damasio, H.; Leahy, R. M. Geodesic Curvature Flow on Surfaces for Automatic Sulcal Delineation. In *9th IEEE International Symposium on Biomedical Imaging (ISBI)*, 2012.
27. Ashrafulla, S.; Shattuck, D. W.; Damasio, H.; Leahy, R. M. An Invariant Shape Representation Using the Anisotropic Helmholtz Equation. *Med. Image Comput. Comput. Assist. Interv.* **2012,** *15* (3), 607–614.
28. Haldar, J. P.; Leahy, R. M. New Linear Transforms for Data on a Fourier 2-Sphere with Application to Diffusion MRI. In *9th IEEE International Symposium on Biomedical Imaging (ISBI)*, 2012; pp 402–405.
29. Bhushan, C.; Haldar, J. P.; Joshi, A. A.; Leahy, R. M. Correcting Induced Distortion in Diffusion-Weighted MRI Using Constrained Non-Rigid Registration. In *Signal & Information Processing Association Annual Summit and Conference (APSIPA ASC)*, 2012; pp 1–19.

CHAPTER 5

DESIGN AND ANALYSIS OF CLASSIFIER FOR ATRIAL FIBRILLATION AND DEEP NEURAL NETWORKS WITH ECG

M. SIVAKUMAR,[1] S. OMKUMAR,[2] B. ARPUTHAMARY,[3] and YOHANNES BEKUMA BAKARE[4]

[1]*Department of ECE, Mohamed Sathak A.J. College of Engineering, Chennai, India*

[2]*Department of ECE, SCSVMV (Deemed University), Kanchipuram, India*

[3]*Department of Computer Applications, Bishop Heber College, Tiruchirappalli, India*

[4]*Department of Electrical and Computer Engineering ,College of Engineering and Technology, Wollega University, Ethiopia*

ABSTRACT

Atrial fibrillation is the popular factor of risk in evaluating the coronary and heart diseases. Deep learning is a quite common area of interest in various medical applications connected with classification of heartbeats with ECG signals. ECG can be employed for the detection of AF. This paper describes constructing a classifier for the detection of atrial fibrillation in signals of ECG using deep neural networks. Generally, neural network structures of convolution (CNN) and recurrent types (RNN) are employed for classification of heartbeats. The proposed method in this paper employs memory networks and analysis using both time and frequency. If raw signals are trained with the LSTM network, the classification accuracy obtained is

Computational Imaging and Analytics in Biomedical Engineering: Algorithms and Applications.
T. R. Ganesh Babu, PhD, U. Saravanakumar, PhD & Balachandra Pattanaik, PhD (Eds.)

poor. Instead if network is trained with features of time-frequency-moment for each signal, the classification performance gets improved and time for training will be less. The simulation results show significant improvement in sensitivity and efficiency for heartbeat classifications. It is very much clear from observations that models of LSTM DL's are suitable for heartbeat classifications.

5.1 INTRODUCTION

The behavior of heartbeat of a person can be observed generally using ECG. ECGs will indicate the heartbeat of a person to be either normal or not. AF can be defined as inappropriate mannerism of heartbeat which prevails if the upper chambers of heart (artia) mismatch when compared with that of chambers of lower region. The data obtained from ECG are available from challenge of PhysioNet[1–3] for the proposed work. The data are sampled at a frequency of 300 Hz and categorized into "4" groups such as Normal, Atrial Fibrillation, Rhythm, and Others. This illustration indicates how the process of classification can be automated using the concept of deep learning. A binary classifier is employed in the procedure to differentiate normal and signed signals. The concept of long short-term memory (LSTM) networks is discussed in this paper to observe the sequence and series of data with respect to time parameter. A bidirectional LSTM layer can monitor and observe sequence of time in both directions—forward and backward.

5.2 MATERIALS AND METHODS

(i) ECG Data Description

The data obtained from the PhysioNet database comprises the classes of people with cardiac arrhythmia (ARR), failure of heart (CHF), and sinus (NSR). The database collected comprises 96, 30, and 36 observations from the patients with ARR, CHF, and NSR, respectively. The MAT-file comprises test ECG data in the format of time series. An image classifier is employed to differentiate between the above "3" groups in this example.

Figure 5.1 shows the DL Simulink model. A MATLAB Function block inside the pre-processing system is used to execute CWT for getting ECG signal scalogram and finally the image is obtained from it. The image classifier will do prediction for classification of images with respect to deep learning CNN.

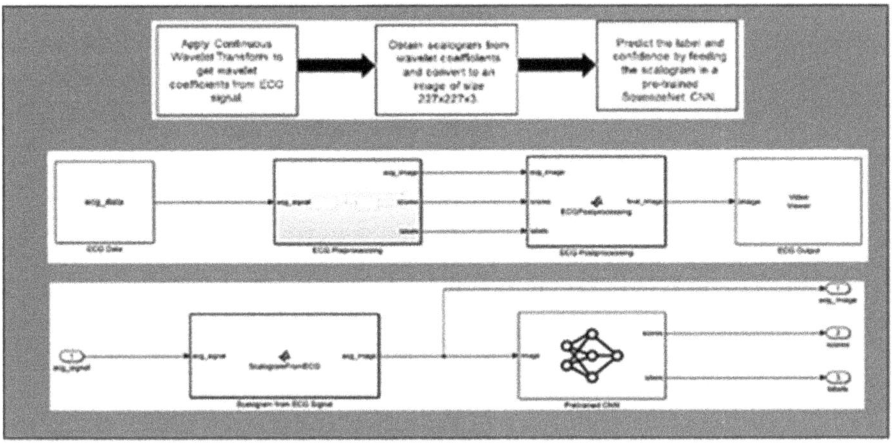

FIGURE 5.1 ECG deep learning simulink model.

A function named ecg_to_scalogram is defined to execute the following activities:

- The input given is 65,536 samples of ECG data with double precision.
- Input representation is available from ECG if Wavelet transform is applied.
- The wavelet coefficients give the scalogram.
- The scalogram is then converted to image (227 × 227 × 3).

The ECG post processing MATLAB function illustrates the corresponding image function to identify label with respect to highest score obtained from classifier of images.

(a) **Loading of Data and Observation:** A MAT file is generated from the *PhysioNet* which comprises ECG signals in a proper format.
(b) **Add Signals and Labels to Workspace:** A cell array named signal is used to include the signals from ECG. A categorical array named label is used to include the ground-truth labels of the signals accordingly.
(c) **Readings:** The number of AFib and normal signals composite in the data can be monitored by the Matlab summary function.
(d) **Generation of Histograms:** Majority of the data signals are 9000 samples long and shown in Figure 5.2.
(e) One signal segment can be monitored from each and every class.

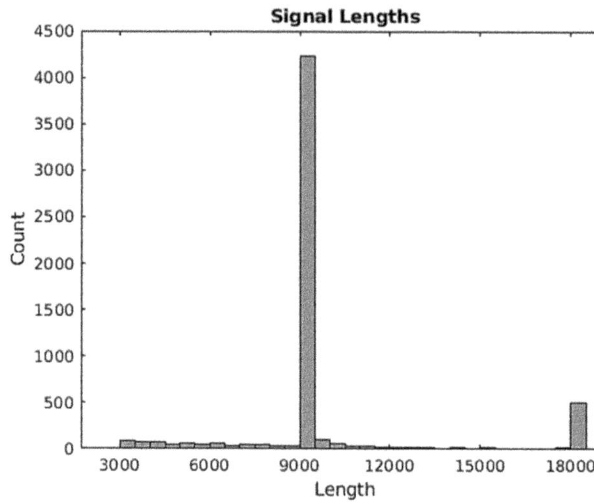

FIGURE 5.2 Histogram plot of ECG signal lengths about 9000 samples.

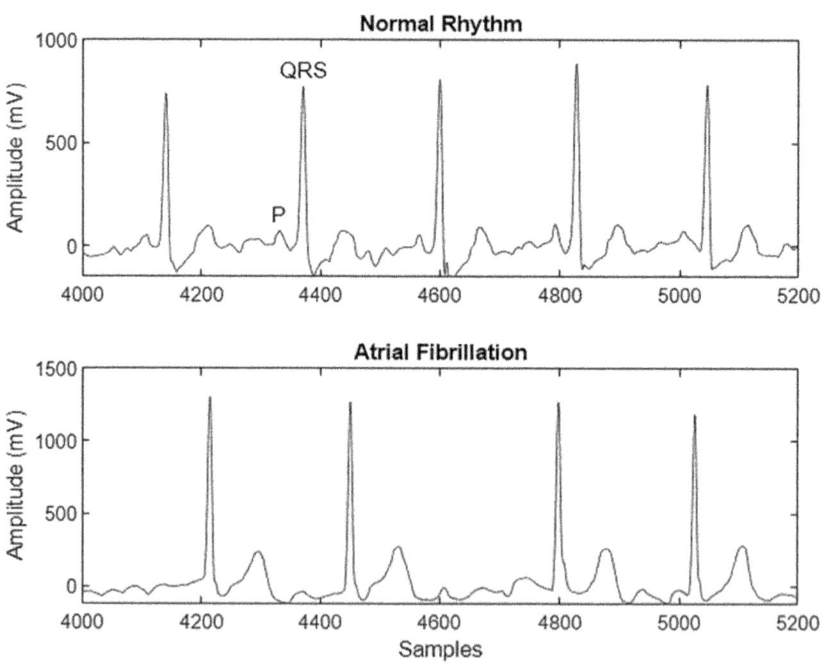

FIGURE 5.3 Normal signal plot—P wave and QRS complex observation.

In Figure 5.3, the irregular and regular intervals of time correspond to AFib heartbeats and normal heartbeats respectively. Generally, the P wave is absent in AFib heartbeats but projects before the QRS complex in case of normal heartbeats.

(f) **Classifier Training with Raw Data:** The raw signals produced earlier can be employed to design the classifier. The set of readings can be used to train and test the classifier accuracy with respect to new data. Finally, the summary function can be applied to indicate ratio of two signals to 1:7.

(ii) Training of LSTM Network Architecture using ADAM Solver

The Matlab *trainNetwork* function can be used to train the LSTM network with various specified options of training and layer architecture. The total process of training will consume some time due to bigger training set. The network training is provided by the ADAM solver.

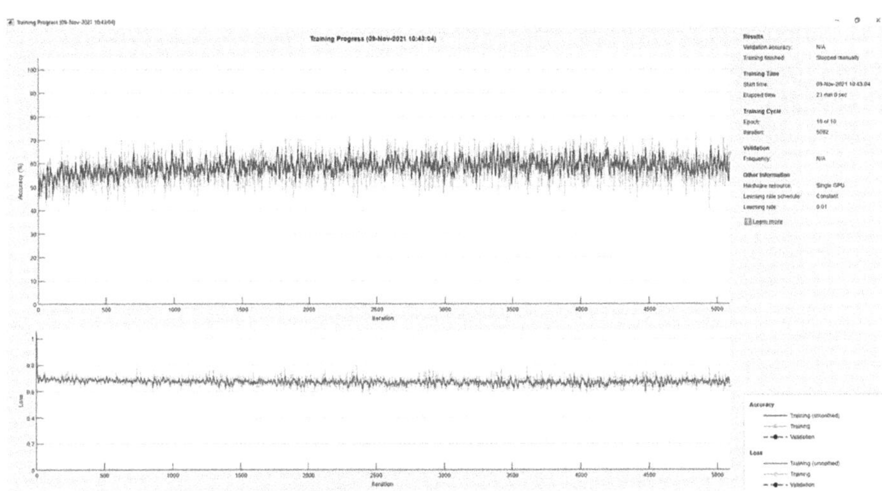

FIGURE 5.4 Plot of training progress.

The accuracy of classification on each batch is represented by training accuracy as shown in Figure 5.4. On successful progress of training, this value reaches 100% at maximum. On successful progress of training, the cross entropy loss value reaches a minimum value of zero. Figure 5.5 illustrates accuracy of classifier's training oscillating between 50% and 60%.

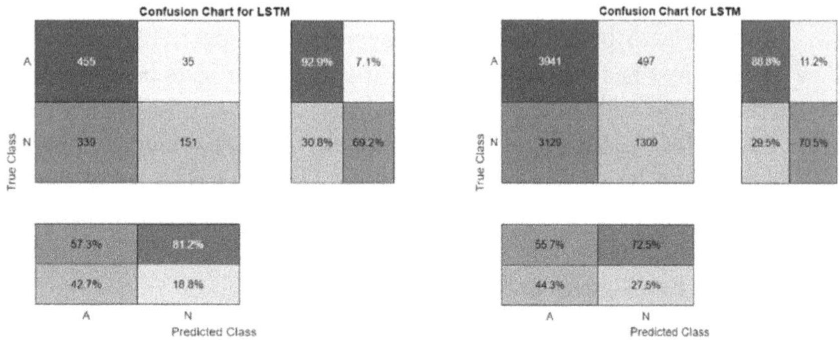

a) LSTM Accuracy = 59.1483 b) LSTM Accuracy = 61.8367

FIGURE 5.5 Confusion matrix chart.

(iii) Proposed Classifier Performance Enhancement with Extraction of Features

The parameters from given data will enhance the classifier accuracies with respect to both training and testing. To conclude on the extraction of features, this paper implements a methodology which computes time/frequency-based images which can be utilized for training the neural networks.

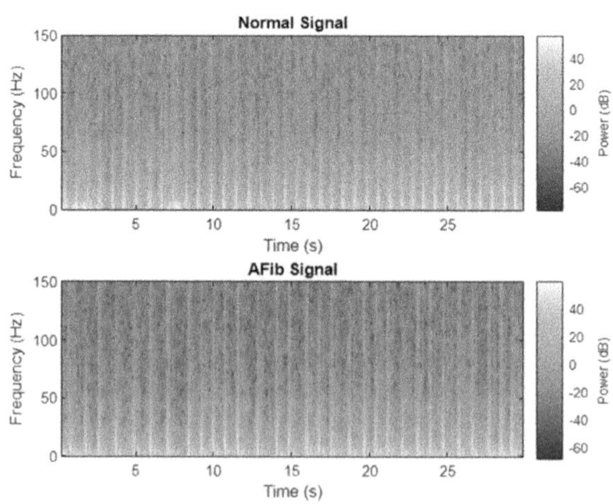

FIGURE 5.6 Spectrogram plot of normal ECG vs AFib ECG.

The LSTM technique is applied instead of CNN and more applicable for signals of single dimension. The information from spectrograms can be extracted by applying time-frequency based (TF) moments. These moments are actual inputs to LSTM. Then it computed by applying Fourier transforms of short time over time windows as shown in Figure 5.6.

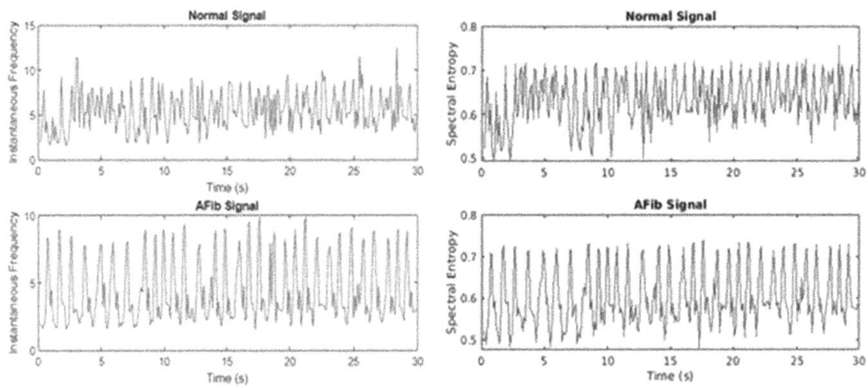

a) Plot of instantaneous frequency of ECG Signal b) Plot of instantaneous frequency of ECG Signal

FIGURE 5.7 Spectrograms for TF moments.

The spectrum of a signal either high or low can be determined by the spectral entropy. If the signal has a spectrum of spiky type, then it has a low value of spectral entropy. On the other hand, if the signal has a spectrum of flat type, then it has a high value of spectral entropy. The spectral entropy is then estimated by Matlab *pentropy* function. The number of time windows used is 255 in this context. The centers of time windows and function time outputs correspond to each other. Figure 5.7 shows the TF moments extracted from the spectrograms.

(iv) Training the LSTM Network with Time-Frequency Features

It is essential to change the architecture of LSTM for two-dimensional signals by mentioning the size of input sequence equal to 2.

(a) The LSTM layer must be bidirectional with output size equal to 100.
(b) Two classes must be specified by involving 2 layers—softmax and classification layers.
(c) The options of training must be specified by allotting the number of epochs to a maximum value of 30.
(d) The Matlab *trainNetwork*. Function can be used for training the LSTM network overall.

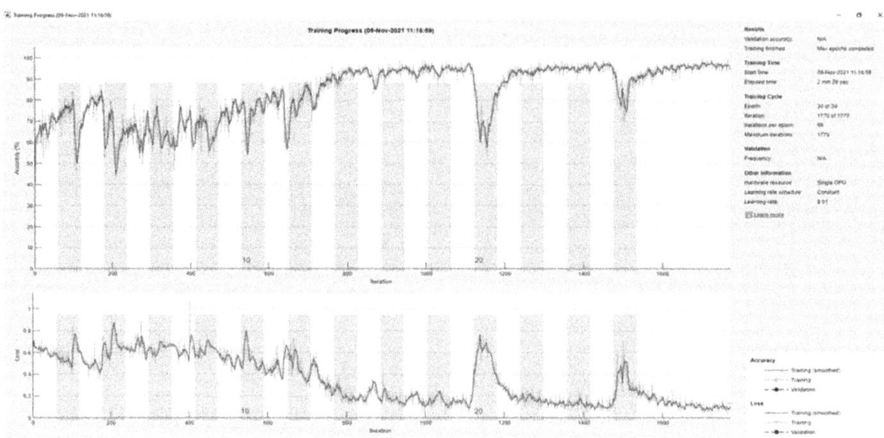

FIGURE 5.8 Performance improved training-progress plot of raw ECG sequences.

Figure 5.8 indicates the enhancement in accuracy for training. The value of cross-entropy loss moves toward the value of 0. Due to shorter TF moments, the training time decreases when compared to that of raw sequences. The confusion chart for LSTM, as shown in Figure 5.9, can be employed to have a clear picture on various concepts such as accuracy in testing, quality of training, etc.

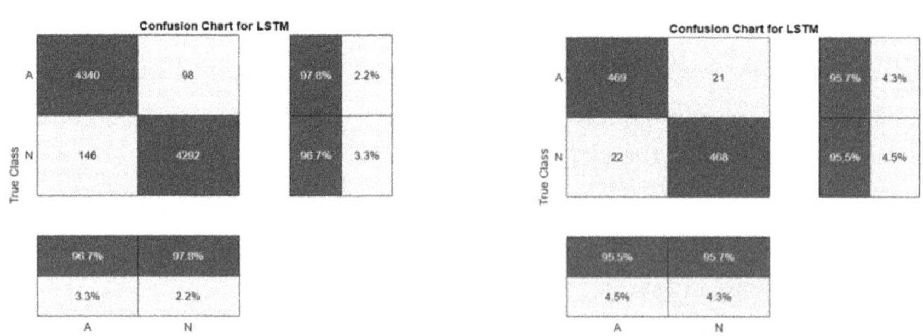

a) LSTM Accuracy = 95.6122 **b) LSTM Accuracy = 97.2510**

FIGURE 5.9 Visualize the improved training and testing accuracy using confusion matrix chart.

5.3 CONCLUSION

This paper illustrates the construction of a classifier for observing atrial fibrillations in ECG signals by employing the network of LSTM. The methodology applies the concept of oversampling to override the bias in classification which prevails in regions where more number of patients with good health are available. Instead of training LSTM network with data of raw signals, training can be provided using features of TF moments for each signal. Thereby, performance of classification is improved and the training time is decreased.

KEYWORDS

- **atrial fibrillation**
- **deep learning**
- **heartbeat**
- **dataset**
- **ECG**
- **heart disease**

REFERENCES

1. AF Classification from a Short Single Lead ECG Recording: The PhysioNet/Computing in Cardiology Challenge, 2017 [Online]. https://physionet.org/challenge/2017/
2. Clifford, G.; Liu, C.; Moody, B.; Lehman, L. W.; Silva, I.; Li, Q.; Johnson, A.; Mark, R. G. AF Classification from a Short Single Lead ECG Recording: The PhysioNet Computing in Cardiology Challenge 2017. *IEEE Comput. Cardiol.* **2017,** *44*, 1–4.
3. Goldberger, A. L., Amaral, L. A.; Glass, L.; Hausdorff, J. M.; Ivanov, P. C.; Mark, R. G.; Mietus, J. E.; Moody, G. B.; Peng, C. K.; Stanley, H. E. PhysioBank, PhysioToolkit, and PhysioNet: Components of a New Research Resource for Complex Physiologic Signals. *Circulation* **2000,** *101* (23), e215–e220. http://circ.ahajournals.org/content/101/23/e215.full
4. Pons, J.; Lidy, T.; Serra, X. In *Experimenting with Musically Motivated Convolutional Neural Networks*, 14th International Workshop on Content-Based Multimedia Indexing (CBMI), June 2016.
5. Wang, D. Deep Learning Reinvents the Hearing Aid. *IEEE Spectr.* **2017,** *54* (3), 32–37. DOI: 10.1109/MSPEC.2017.7864754.

6. Brownlee, J. How to Scale Data for Long Short-Term Memory Networks in Python [Online] July 7, 2017. https://machinelearningmastery.com/how-to-scale-data-for-long-short-term-memory-networks-in-python/.
7. Shen, M.; Zhang, L.; Luo, X.; Xu, J. In *Atrial Fibrillation Detection Algorithm Based on Manual Extraction Features and Automatic Extraction Features*, Proceedings of the IOP Conference Series: Earth and Environmental Science; Hulun Buir, China; August 28–30, 2020.
8. Wang, J.; Wang, P.; Wang, S. Automated Detection of Atrial Fbrillation in ECG Signals Based on Wavelet Packet Transform and Correlation Function of Random Process. *Biomed. Signal Process. Control* **2020,** *55*, 101662.
9. Shankar, M. G.; Babu, C. G. An Exploration of ECG Signal Feature Selection and Classification Using Machine Learning Techniques. *Int. J. Innov. Technol. Explor. Eng. Regul.* **2020,** *9*, 797–804.
10. Shi, H.; Wang, H.; Qin, C.; Zhao, L.; Liu, C. An Incremental Learning System for Atrial Fibrillation Detection Based on Transfer Learning and Active Learning. *Comput. Methods Prog. Biomed.* **2020,** *187*, 105219.
11. Hurnanen, T.; Lehtonen, E.; Tadi, M. J.; Kuusela, T.; Kiviniemi, T.; Saraste, A.; Vasankari, T.; Airaksinen, J.; Koivisto, T.; Pänkäälä, M. Automated Detection of Atrial Fibrillation Based on Time–Frequency Analysis of Seismocardiograms. *IEEE J. Biomed. Health Inf.* **2016,** *21*, 1233–1241.
12. Andersen, R. S.; Peimankar, A.; Puthusserypady, S. A Deep Learning Approach for Real-Time Detection of Atrial Fibrillation. *Expert Syst. Appl.* **2019,** *115*, 465–473.
13. Muthukumaran, D.; Sivakumar, M. Medical Image Registration: A Matlab Based Approach. *Int. J. Sci. Res. Comput. Sci. Eng. Inf. Technol.* **2017,** *2*, 29–34.
14. Ahalya, S.; Umapathy, K.; Sivakumar, M. Image Segmentation–A MATLAB Based Approach. *Int. J. Eng. Res. Comput. Sci. Eng.* **2017,** 10–14, ISSN: 2394–2320.
15. Lehari, M. N. S.; Umapathy, K.;Sivakumar, K. M. SVD Algorithm for Lossy Image Compression. *Int. J. Eng. Res. Comput. Sci. Eng.* **2017,** 44–48, ISSN: 2394–2320.

CHAPTER 6

DESIGN AND ANALYSIS OF EFFICIENT SHORT TIME FOURIER TRANSFORM BASED FEATURE EXTRACTION FOR REMOVING EOG ARTIFACTS USING DEEP LEARNING REGRESSION

M. SIVAKUMAR,[1] K. UMAPATHY,[2] P. SRINIVASAN,[3] and S. PUNITHA[4]

[1]*Department of ECE, Mohamed Sathak A.J. College of Engineering, Chennai, Tamil Nadu, India*

[2]*Department of ECE, SCSVMV (Deemed University), Kanchipuram, Tamil Nadu, India*

[3]*Department of CSE, Muthayammal Engineering College, Rasipuram, Tamil Nadu, India*

[4]*Department of ECE, Muthayammal Engineering College, Rasipuram, Tamil Nadu, India*

ABSTRACT

This paper enunciates how to eliminate electro-oculogram (EOG) noise from the signals of electroencephalogram (EEG) by employing the benchmark dataset of EEG DenoiseNet. This work involves EOG-contaminated signals of EEG signals for training a Regression model based on deep learning to eliminate the artifacts of EOG. Initially, original input signals were used to

Computational Imaging and Analytics in Biomedical Engineering: Algorithms and Applications.
T. R. Ganesh Babu, PhD, U. Saravanakumar, PhD & Balachandra Pattanaik, PhD (Eds.)

train the model of regression and then transformed signals using short time Fourier transform (STFT) were employed. The proposed model based on STFT enhances the total especially at lower values of SNR. The mean square error between actual EEG and denoised EEG signals is considered as the metric for performance. The MSE value is also calculated between actual signal and noisy EEG signals to indicate the worst case MSE if no denoising is implemented. The results were simulated using MatLab and they indicate that greater improvement in performance can be achieved by using the sequences of STFT even at worst values of SNRs.

6.1 INTRODUCTION

Nowadays, deep learning networks are gaining more importance in different domains including electroencephalography (EEG) signal processing. They provide enhanced performance when compared to techniques available in tradition. But currently the constraint for implementing solutions of deep learning for EEG denoising is nonavailability of standard datasets with proper benchmark. Hence, benchmark called EEGdenoiseNet is presented here for training and testing denoising models based on deep learning.

This dataset comprises about 4514 clean segments, 3400 ocular artifacts, and 5598 muscular artifacts by which contaminated EEG segments can be synthesized with that of clean EEG segments. By applying the benchmark, the performance of denoising can be evaluated for different types of classical networks. Moreover, the methods of deep learning will have more significance for denoising of EEG even under contamination of high noise.

There was a consideration for wavelet transformation, but there exists a compromise in classification accuracy in various time/frequency decomposing methods. STFT is a method by which time and phase-based data can be obtained from a signal. This method shall be employed for training of EEG signals.

6.2 MATERIALS AND METHODS

(i) Feature Extraction using STFT Implementation

In the place of FFT, STFT can be applied to obtain the structure of frequency for entire EEG signal in terms of time windows. Generally, applying extracted features in lieu of original signal data is the traditional strategy for enhancing the performance of a model of deep learning. The extracted features will give a good means of representation of input data easily accessible and understandable by the network. This transformation with 64 windows and 63 overlap samples will yield 33 features of complexity with 449 samples each. The complex features must be segregated into two components of real and imaginary by placing the real features over the imaginary features to obtain real features of 66 with each one length of 449 samples.

Figure 6.1 illustrates the complete activities of the process. A small signal segment with a time window of few ms is taken for consideration and then tapered to avoid edge artifacts. In this example, the hann window is applied. FFT transformation is applied on the data segment tapered at the rate of Nyquist frequency. The spectrum of the particular segment is then placed in time space corresponding to that of FFT.

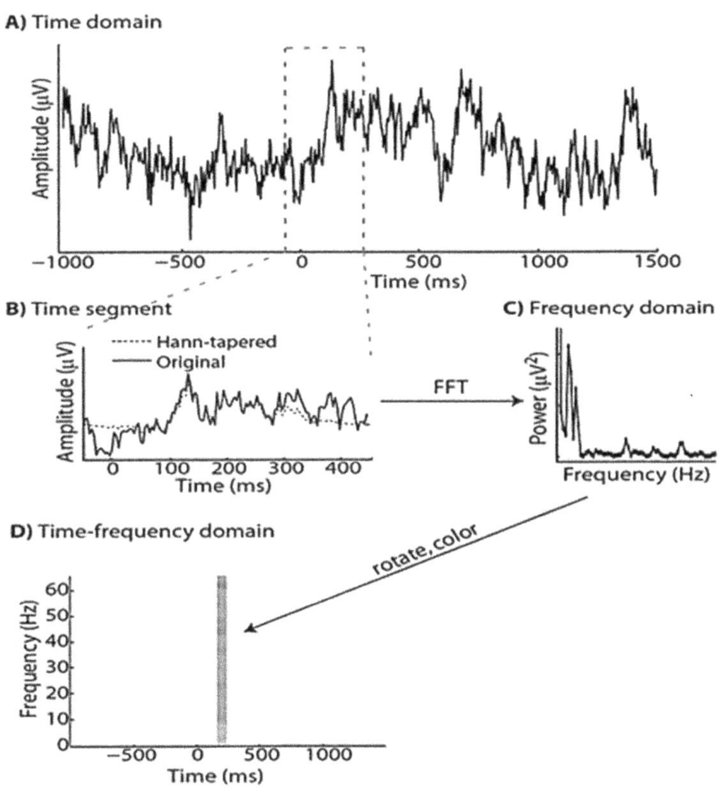

FIGURE 6.1 (A) Time segment extracted; (B) signal tapered; (C) FFT on tapered signal; (D) FFT into time-frequency domain.

(ii) Removal of EOG Artifacts:

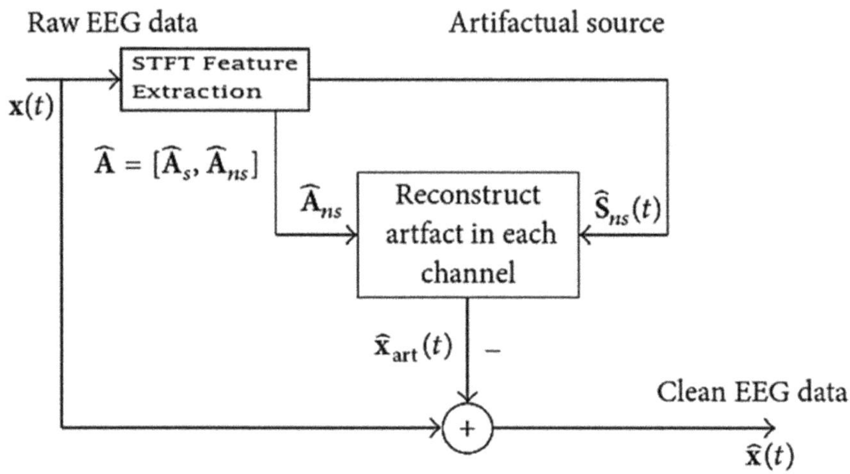

FIGURE 6.2 Block diagram of proposed system.

Figure 6.2 shows the block diagram of the proposed approach for the removal of EOG artifacts. It includes extraction of features using STFT and reconstruction of EOG artifact signal. The Matlab *transformSTFT* helper function normalizes the input signal and then computes its STFT-based clean EEG data.

(iii) Implementation Using MatLab:

Training, validation, and testing datasets can be generated by applying *createDataset* helper function of MatLab. This helper function produces pairs of clean and noisy segments of EEG with different values of SNR by integrating both clean EEG and EOG signals. The following expression can be used to get a noisy segment with a given value of SNR:

$$\text{noisyEEG} = \text{EEG} + \lambda \cdot \text{EOG}$$

The value of λ can be varied to control the artifact power with a specific value of SNR. The following variables are included in the data segment for implementation using MatLab.

- EEG for a clean EEG segment
- EOG for an EOG segment

- noisyEEG for a noisy EEG segment
- SNR for signal-to-noise ratio of noisy segment
- Fs for sample rate of signal segments

Initially, the training file is read for data and both clean and noisy EEG signals are plotted. Then call to methods of data store is done. It provides an array of size of 1 × 2 with the first element and the second element corresponding to noisy EEG segment and clean EEG segment, respectively. The input and output signals are normalized for improving the performance of the system.

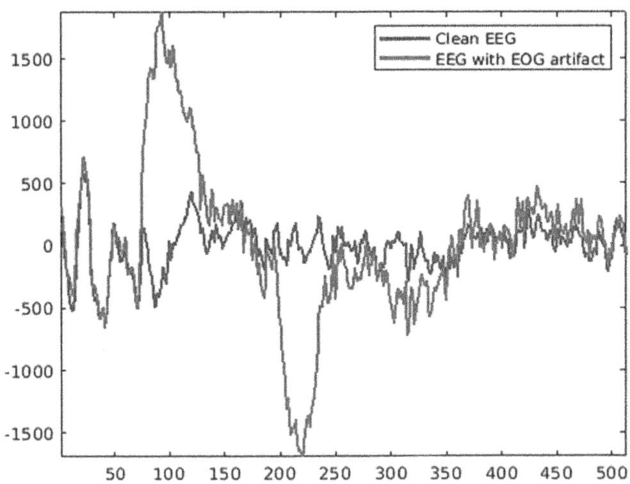

FIGURE 6.3 Training plot of the clean and EEG with EOG artifact segments.

Figure 6.3 shows the training plot of the clean and EEG with EOG artifact segments. The purpose of the proposed system is to train a network so that it can yield STFT denoised signal representations with respect to input of STFT corresponding to that of noisy signals. Finally, inverse STFT (ISTFT) is applied to recover the denoised signal as shown in Figure 6.4.

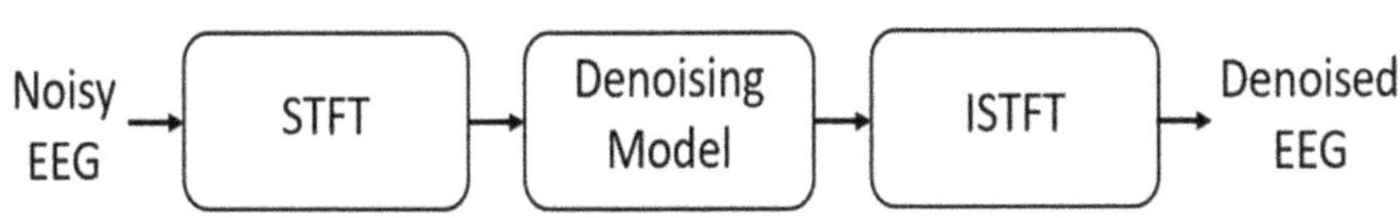

FIGURE 6.4 Recovering the denoised EEG STFT segment for deep learning regression.

Thus, the enhancement in performance can be obtained by using *transformSTFT* helper function.

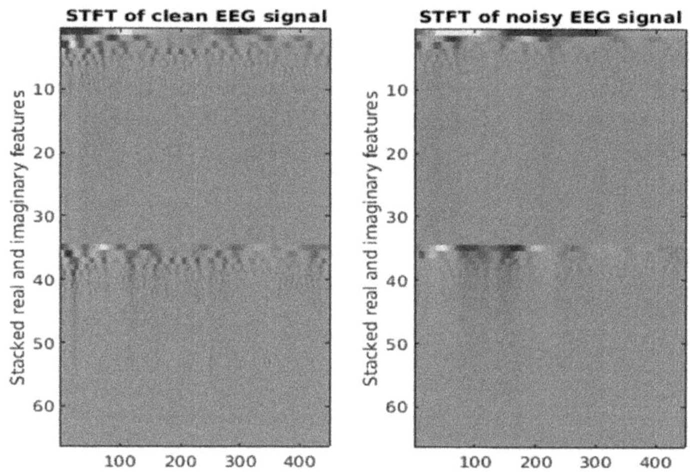

FIGURE 6.5 Resultant plot of STFT-based regression-based signal representation.

Figure 6.5 shows the resultant plot of STFT-based regression-based signal representation with enhancement in extracted features in lieu of the original raw signal data. The *getRandomEEG* helper function obtains EEG signal of random with a particular value of signal-to-noise ratio from the dataset in test.

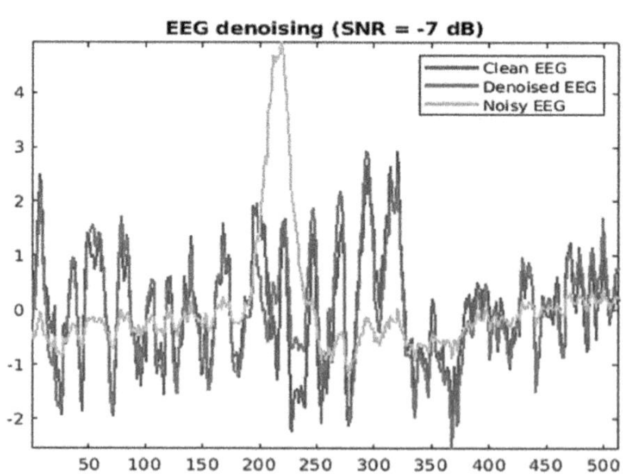

FIGURE 6.6 Resultant plot of EEG denoising noisy signals for different SNRs.

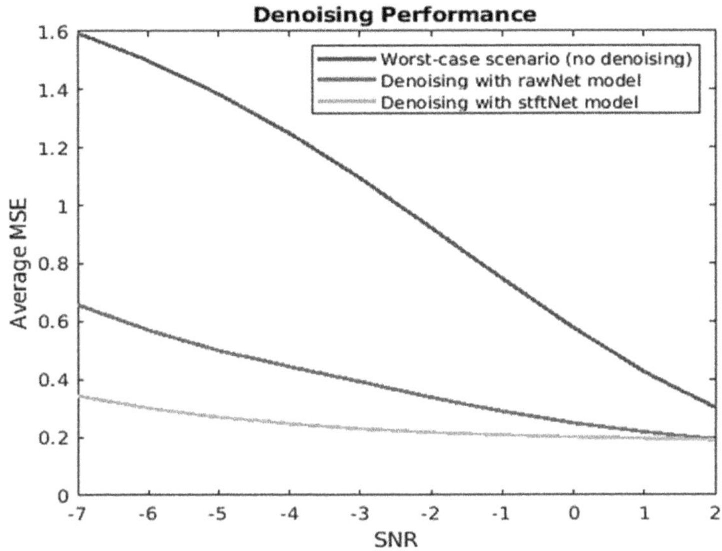

FIGURE 6.7 Improved SNR performance plot of average MSE without denoising.

Figures 6.6 and 6.7 show the plot of average value of MSE obtained without denoising and denoising network trained with raw input signals and STFT transformed signals, respectively. The performance of the system got improved by using STFT with respect to decreased values of SNR.

6.3 CONCLUSION

This paper explains how a deep network shall be trained with EEG signals to perform regression for signal denoising and removing artifacts of EOG using the feature extraction model. Additionally, a comparison between the two models trained with raw clean and noisy signals of EEG is also done.

The other technique applied is short-time Fourier transform to train, validate, and test datastores using the Matlab *transformSTFT* function. The complex features of EEG signals are treated as independent real features by using Matlab implementations. From the simulation results, it is clear that using STFT-based sequences yields greater performance enhancement at worst values of SNRs and both approaches converge in performance with the improvement in SNR.

KEYWORDS

- **regression model**
- **short-time Fourier transform**
- **mean-squared error**
- **denoising**
- **deep learning networks**

REFERENCES

1. Mashhadi, N.; et al. In *Deep Learning Denoising for EOG Artifacts Removal from EEG Signals*, IEEE Global Humanitarian Technology Conference (GHTC), 2020.
2. Yang, B.; Duan, K.; Fan, C.; Hu, C.; Wang, J. Automatic Ocular Artifacts Removal in EEG Using Deep Learning. *Biomed. Signal Process. Control* 2018, *43*, 148–158. DOI: 10.1016/j.bspc.2018.02.021.
3. Zhang, H.; Zhao, M.; Wei, C.; Mantini, D.; Li, Z.; Liu, Q. A Benchmark Dataset for Deep Learning Solutions of EEG Denoising, [Online] 2019. https://arxiv.org/abs/2009.11662
4. Gandhi, T.; Panigrahi, B. K.; Anand, S. A Comparative Study of Wavelet Families for EEG Signals Classification. *Neurocomputing* 2011, *74* (17), 3051–3057.
5. Bruns, A. Fourier-, Hilbert- and Wavelet-Based Signal Analysis: Are They Really Different Approaches. *J. Neurosci. Methods* 2004, *137* (2), 321–332.
6. Zeng, H.; Song, A. Removal of EOG Artifacts from EEG Recordings Using Stationary Subspace Analysis. Hindawi Publishing Corporation. *Sci. World J.* 2014, *2014*.
7. Hussin, S. S.; Sudirman, R. R. EEG Interpretation through Short Time Fourier Transform for Sensory Response Among Children. *Australian J. Basic Appl. Sci.* 2014, *8* (5), 417–422.
8. Mowla, Md. R.; Ng, S. C.; Zilany, M. S. A.; Paramesran, R. Artifacts-Matched Blind Source Separation and Wavelet Transform for Multichannel EEG Denoising. *Biomed. Signal Process. Control.* 2015, *22*, 111–118. DOI: 10.1016/ j.bspc. 2015.06.009.
9. Nguyen, H. A. T.; et al. EOG Artifact Removal Using a Wavelet Neural Network. *Neurocomputing* 2012, *97*, 374–389.DOI: 10.1016/j.neucom.2012.04.016.
10. He, C.; Xing, J.; Li, J.; Yang, Q.; Wang, R. A New Wavelet Threshold Determination Method Considering Interscale Correlation in Signal Denoising. *Math. Probl. Eng.* 2015, *2015*, 280251. DOI: 10.1155/2015/280251.
11. Klados, M. A.; Bamidis, P. D. A Semi-Simulated EEG/EOG Dataset for the Comparison of EOG Artifact Rejection Techniques. *Data Br.* 2016, *8*, 1004–1006. DOI: 10.1016/j.dib.2016.06.032.
12. Muthukumaran, D.; Sivakumar, M. Medical Image Registration: A Matlab Based Approach. *Int. J. Sci. Res. Comput. Sci. Eng. Inf. Technol.* 2017, *2*, 29–34.
13. Ahalya, S.; Umapathy, K.; Sivakumar, M. Image Segmentation–A MATLAB Based Approach. *Int. J. Eng. Res. Comput. Sci. Eng.* 2017, 10–14, ISSN: 2394–2320.
14. Lehari, M. N. S. Umapathy, K.; Sivakumar, K. M. SVD Algorithm for Lossy Image Compression. *Int. J. Eng. Res. Comput. Sci. Eng.* 2017, 44–48, ISSN: 2394–2320.

CHAPTER 7

MACHINE LEARNING FOR MEDICAL IMAGES

N. GOPINATH,[1] M. SURESH ANAND,[1] and M.V. ISHWARYA[2]

[1]*Department of Computer Science and Engineering, Sri Sairam Engineering College, Chennai, India*

[2]*Department of Computer Science and Engineering, Agni College of Technology, Chennai, India*

ABSTRACT

Research in medical imaging is increasingly turning to machine learning approaches. To recognize patterns in medical images, this method relies on pattern recognition. To enhance medical imaging applications, these algorithms are now being used to detect faults in the sickness diagnosing system, which may lead to significantly ambiguous medical therapy. Medical imaging relies heavily on these algorithms to detect early indications of sickness. To make predictions, it employs either supervised or unsupervised algorithms based on a predetermined dataset. Image categorization, object identification, pattern recognition, and reasoning are just a few of the principles that are examined in medical imaging. By extracting meaningful patterns for a specific condition, they are employed in medical imaging to improve accuracy. They also aid in the process of deciding on a course of action. Using Machine Learning to improve medical imaging is a trend that will continue in the future. Machine Learning approaches in medical imaging are the focus of this chapter's discussion. Machine Learning is a must-know for medical imaging professionals. We intended to give researchers a framework for

Computational Imaging and Analytics in Biomedical Engineering: Algorithms and Applications.
T. R. Ganesh Babu, PhD, U. Saravanakumar, PhD & Balachandra Pattanaik, PhD (Eds.)

understanding existing medical imaging approaches, highlighting the benefits and shortcomings of these algorithms, and discussing future prospects.

7.1 INTRODUCTION

"It is thrilling to see what machine learning can do for the area of computer science and engineering. As a subfield of artificial intelligence, it promotes the extraction of meaningful patterns from instances, which is an essential component of human intelligence."[1] To educate a computer system to think like an expert, machine learning methods are essential.[2] The goal of machine learning research is to give computers the ability to learn on their own.[3] For example, a computer may collect patterns from data and then analyze them on its own for autonomous reasoning in this domain.[4] Medical imaging is a fast expanding study field that is used to identify and cure diseases early. Based on certain expectations, digital image processing has a major impact on decision-making procedures. It improves the accuracy and extraction of features. The process of functional evaluation is complex and contains a wide range of attributes.[5,6] There are several computer systems available that use digital image processing methods. Image processing technologies must be validated before they can be used to implement specific operations that impact their performance. Medical imaging techniques are used for making choices and taking action. Basic and advanced functionality are available for image analysis and visualization. There are two subdomains of Artificial Intelligence: Machine Learning and Deep Learning.[8] To achieve Machine Learning (ML), AI, and Deep Learning (DL) all require the same methodologies, which in turn are used to achieve AI.[9]

Medical pictures go through a series of processes before they may be used to identify an output signal. It initially goes via the machine learning and deep-learning algorithms. To focus on a certain area, the picture is divided up into smaller sections. From these segments, attributes may be found utilizing information retrieval methods. We then filter out any extraneous noise before selecting the suitable quality. There are many ways to categorize the data and generate predictions based on it in the end. The following stages are always followed in a machine learning experiment. The most prevalent types of machine learning algorithms include supervised, semisupervised, unsupervised, reinforcement, and active learning. As a result, deep learning is a kind of advanced machine learning that uses neural networks to categorize and predict more precisely.[10,11]

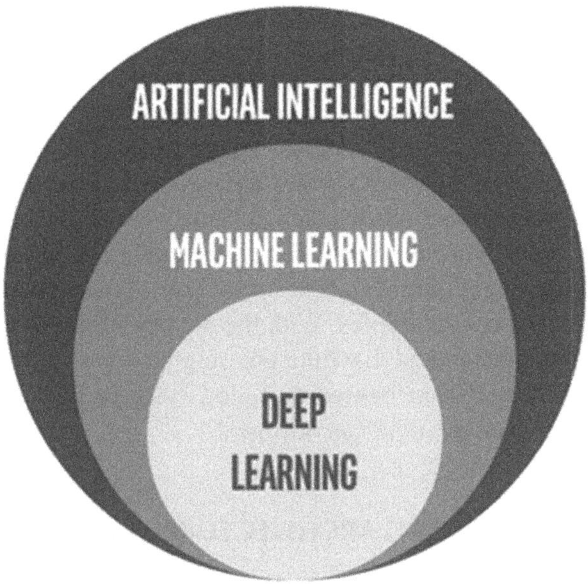

"**FIGURE 7.1** Artificial intelligence, machine learning, and deep learning domain.

7.2 MEDICAL IMAGING AND ITS TYPES

Without any injury to the patient, we would be able to identify and treat illnesses. Surgery and other invasive treatments are not necessary with medical imaging, which allows us to observe what is going on within the body without causing any harm. Possibly, we have taken it for granted at one point or another. Medical imaging is one of the most powerful tools we have for efficiently caring for our patients because it may be utilized for both diagnosis and treatment.

In terms of diagnosis, common imaging types include:

- CT (computer tomography)
- MRI (magnetic resonance imaging)
- Ultrasound
- X-ray
- Nuclear medicine imaging (including positron-emission tomography (PET)).
- Single photon emission computed tomography (SPECT), etc.

7.3 HISTORY OF MEDICAL IMAGE ANALYSIS

Low-level photo processing and mathematical modeling were used successively to construct a rule-based system that could only perform a limited set of tasks in the early days of medical image analysis. GOFAI (Good Old Fashioned Artificial Intelligence)[12] laws also existed in the field of artificial intelligence. As supervised approaches became increasingly popular in the 1990s, they became more prevalent in the area of medical image analysis, notably in the use of training data to train models. Active shape models and atlas approaches are two examples. With the development of certain creative ideas, pattern recognition and machine learning have become more popular. As a result, the move from human-designed systems to computer-trained systems based on example data has occurred.

7.4 MACHINE LEARNING ARCHITECTURE

The field of machine learning architecture has moved from the realm of possibility to the realm of evidence. An early machine learning strategy for identifying patterns established the framework for a future huge AI project. There are three kinds of machine learning architecture: In addition to the data collection, data processing, model engineering, exploration, and deployment that go into this architecture, there are also three types of unsupervised learning: reinforcement learning, unsupervised learning, and self-paced learning.[13]

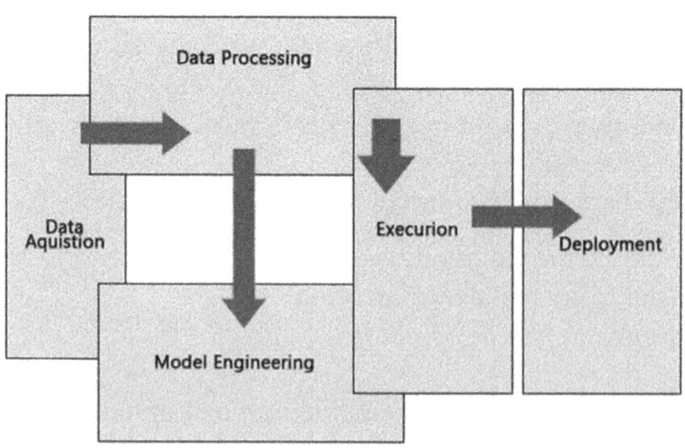

FIGURE 7.2 Block diagram of decision flow architecture for machine learning systems

7.5 TYPES OF MACHINE LEARNING ARCHITECTS

7.5.1 SUPERVISED LEARNING

It gives a computer system a collection of realistic training situations with clear goals. This training set system provides accurate responses to given probable inputs. Supervised Learning is divided into two categories: regression and classification. Classifying inputs using different classification algorithms is an important part of a taught system's operations. Experiments like this one are known as "multi-labeling." It is a spam purification procedure that categorizes emails into one of two categories: Spam or Not Spam. There are no discrete outcomes in the regression, which is a supervised approach that returns continuous findings. Regression predictions are evaluated using the Root Mean Squared Error (RMSE), while classification predictions are evaluated using accuracy (RMSE).

7.5.2 UNSUPERVISED LEARNING

Rather than training on a dataset, the system will make the decision on its own. The system receives no labeling that may be utilized to make predictions. Unsupervised learning may be used to discover the hidden pattern by learning features from the data. An unsupervised learning technique known as clustering is used to classify inputs into distinct clusters. These clusters had previously gone unnoticed. It creates groupings based on similarities.

7.5.3 SEMISUPERVISED LEARNING

The system is designed to be a training dataset in semisupervised learning. Using previously trained data, this technique is used to uncover any potential outcomes that may be missing from the final model. This method is used to train commitment on data that is not labeled. The semisupervised learning method was developed using both unlabeled and labeled data, and it demonstrated characteristics of both unsupervised and supervised learning.

7.5.4 ACTIVE LEARNING

It only obtains training tags for a small number of active learning events. A substance's optimality is enhanced to achieve the desired tag,. when it comes to an organization's budgeting tasks, for instance.

7.5.5 REINFORCEMENT LEARNING

As long as the learner is motivated by the program's actions, such as driving a vehicle or playing video games, the learnt data are only supplied in response to the program's activities.

7.5.6 EVOLUTIONARY LEARNING

Biological organisms and their progeny's survival and fatalities are mostly studied and predicted using this technique in biology. This model may be used to predict how fitness information can be used to correct the outcome.

7.5.7 DEEP LEARNING

When neural networks are utilized for data learning and prediction, this is the highest level of machine learning. Various algorithms make up this system. Systems that can solve any issue and anticipate the result are built using them. An extensive graph with various processing layers is used, which is composed of numerous linear and nonlinear conversions.[14]

There are several factors that go into diagnosing an illness in today's medical environment. In order to accurately diagnose patients, doctors must conduct thorough examinations and assessments. Health care data include everything from medical assessments to patient comments to treatment to supplement and prescription use. The healthcare business creates a lot of data. Poor data management may lead to an association effect in these reports.[15] This is the main worry. If these medical records are to be mined and processed effectively, this data has to be restructured. Different machine learning methods may be used to distribute data based on its properties using classifiers that are tailored to the specific needs of each project. Multiple categories are available for sorting the data. Data from the medical field is examined using these kinds of classifiers. It was for recognizing medical data sets that machine learning technologies were first developed and put to use. A wide range of options are now available to organize and analyze medical data using machine learning. Most new hospitals include data collecting and inspection systems that are used to gather and share data. In the medical field, it is utilized to rectify numerous diagnoses. There must be an accurate diagnosis in order for an algorithm to work, and findings may be drawn from earlier instances that have already been cracked. It is a machine learning concept that uses patterns in medical pictures to predict and draw inferences about the health of a patient.[16,17]

7.6 MEDICAL IMAGING AND MACHINE LEARNING TECHNIQUES

The use of machine learning algorithms in medical imaging is quite advantageous in the examination of specific illnesses. In medical image processing, lesions and organs may be too complicated to depict correctly by a simple mathematical solution. Image analysis using pixel-based analysis was employed by the author of Ref. [18]. In medical image processing, machine learning has been used to analyze pixel values rather than extracting features from chunks of data. In certain circumstances, the implementation of this technique may be preferable than the development of feature-based classifiers. The lack of contrast between individual pixels in high contrast photos makes them difficult to analyze. Pixel-based machine learning does not need feature computation or segmentation, unlike traditional classifiers that defend against erroneous segmentation and feature computation problems. Pixel analysis takes a long time to master because of the high dimensionality of the data (a large number of pixels in a picture). The author used low-contrast images for analyzing medical pictures. Histogram Equalization (HE) is the most effective technique for improving contrast. "Modified Histogram-Based Contrast Enhancement with Homomorphic Filtering" (MHFIL) is proposed by the authors. It used a two-phase handling procedure, the first of which boosted global contrast by modifying the histogram. For picture sharpening, second-phase homomorphic filtering is also planned. The investigation looks at 10 medical images of low-contrast chest X-rays. In comparison to other approaches, the MHFIL has minimum values in all ten photographs. Radiologists have the highest responsibility for medical image clarification, with assignments comprising both high-quality images and their analysis. CAD (computer-aided design) has been created for a variety of purposes. Linear discriminant analysis, support vector machines, decision trees, and other machine learning methods are used to analyze medical pictures. The author of Ref. [20] employed machine learning algorithms to evaluate medical images. They employed local binary patterns widely among texture descriptors in particular. In addition, a new trial using many low binary pattern descriptors of biomedical images was investigated. Beginning with face descriptions, the dataset of neonatal facial photos was used to categorize pain syndromes. Results on a painstakingly prepared 2D HeLa dataset and the recommended descriptor, which incorporates all of the different texture descriptors, achieve maximum implementation. A linear support vector machine classifier is used on the 2D-HeLa and PAP datasets. Accuracy of 92.4% was the highest among all other dataset descriptors. Neuronal network techniques are used in medical pictures to study disease characteristics.[21] The neural network groupings are retained for cancer research. It is used to criticize scenarios in which a cell is considered to be normal with a high degree of assurance, and when each single network has just two conceivable outcomes: either a normal cell or a cancer cell. To integrate the predictions from the network of cells, a common approach known as plurality voting is employed. There was a high rate of accuracy and low false-negative analytical value for the neural network as a whole, according to these findings: Expert systems that employ machine learning may generate hypotheses based on patient data. An expert system is built using rules gleaned from the knowledge of specialists. It is possible to apply machine learning methods to develop a systematic description of the clinical traits that are notably described in a collection of clinical issues that may be utilized as examples. Decision trees and other decision-tree-like diagrams may be used to represent this information. KARDIO, which was developed to interpret ECGs, is an excellent example of this method. The use of statistics to examine the properties of images in medical image analysis is a sound criterion. Nuclear medicine imaging equipment is known as a "channeled Hotelling observer" (CHO). Amenable themes in human visual structure activate the channels. Image quality assessment may be done using this method, and the results are positive for medical imaging. It is called a channelized SVM because of the way it is implemented in the algorithm (CSVM). Two medical physicists rated fault discernibility in 100 noisy pictures on a six-point scale and then assessed the score certainty of a lesion's existence in the current time period. After that, an additional 60 photographs are included in a training session. The human observers were able to complete this task for six different flattening filter selections and two different numbers of repetitions in the OS-EM rebuilding method.[23]

7.7 CONCLUSION

Since this technique was first conceived 50 years ago, machine learning technology has advanced tremendously. Initially, the models were simplistic and "brittle," which meant that they could not handle any deviations from the examples supplied in training. Due to the fast advancements in technology, machine learning systems will soon be taking on tasks formerly reserved for human beings. During the previous several years, advances in machine learning have been made. Currently, machine learning algorithms are quite resilient in actual circumstances, and the frameworks make the most of the learning process. It was formerly used to describe the practice of medical imaging, and it is predicted to increase quickly in the near future. The use of machine learning in medical imaging has substantial implications for medicine. It is imperative that our research leads to improving patient care. Machine learning's benefits must be taken seriously to make the most use of them feasible.

KEYWORDS

- **medical imaging**
- **machine learning**
- **image enhancement**
- **information retrieval**
- **supervised learning**
- **unsupervised learning**

REFERENCES

1. Erickson, B. J.; Korfiatis, P.; Akkus, Z.; Kline, T. L. Machine Learning for Medical Imaging. *Radiographics* **2017,** *37* (2), 505–515 [Online]. DOI: 10.1148/rg.2017160130 PMCID: PMC5375621

2. Latif, J.; Xiao, C.; Imran, A.; Tu, S. In *Medical Imaging Using Machine Learning and Deep Learning Algorithms: A Review*, 2019 International Conference on Computing, Mathematics and Engineering Technologies – iCoMET 2019, March 2019. DOI: 10.1109/ICOMET.2019.8673502
3. Valiant, L. G. A Theory of the Learnable. *Commun. ACM* **1984,** *27* (11), 1134–1142.
4. Robert, C.; In *Machine Learning, A Probabilistic Perspective*; Taylor & Francis, 2014.
5. Doi, K. Computer-Aided Diagnosis in Medical Imaging: Historical Review, Current Status and Future Potential. *Comput. Med. Imag. Graph.* **2007,** *31* (4-5), 198–211.
6. Mahesh, M. Fundamentals of Medical Imaging. *Med. Phys.* **2011**.
7. Jannin, P.; Grova, C.; Maurer, C. R. Model for Defining and Reporting Reference-Based Validation Protocols in Medical Image Processing. *Int. J. Comput. Assist. Radiol. Surg.* **2006,** *1* (2), 63–73.
8. https://www.intel.com/content/www/us/en/artificial-intelligence/posts/difference-between-ai-machine-learning-deep-learning.html
9. Michalski, R. S.; Carbonell, J. G.; Mitchell, T. M. Eds.; In *Machine Learning: An Artificial Intelligence Approach*; Springer Science & Business Media, 2013.
10. Norris, D. J. Machine Learning: Deep Learning. In *Beginning Artificial Intelligence with the Raspberry Pi.*; Springer, 2017; pp 211–247.
11. Jankowski, N.; Grochowski, M. In *Comparison of Instance Selection Algorithms I. Algorithms Survey*, International Conference on Artificial Intelligence and Soft Computing; Springer, 2004.
12. Wernick, M. N.; Yang, Y.; Brankov, J. G.; Yourganov, G.; Strother, S. C. Machine Learning in Medical Imaging. *Process. Mag.* **2010,** *27*, 25–38.
13. https://www.educba.com/machine-learning-architecture/
14. Schmidhuber, J. Deep Learning in Neural Networks: An Overview. *Neural Netw.* **2015,** *61*, 85–117.
15. Warwick, W.; et al. A Framework to Assess Healthcare Data Quality. *Eur. J. Soc. Behav. Sci.* **2015,** *13* (2), 1730.
16. Ghassemi, M.; et al. Opportunities in Machine Learning for Healthcare, 2018. arXiv preprint arXiv:1806.00388.
17. Dua, S.; Acharya, U. R.; Dua, P. In *Machine Learning in Healthcare Informatics*, 2014.
18. Suzuki, K. Pixel-Based Machine Learning in Medical Imaging. *J. Biomed. Imag.* **2012,** *2012*, 1.
19. Agarwal, T. K.; Tiwari, M.; Lamba, S. S. In *Modified Histogram Based Contrast Enhancement Using Homomorphic Filtering for Medical Images*, Advance Computing Conference (IACC); 2014 IEEE International; IEEE, 2014.
20. Nanni, L.; Lumini, A.; Brahnam, S. Local Binary Pattern Variants as Texture Descriptors for Medical Image Analysis. *Artif. Intell. Med.* **2010,** *49* (2), 117–125
21. Shi, Z.; He, L. In *Application of Neural Networks in Medical Image Processing*, Proceedings of the Second International Symposium on Networking and Network Security; Citeseer, 2010.
22. Bratko, I.; Mozetič, I.; Lavrač, N. KARDIO: A Study in Deep and Qualitative Knowledge for Expert Systems; MIT Press, 1990.
23. Narasimhamurthy, A. An Overview of Machine Learning in Medical Image Analysis: Trends in Health Informatics. In *Classification and Clustering in Biomedical Signal Processing*; IGI Global, 2016; pp 23–45.

CHAPTER 8

INNOVATIONS IN ARTIFICIAL INTELLIGENCE AND HUMAN COMPUTER INTERACTION IN THE DIGITAL ERA

M. V. ISHWARYA,[1] M. SURESH ANAND,[2] A. KUMARESAN,[3] and N. GOPINATH[4]

[1]*Department of Artificial Intelligence and Data Science, Agni College of Technology, Chennai, India*

[2]*Department of Computing Technologies, School of Computing, SRM Institute of Science & Technology, Kattankulathur, India*

[3]*School of Computer Science and Engineering, Vellore Institute of Technology, Vellore, India*

[4]*Department of Computer Science and Engineering, Sri Sairam Engineering College, Chennai, India*

ABSTRACT

The artificial intelligence (AI) and human–computer interaction (HCI) people group have frequently been described as having contradicting perspectives on how people and PCs ought to interface. As the two of them advance, there is a more profound differentiation that cuts across these networks, in how scientists imagine the connection among information and plan. By inspecting the rationalistic and plan directions fundamental groups of work in the two disciplines, we feature applicable contrasts and opportunities for

Computational Imaging and Analytics in Biomedical Engineering: Algorithms and Applications.
T. R. Ganesh Babu, PhD, U. Saravanakumar, PhD & Balachandra Pattanaik, PhD (Eds.)

powerful communication with PCs. The new ascent of profound learning has changed AI and has prompted a heap of pragmatic strategies and apparatuses that fundamentally sway regions outside of center AI.

Specifically, current AI strategies presently power new ways for machines and people to communicate. Accordingly, it is convenient to examine how current AI can drive HCI research recently and how HCI exploration can assist with coordinating AI advancements. This study offers a gathering for specialists to examine new freedoms that lie in bringing current AI strategies into HCI research, recognizing significant issues to explore, exhibiting computational, and logical techniques that can be applied, and sharing datasets and apparatuses that are now accessible or proposing those that ought to be additionally evolved. The themes we are keen on including profound learning techniques for comprehension and demonstrating human practices and empowering new association modalities, cross breed knowledge that consolidate human and machine insight to address troublesome errands, and apparatuses and strategies for collaboration information duration and enormous scope information driven plan. At the center of these points, we need to begin the discussion on how information driven and information driven methodologies of present day AI can affect HCI.

With regards to these and different difficulties, the jobs of people working couple with these frameworks will be significant, yet the HCI people group has been just a tranquil voice in these discussions to date. This article diagrams a background marked by the fields that recognize a portion of the powers that avoided the fields as much as possible. Computer-based intelligence was by and large set apart by an exceptionally aggressive, long haul vision requiring costly frameworks, albeit the term was infrequently imagined as being the length of it end up being, while HCI centered more around advancement and improvement of broadly utilized equipment inside a brief time frame scale. These distinctions prompted various needs, techniques, and appraisal draws near.

We propose a human-PC communication stage for the hearing weakened, that would be utilized in medical clinics and banks. To grow such a system, we gathered Bosphorus Sign, a Turkish Sign Language corpus in wellbeing and finance spaces, by counseling communication through signing etymologists, local clients, and area specialists. Utilizing a subset of the gathered corpus, we have planned a model system, which we called HospiSign that is expected to help the Deaf in their clinic visits. The HospiSign stage directs its clients through a tree-based movement outline by asking specific questions and requiring the clients to reply from the given alternatives. In order to perceive signs that are offered as responses to the cooperation stage, we propose during hand position, hand shape, hand development, and chest area present highlights to represent signs. To demonstrate the fleeting part of the signs we utilized dynamic time warping and temporal templates. The order of the signs is finished utilizing k-Nearest Neighbors and Random Decision Forest classifiers. We led experiments on a subset of Bosphorus Sign and assessed the adequacy of the framework in terms of highlights, transient demonstrating strategies, and characterization techniques. In our experiments, the mix of hand position and hand development highlights yielded the highest acknowledgment execution while both of the fleeting demonstrating and classification techniques gave serious outcomes. In addition, we researched the impacts of utilizing a tree-based action chart and discovered the way to deal with increment the recognition performance, yet additionally facilitate the transformation of the clients to the framework. Furthermore, we researched area variation and facial milestone limitation methods and examined their appropriateness to the motion and communication through signing acknowledgment errands.

8.1 INTRODUCTION

Human–computer interaction (HCI) centers around the plan, assessment, and utilization of data and correspondence advances with an express objective to further develop client encounters, task execution, and personal satisfaction. HCI is right now being formed and molding the utilizations of man-made reasoning (artificial intelligence [AI]) and keen increase (IA). This is prompting the quick development of new and invigorating examination themes. These subjects and the inquiries got from them are broadening and testing our ebb and flow hypothetical establishments and exploration philosophies. This calls for both reflection and conversation among IS researchers.

This track plans to give a gathering to talk about, create, and advance a scope of examination investigating novel speculations, systems, and exact bits of knowledge identified with marvel relating to HCI, AI, or potentially IA. We invite entries mirroring a breath of examination customs and approaches, including conduct, financial aspects of IS, econometrics, plan science, and information science, that address new and arising issues in these fields.

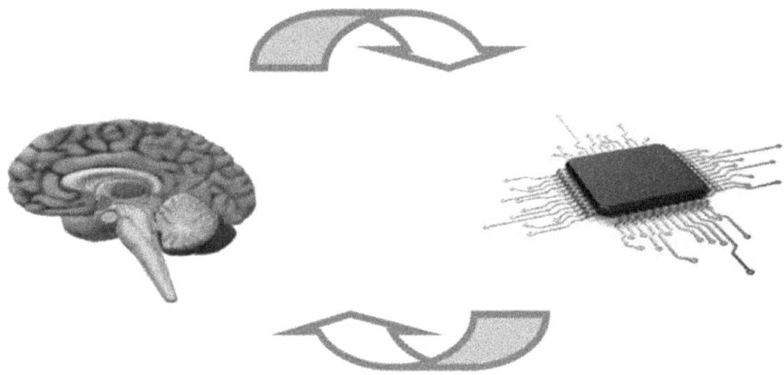

FIGURE 8.1 HCI and AI interaction.

Exploration applicable to the track might offer novel hypothetical points of view on HCI, AI, and IA in an assortment of settings including, however, not restricted to web-based business, m-trade, work, associations, human collaborations with shrewd advances, human robot cooperations, novel interface plans for computer-generated reality/expanded reality (VR/AR) innovations, and assessment of HCI issues through neurophysiological devices and gadgets (e.g., EEG, fMRI, GSR, and eye-trackers). Significant exploration may likewise incorporate examinations analyzing the effect of AI/IA on business technique improvement, applications in open versus private associations, the impact of AI available, economy and society, and the business esteem and unexpected results of AI. The logical and designing disciplines that make up or are firmly aligned with AI and HCI are of somewhat ongoing beginning. They date back to the 1950s at the soonest, with the main reference book article on AI showing up in 1967. AI and HCI have been quickly changing and growing both in scholarly substance and in application, particularly somewhat recently. The new sped up speed of progress has been expected in no little part to the practically amazing advancements and item improvements in supporting microelectronics, elect optics, and show advances—all equipment escalated regions.

Development in the measure of new wording and specialized language consistently goes with such changes and advances in innovation. Perceiving the uselessness of endeavors to create extensive definitions considering these quick changes, the board has picked to give brief portrayals of four regularly utilized terms: computerized reasoning, human–PC interface, virtual universes, and manufactured conditions. Man-made consciousness is the assortment of calculations that whenever make it conceivable to help clients to see, reason, and act. Since it is calculations that make up AI, the elements of seeing, thinking, and acting can be cultivated heavily influenced by the computational gadget (e.g., PCs or mechanical technology) being referred to.

Simulated intelligence at least incorporates:

- Representations of "reality," insight, and data, alongside-related techniques for portrayal;
- Machine learning;
- Representations of vision and language;
- Robotics; and
- Virtual reality (characterized underneath)

Human–PC interface (HCI) comprises of the accompanying:

- The machine combination and understanding of information and their show in a structure advantageous to the human administrator or client (i.e., shows, human insight copied in computational gadgets, and reproduction and engineered conditions).

Availability and, specifically, versatility, and ease of use are helpful highlights for a PC – equipment or programming asset, since they universalize the advantages of its legitimate use. Openness implies the joining of properties to permit utilizing a PC asset by individuals with some sort of scholarly incapacity or sensorial/actual hindrance, just as the incorporation and expansion of its utilization to all areas of society. Versatility is characterized as the limit of a PC asset to work in various conditions by changing its setup which could be made by end clients, or more than likely its ability to learn with client collaboration and adjust to accommodate its clients. At long last, ease of use is identified with the office to learn and utilize a PC asset. These ideas support the full advancement of availability, flexibility, and convenience highlights as a client alternative, and not as an inconvenience of the asset configuration measure. To advance a superior readability, the term openness is utilized in future as an overall term for availability, versatility, and convenience. Def physical weakness, just as the consideration and augmentation of its utilization to all areas of society. Flexibility is characterized as the limit of a PC asset to work in various conditions by changing its arrangement—which could be made by end clients, or probably its ability to learn with client collaboration and adjust to accommodate its clients. At last, convenience is identified with the office to learn and utilize a PC asset.

These ideas build up the full advancement of availability, versatility, and ease of use highlights as a client choice, and not as a burden of the asset configuration measure. To advance a superior decipherability, the term availability is utilized from now on as an overall term for openness, versatility, and convenience. Characterizing what, how, when and for what reason to fabricate PC assets with availability highlights is one of the primary worries of the HCI region. The association between a client and the asset is portrayed by a two-way course, wherein each incidental affects the conduct or the outcomes created by the other one. The PC asset configuration measure needs to consider the pertinent impacts delivered by such connection, and meeting openness prerequisites includes the advancement of exceptional systems to further develop this client collaboration what, how, when, and for what reason to fabricate PC assets with availability highlights is one of the fundamental worries of the HCI region.

The connection between a client and the asset is described by a two-way course, where each incidental affects the conduct or the outcomes delivered by the other one. The PC asset configuration measure needs to consider the applicable impacts delivered by such cooperation, and meeting openness necessities includes the advancement of exceptional procedures to further development of this client association. Simulated intelligence instruments used to execute it are momentarily examined; at long last, the future examination bearings and ends are given and the connected book indices are recorded. The main methodology, which I will call "rationalistic" (see conversation of this term in tries to show individuals as intellectual machines, whose inside instruments equal to those we incorporate into computerized PCs. The clearest articulation of this view is Newell and Simon's Physical Symbol System Hypothesis,[9] which impacted an age of scientists both in AI and HCI. It was the hypothetical articulation of what Haugeland[5] calls "Ordinary AI" (GOFAI), which was the prevailing worldview in the period of elevated requirements for the close term making of human-like AI. Newell's origination was additionally the vital motivation for the establishing text on HCI as a discipline of psychological designing, which stays powerful in the HCI people group today.

A speedy look at the papers in the yearly ACM SigCHI meeting shows numerous papers that address communication issues from an exact quantitative point of view. The critical suspicions of the rationalistic methodology are that the fundamental parts of thought can be caught in a formal emblematic portrayal. Regardless of whether it relates straightforwardly to a conventional rationale, it works like a rationale in that clear cut algorithmic guidelines can be applied to models (cycles and information) as image structures. Furnished with this rationale, we can make shrewd projects and we can plan frameworks that improve human cooperation. The subsequent methodology is more enthusiastically to mark. It has fondness to the individuals who call their methodology "phenomenological," "constructivist," and "natural," and I will allude to it as a "plan" approach. In the plan approach, the attention isn't on demonstrating canny inward operations, yet on the connections between an individual and the encompassing climate. Alongside this shift of center goes a change in the sort of understanding that is sought after. A critical piece of the thing that matters is in the job of formal demonstrating and clarification. In plan, we regularly work in spaces of human translations and practices for which we don't have prescient models. The topic of "Accomplishes it work?" isn't drawn nearer as an estimation before development, however, as an iterative cycle of model testing and refinement. David Kelley, author of the plan firm IDEO and the head of Stanford's Institute for Design (dschool. stanford.edu) is regularly referred to his explanation that "Edified experimentation beats the arranging of impeccable acumen." This isn't an assertion against astuteness, yet rather an affirmation of the impediments of knowing and demonstrating the intricacies of the genuine human world. Inside the HCI people group, there has been a continuous movement away from Newell's intellectual designing methodology, toward a plan situated position that draws significantly more from the experience of other plan disciplines.[14] The present CHI people group is occupied with progressing banters about the jobs of investigation and configuration in making new interfaces and understanding the manner in which individuals associate with them (see, for instance, Ref. [15]). Obviously, "plan" is deciphered by various individuals in many related, however, non-indistinguishable manners.

A speedy web look for the terms HCI and design gives a vibe for the variety and enthusiasm of the conversation. Over a similar period, AI has followed a comparable to direction. We have seen the overall surrender of GOFAI and the ascendance of measurable, encapsulated, and constructivist draws near. It is difficult to do equity to the significant ideas and contrasts among these in a short remark, yet a string goes through them that resounds with the 1258 T. Winograd/Artificial Intelligence 170 (2006) 1256–1258 plan approach in HCI. Maybe than putting together AI skill with respect to a coherent portrayal of the setting and the specialist's information, there is an exchange between broad versatile systems and world experience, which leads after some time to shrewd practices, frequently as the result of broad models and preparing. Numerous pundits of AI, starting with Dreyfus[2] have underlined the indistinguishability of human idea and human actual epitome. Taking on a similar mindset as a human isn't simply a question of the right cerebrum engineering or quality grouping, yet of a complex formative communication between the entire creature and its current circumstance, including others. He and others have contended that for a gadget to have human-like knowledge would require its encountering something similar to human turn of events and childhood. The rationalistic reaction to this is to imitate the outcome of actual world involvement in an assortment of suggestions, as in CYC,[6] a task which has neglected to satisfy its hopes after numerous long periods of exertion (see Ref. [13]). Obviously, there is a significant job for a rationalistic methodology in making and understanding the versatile components. Work in measurable language understanding, neural organizations, or AI depends on profound investigation and quantitative models of the various components and strategies that structure the reason for transformation. In any case, the analyst isn't needed (or ready) to unequivocally address the information or rules of conduct for the smart framework.

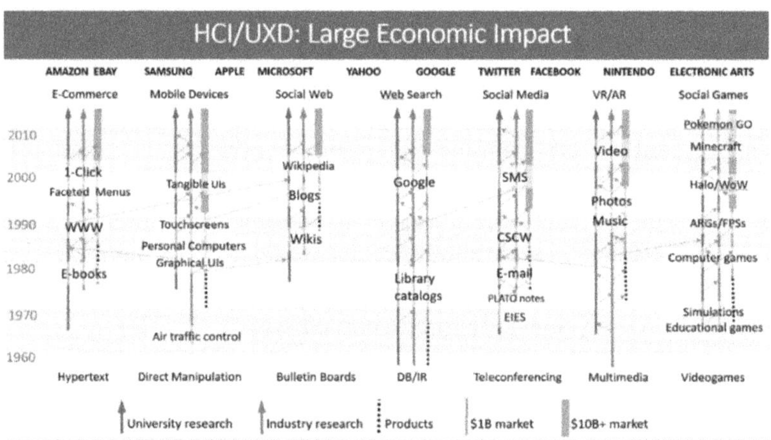

FIGURE 8.2 Growths of HCI and AI.

Eventually, "Edified experimentation beats the arranging of perfect mind. Human and man-made brainpower collaboration are introducing huge and energizing mechanical freedoms for common advancement even in the present current innovative environment, however, the genuine potential for shared improvement is within a reasonable time-frame and past which can possibly be astounding. With the proceeded with fast advancement of innovation as well as in quantum research, the potential for AI to develop at a terrifying rate is inside our grip. The structure squares to accomplish this incredible jump in innovation comprises of Quantum Computing, Big Data, and mind PC interface (BCI), each becoming venturing stones and progressing to such a phase where when innovation just considered in sci-fi will be an ordinary component of present day human and AI innovation. Man-made intelligence and people carry their own one of a kind characteristics with each gaining from different, people bring ascribes including experience, qualities, and judgment which can weave along with the gigantic traits that AI can contribute. Simulated intelligence and human association can possibly create and push the limits of innovations on earth as well as space investigation. For instance, with this coordinated effort comes the potential for the fruitful turn of events and business utilization of nuclear fusion to supply a clean new tremendous fuel source. Add this along with the capacity of AI to nullify the need to rest, to take care of or the need to breath oxygen, giving AI the capacity for huge space travel while examining conceivable future human investigation and in any event, driving the limits further with possible human colonization. Tremendous measures of large information and the copious stock of organized information are fundamental for the future headway of AI, the preparation of learning calculations for use in deep learning and neural networks is vital for make AI more shrewd.

The preparing force of the present PCs is now working related to enormous information ceaselessly propelling the capacity to handle huge information, all the more productively developing close by the appearance of equal preparing and Hadoop with both progressing at a frantic rate which thus is empowering AI to turn out to be considerably more astute. Envision the advancement of AI if the force of quantum computing was promptly accessible with potential preparing power multiple times more impressive than the present old style PCs with every producer endeavoring to arrive at quantum supremacy as well as far and past this progressive advancement Humanity can outfit this astounding innovation with AI and BCI creating another innovation transformation, this is the future, yet what's to come is quick drawing closer. With quantum computing carrying the capacity to handle enormous information in plenitude, this will thus soar AI improvement prompting the progression of BCIs, with AI being able to pinpoint as well as to guide and increment for instance DNA planning distinguishing qualities in strength and knowledge. The opportunities for super strength and incredibly smart people to coordinate with the enormous progression in AI insight are achievable. With the peculiarity sure to be outperformed later on people and AI will take a stab at artificial general intelligence (AGI).

This potential for humans and AI to develop constantly together is faltering, however, the topic of security, regulations, and ethics close by AI and future human BCI headways features the requirement for standard security and guidelines to be set up. From man-made brainpower and choice emotionally supportive networks and their utilization in a scope of settings, the following two articles research non-traditional HCI and moral contemplations in regards to explicit socio-specialized frameworks with regards to crisis and catastrophe the board. For example, Shipman investigates the moral difficulties related to the plan of foundation that is educated by human conduct and person on foot elements. While the creator distinguishes the chances this presents, from supporting individualized encounters and personalization in brilliant urban areas, to working with improved departures during threatening or regular crises, the moral dangers are likewise underlined. The requirement for industry self-guideline is obvious, just like the significance of analysts and modelers displaying familiarity with the pertinent moral difficulties.

8.2 HEALTHCARE SYSTEMS

This paper presents a continuous investigation pointed toward EMR quality evaluation and improvement as a piece of general calculated and mechanical reason for CDSS building.[16] The considered issues show up and are settled inside an extent of advancement for dynamic in medication and medical care which utilizes EMR as a center wellspring of data on patients' curation in clinic. This makes the EMR quality improvement to be considered as a significant issue for CDSS advancement. The appropriate plan of EMR GUI isn't the lone issue for aversion of slip-ups in EMR. As a matter of fact, just slip-ups of 2–4 classes portrayed in 2.2 could be tackled thusly. Different slip-ups classes need more clever arrangements, while the framework ought to give hints dependent on recently entered data during the current client meeting. This assignment can be settled utilizing CDSS coordinated with EMR. Class 1 errors might prompt befuddle in drugs buy that prompt inaccurate medication consumption.[23] Creators in Ref. [24] propose the rundown of activities for clinical specialists to keep away from such circumstance, in this way, we accept that the CDSS can help by mistakes checking with respect to finding and select that error amendment that fits to the analysis. Another approach to further develop drug solution is to naturally change drug names to utilize Tall Man lettering, while it can diminish drug name disarray errors.[25]

Computerized diagnostics of certain sicknesses dependent on clinical trial results examination could assist with remedying class 4 errors. Different account strategies for a similar data are a more muddled issue and there are various approaches to settle it. One way is perceive current info explanation and give autocomplete hint for the client. This clue might be founded on most normal structure for such explanations recovered from EMR. What direction we ought to choose is the theme for future examination. The CDSS can perform such undertakings as checking conceivable medication drug collaborations (DDI) in doctor prescriptions[26] or checking the consistency of field "analysis," clinical trials, and endorsed drugs. In Ref. [27], creators led meta-examination of a few investigations on CDSS. Creators notice that advanced CDSS ought to be not difficult to utilize, and yet, ought to keep away from the chance of disregarding proposals in the framework. With everything taken into account, it is an uncommon test to foster HCI in CDSS. Our flow project is committed to the improvement of a CDSS for cardiovascular infections coordinated with EMR. The principal point of this CDSS is to further develop HCI to diminish the quantity of missteps in EMR. Our framework coordinates with the as of now utilized EMR (see Fig. 8.1 for theoretical engineering), utilizes the records put away in EMR, reuses and broadens the accessible EMR (UI). The second permits us to save existing client experience, stay away from most mix-ups associated with curiosity and acquire fair assessment of the progressions in the medical care quality. Specialized issues of the coordination interaction are settled by the UI expansion. That UI augmentation screens current client meeting in EMR, cooperates with worker side of DSS for meeting investigation and shows cautions and ideas when a few occasions are set off.

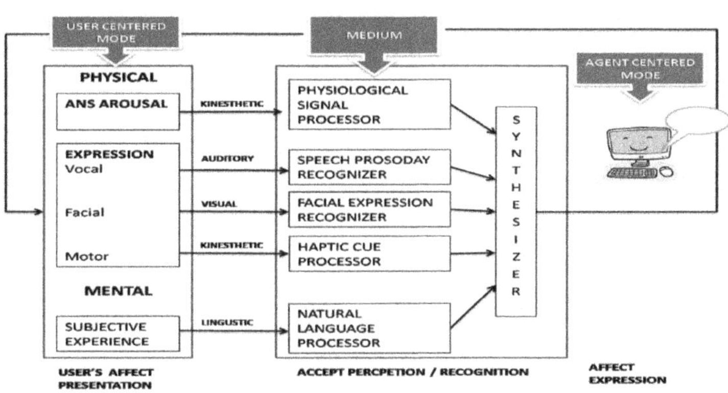

FIGURE 8.3 HCI and AI for healthcare systems.

As of now the proposed engineering is effectively being worked out to give (1) the adaptable incorporation with MIS; (2) gradual update HCI. To play out certain slip-ups adjustment errands normal language preparing techniques ought to be utilized and distinctive AI and information recovering strategies, including counterfeit neural organization, choice trees, and other characterization calculations could be utilized to make the information base. Work on this task is as yet in progress and we are in the beginning phase of advancement including the decision of explicit calculations (Fig. 8.1). Reasonable engineering of the CDSS incorporated in EMR. To further develop client association of the clinical data framework, we utilize the cycle DMAIC (define, measure, analyze, improve, and control).[28] Currently, we have characterized the issue and distinguished potential causes as per the primary period of the DMAIC cycle. The examination was led in genuine settings and gives rich subjective information to foster the CDSS and the vision of things to come framework. In the pattern of measure–analyze–improve, we will make another adaptation of CDSS. Because of the way that clients conjure the MIS consistently, we will apply a technique for Subjective Mental Effort Question (SMEQ)[29] for estimating fulfilment on each phase of cycle. A seven-digit size of reactions gives a decent harmony between the quantity of alternatives for settling on a powerful decision by the respondent and special cases blunder of estimation. Culmination and exactness of information in EMR structures are the reason for generous advantages, remembering better consideration, and decline for medical services costs.

The significant piece of the EMR is the plan of HCI. Improvement of human–PC communication influences definitely a larger number of cycles than it may appear at the main look. Remembering that HCI influences doctor appropriation of EMR, it likewise contacts the course of therapy, medical care quality estimation, clinical information investigation, and numerous different viewpoints that are not shrouded in this paper. This is the motivation behind why EMR frameworks ought to be planned, carried out, and utilized fittingly. Something else, clinical frameworks will bring about accidental unfavorable results, for example, misdiagnosis, underestimation of infection seriousness of associative sicknesses or medication drug communications. The entirety of this might influence patient wellbeing.

Use of coordinated CDSS pointed toward further developing HCI gives certain advantages and diminishes the quantity of mix-ups in EMR. It likewise animates the appropriation of EMR by doctors. These progressions ought to work on the nature of clinical consideration. Great nature of EMR furnishes information researchers with more information helpful for investigation and more certainty about the pertinence of the outcomes acquired from EMR information examination. In the meantime, as referenced above, progress of human–PC association in EMR requests specialized arrangements, yet in addition assistance of doctors' comprehension of the significance of such frameworks for their normal practice and further utilization of information put away in such frameworks. CDSS engineers will work intimately with doctors and information researchers to comprehend their requirements and hardships.

Along these lines, we will lead ethnographic meetings with doctors, as we did with information researchers, to comprehend the conduct and customs of individuals interfacing with the framework. We intend to utilize DMAIC cycle to work on the ease of use of the framework being worked on. Shrewd Health with the expanding application and acknowledgment of digitalization, especially in agricultural nations like India, the Internet turns into a fundamental factor for medical services. Savvy wellbeing like e-Health administrations has the critical potential to upgrade patient analysis and to work on the availability and nature of treatment using Internet applications and cloud stage.[28] With the progression of ICT, Web 2.0 innovation has carried the degree to expand these administrations of by permitting patients, patient's families, and local area everywhere to partake all the more effectively during the time spent wellbeing advancement.

What's more, schooling through long-range interpersonal communication measure,[15] many agricultural nations are receiving e-Health for its adaptability and speed to defeat hindrances like the comfort of arrangement, diminished driving expenses and time, decreased patient holding up period, further developed admittance to clinical subject-matter experts and wellbeing data alongside directed plans and subsequent meet-ups. Such highlights open another window for cutting edge and innovation empowered treatment applications. This likewise makes the consideration cycle a worked on personal satisfaction for a great many country patients experiencing clinical problems which can be better overseen.[28] Because of the progression of innovation alongside responsive highlights, a large number of the online applications are currently available through cell phones and other shrewd gadgets, which is carrying adaptability and solidarity to the client and saving his time and cost. To perceive the future extent of cell phone in medical services, Joshi and Pant have recommended from their exploration that execution of versatile applications in wellbeing administrations is in the underlying stage yet its further use in both giving clinical instruction and expert medical care will change the medical care industry.

HCI professionals give off an impression of being staying away from a significant region; similarly, individuals working in the space appear to be ignorant of HCI. For instance, are software engineering divisions graduating designers ignorant of HCI? Maybe plan/makers/controllers/obtainment may not be utilizing qualified individuals. Assuming we are to work on the nature of wellbeing basic regions, HCI must be an innovator in uncovering issues and showing it can keep away from or oversee abandons. There is a minuscule writing on HCI in safety critical regions, especially in medical care. In the medical care space, the majority of the writing focuses on clinical sequelae of antagonistic occasions (e.g., by directing CPR and a remedy, the patient recuperated from a medication glut), not on the plan or HCI issues (e.g., the ergonomics of the keypad urged the attendant to enter some unacceptable medication portion, and the UI configuration gave no dynamic survey of the portion) that made the idle conditions for the blunder. In the USA, the obligatory clinical gadget mistake reports frequently fault the client at whatever point no specialized breakdown could be recognized.

HCI and dormant conditions prompting the occurrence are by and large overlooked. One of the conceivable outcomes is that HCI is intricate and that clinical gadgets and conditions are perplexing. One may likewise add further reasons like the exceptionally cut-throat market, quickly evolving advancements, and issues undertaking dependable environmental client contemplates (e.g., patient classification, or generally unessential clinical exploration guidelines, for example, randomized control preliminaries). In actuality, this paper showed that fundamental HCI can add to the medical care. Obviously, a few issues of security basic frameworks lie outside HCI itself; maybe plans of action inside an administrative system that disregards HCI. Consider how we prepare and spur HCI experts to draw in—or stay away from—significant regions like medical care. The turn of events and take-up of various advancements might appear to be exceptionally unique in various pieces of the world, due to accessibility, expenses, and specialized availability. The Western world has seen an improvement from fixed PCs, through PCs tablets, and versatile PCs, presently moving toward the Internet of things, sensors, and universal advancements. It has gone from proficient business related use to more private and recreation use. Numerous nations from the creating scene skirt at least one stage in this interaction, furnishing admittance to versatile advances with no or little utilization of fixed or PCs.

Portable innovations are less expensive, more secure, and simpler to use in a more customized way.

The methods of utilizing the computerized innovations fluctuate a ton in various pieces of the world because of culture, yet in addition to accessibility. In the Western world, a cell phone might be an extremely private device that one infrequently provides for another person; yet in the country Africa, a cell phone might be shared by numerous individuals in a more prominent social setting.

Through the expanding digitalization of advanced education and the expanding global accessibility through MOOCs and other carefully accessible innovations for instructing and learning, the chances for building abilities and occupations in the agricultural nations also increment immensely. A particularly mechanical improvement will, if appropriately utilized, assume a significant part in democratization and fair freedoms for advancement and in adding to a few of the maintainability advancement objectives.

It is important to create and adjust the techniques and cycles for plan and improvement of computerized change to the particular territorial conditions in each separate region. By opening up advancement, publicly supporting improvement to take into account numerous to contribute, by being more deft and considering changing prerequisites improvement could be made considerably more applicable for different pieces of the world. Changing conduct and mentalities to permit a more client focused methodology, notwithstanding, has so far ended up functioning admirably in various pieces of the world, regardless of whether the manners by which it has been embraced change in various districts.

There is likewise a danger for "imperialistic colonialization" of programming improvement in which "advanced superpowers" force their advancement societies without recognizing the neighborhood perspectives in light of a legitimate concern for digitalization. Supporting global turn of events and digitalization requires a reasonable cycle of quietude and regard for nearby customs and societies to deal with a solid presentation of new improvement measures, instruments, practices, and strategies. In addition, research has shown that the presentation of new innovation in medical care is once in a while met with opposition and that dormancy is solid with regards to the foundation of innovation. At the point when online clinical records were presented in Sweden doctors emphatically disliked the framework as they were stressed over their workplace and about patients perusing clinical records on the web and becoming stressed over what they read.[25]

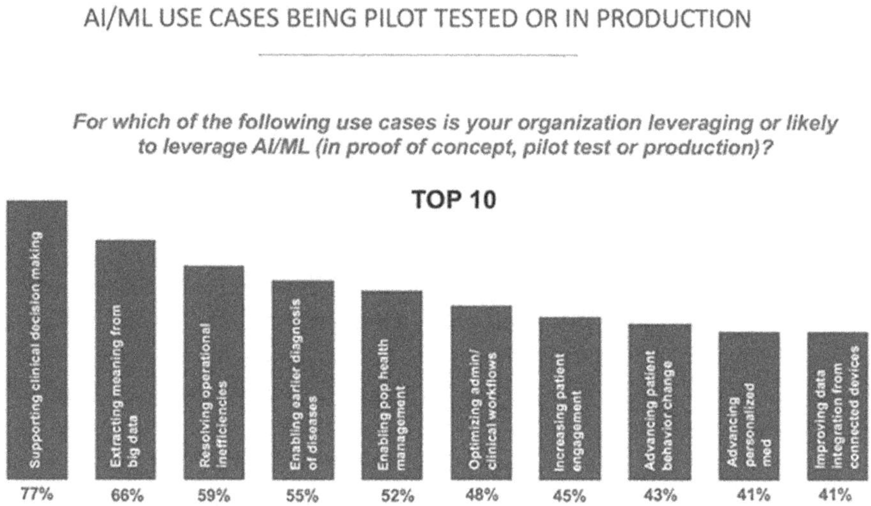

FIGURE 8.4 Healthcare growth with HCI and AI.

All in all, digitalization is in a general sense changing society through the turn of events and utilization of computerized innovations. It might profoundly affect the advanced improvement of each country on the planet. Be that as it may, it should be created dependent on nearby practices, it needs worldwide help and not to be restricted by any mechanical requirements. There is a requirement for a dependable plan measure that includes a comprehension of neighborhood societies and customs for sound global frameworks advancement and digitalization. Especially, digitalization to help worldwide wellbeing requires a significant comprehension of the clients and their specific circumstance and the whole medical care framework and accordingly requires a client focused frameworks configuration approach.

8.3 DIGITAL HEARING

It is to some degree secret that sound isn't utilized all the more broadly in human–PC cooperation plan as we in reality can hear a ton in any case, maybe, the hear-able truth is as well "undetectable" and hence hear-able interface configuration seems, by all accounts, to be troublesome. There is actually no explanation, aside from availability for outwardly weakened clients, to add sound to human–PC interfaces that are as of now advanced for the visual methodology, as it doesn't actually add to framework convenience.

In original human–PC cooperation ideal models,[1–3] like universal, inescapable, wearable, and vanishing figuring, intuitive sonification may offer potential options to the generally predominant visual presentations, opening up our eyes to see the encompassing scene or do what little visual showcases don't progress nicely. In short, utilizing sound for portrayal in collaboration configuration is valuable for standing out to occasions or areas, for non-visual correspondence as a general rule, including discourse, cautions, warning, and criticism. Sound is less helpful for consistent portrayal of items (as we have no earlids), for outright readings (as the vast majority see hear-able measurements like pitch, din, and tone) as this work was upheld to a limited extent by Enterprise Ireland under the Basic Research Program Multimodal Browsing project and by Mikael Fernström who is associated with the Interaction Design Center, Department of Computer Science and Information Systems, University of Limerick, Limerick, Ireland and Eoin Brazil, who is associated with the Interaction Design Center, Department of Computer Science and Information Systems, University of Limerick, being relative), for fine-nitty gritty spatial portrayal, and furthermore is tricky in loud or clamor delicate conditions. Planning intelligent sonifications for human–PC communication necessitates that various issues are tended to. We need to consider where and how sonification is suitable. As fashioners, we should consider the clients' abilities, while doing errands in genuine conditions, and furthermore consider that encompassing commotion levels may veil the framework's sounds. In case, strong is considered to upgrade collaboration, methods of making and testing hear-able allegories should be investigated and how much the utilization of sound adds to the clients' exhibition and abstract nature of utilization. To have the option to plan with sound, we need an undeniable level comprehension of what and how we hear (for example Gaver[4,5]). While there are a broad measure of studies on the view of melodic sounds and discourse, moderately little is thought about different sorts of non-discourse sounds and specifically purported ordinary sounds. In the new association standards recently referenced, we can interrogate how we think regarding connection and human movement. In past work on hear-able interfaces, going from Gaver's Sonic Finder[6] to Brewster's progressive earcons (for example, in Ref. [7]), human activity has generally been considered in a discrete manner, for example, like kicking a football, where a client activity begins a cycle that then, at that point finishes with no further client control. This view may be fitting when pressing catches or flicking switches. An elective view is activity as a nonstop stream, for example, a pen stroke, where we consistently move a pencil on a surface, depending on our learnt signal through proprioception, just as haptic, visual, and hear-able criticism. This last view is becoming significant as large numbers of the new information gadgets like pens, digitizers, cameras, and so on, are equipped for identifying very unpredictable human activities. In any case, at the center of our plan space, a major issue is the way to group and choose sounds to be reasonable for a specific collaboration plan. Contingent upon our proposed clients, errands and setting there is at first a wide continuum in this plan space, going from cement to extract portrayals, for example, from hear-able symbols to earcons.[8,9] In case, we are planning for easygoing ordinary use, we most likely need to think about substantial structures while in case we are planning for profoundly specific spaces, (for example, cockpit or interaction control applications) where our clients will be chosen and prepared, having elite necessities, we may have to zero in on psychoacoustic issues, for example, while it is accepted that regular sounds have intrinsic importance, gained from our regular exercises, hearing such sounds in segregation without setting can be very befuddling. The sound of a solitary confined stride can for instance be heard as a book being dropped on a table. Curiously, this issue is fairly like how phonetic homonyms work, for example, expressions of a similar spelling or sound as others, however, with various implications.[11] To additionally foster our comprehension of what individuals think they hear, we chose to lead listening tests, a methodology likewise utilized by numerous different scientists for example.[12,13]

Made top caliber (44.1 KHz, 16-digit) chronicles of 104 regular sounds (terms somewhere in the range of 0.4 and 18.2 s) and had 14 postgraduate understudies pay attention to the recorded sounds in irregular request utilizing earphones, reacting in free-text configuration to what exactly each solid was. As a rule, the portrayals given were very rich. In light of Ballas' strategy for computing causal vulnerability,[14–16] the reactions were arranged and sorted just as assessed if right or not. From the reactions, we removed activities and instruments/specialists sections of the writings, for example, what the members thought were the items/materials making the sound and how the articles collaborated. In our starter examination, we tracked down that 77% of the activities and 71% of the instruments/specialists were effectively recognized. We likewise noticed that the subsequent informational index, with all reactions, categorizations, and estimations of causal vulnerability and precision could be utilized for proposing conceivable utilization of sounds in communication plan, to some degree like Barrass' strategy for gathering tales about when sound is valuable in regular day to day existence.[17] According to a fashioner's perspective, it is fascinating to take note of that the informational collection utilized in this manner contains data about how individuals depict regular sounds just as estimations of how precise the distinguishing proof of those sounds is. Taking a gander at the sequencing between various sounds, we can see that occasionally the request for sounds influences the distinguishing proof, especially for sounds with high causal vulnerability. For instance, the sound of falling precipitation is regularly haphazardly recognized either as downpour or fricasseeing bacon. On the off chance that the sound before the downpour sound is a birdcall, the reaction is all the more regularly downpour. In the event that the sound before the downpour sound is conflicting of ceramic plates and cutlery, audience members will in general react that it is the sound of singing and the area is a kitchen. Comparative perceptions are examined by Ballas and Howard[14] proposing that there are a few equals between how we measure language and regular non-discourse sounds with the discoveries of the listening tests, we can begin to recommend conceivable hear-able portrayals and allegories for connection plan. In light of Barrass'[17] TaDa approach, we can do an errand and information investigation that permits us to choose sounds that can impart the measurements and headings that give clients satisfactory criticism about their activities comparable to the framework and the status and occasion of the framework. We then, at that point need to make ways so the framework can deliver the chose sounds lastly assess the subsequent plan with clients.[18] In case we were to simply play sound records for addressing client activities, they would consistently solid the equivalent and never (or sometimes) be, for instance, expressive, for example to be planned to the client's distinguished exertion or the size of the articles in question. This was one of the issues tended to by the sounding object project. The sounding object project investigated new techniques for truly motivated displaying of sounds for sound blend. Our work was at first generally educated by natural acoustics and for instance Gaver's work on hear-able symbols.[4,5,19] We likewise pursued cartoonification of sound models, for example, improving on the models while holding perceptual invariants. The models were carried out in pd2 and tried in various ways going from perceptual analyses to imaginative execution. Contrasted with normal sound records, sound items can give "live" sound models that can be parametrically controlled progressively. The distinction can likewise be depicted as full-fledged articles versus developmental items.[20] For a full grown item, the entirety of its boundaries is realized when actuated while for a developmental article boundaries might change all through a persistent use or initiation of the article.

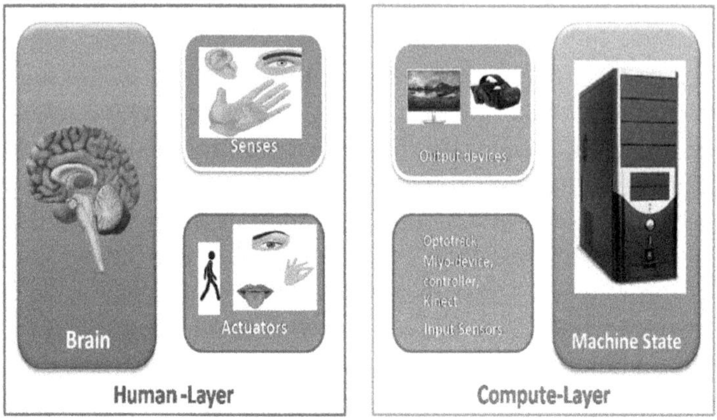

FIGURE 8.5 AI and HCI on digital hearing.

FIGURE 8.6 Digital hearing.

This suggests that sound items, which can be utilized both as full grown and transformative articles, are appropriate for hear-able portrayals for new communication ideal models with non-stop control by client activities. With cartoonification (or as some like, caricaturization), the sound models can be made computationally more proficient than completely sensible models; subsequently, they are bounded to be appropriate for "slender" stages like wearable or handheld PCs.

Applications we need to truly consider if a bunch of sonification natives can be incorporated with a working framework. This can thusly bring about the turn of events and check of solid toolbox for frameworks engineers, like what is accessible for graphical UIs today. There is likewise a need to instruct and uphold collaboration fashioners with the goal that they can open up their innovative intuition toward intelligent sonification, for example, that it is feasible to give persistent criticism continuously for signal-based gadgets. All parts in human–PC interfaces likewise have tasteful properties. It is presumably conceivable to plan sonifications that are psychoacoustically right and very productive yet extremely horrendous to pay attention to. As proposed by Eric Somers, we need to draw upon the information and thoughts of Foley specialists (sound plan for film, radio, and TV) just as exercises gained from different hypotheses of acousmatic music. We have examined some clever ways to deal with intuitive sonification in human–PC collaboration plan, specifically, for relaxed use and pervasive and wearable applications. We recommend that such plans ought to be founded on the outcomes from listening tests with potential similitudes being extricated from clients' depictions of ordinary sounds. The listening tests can likewise give direction in our comprehension of how blends of hear-able symbols can be deciphered by clients. At long last, we portray an instance of utilizing sound objects for communication plan and propose spaces of future exploration. Have executed a programmed interpretation framework, which converts finger-spelled expressions to discourse and the other way around, in a customer worker architecture. The objective of the investigation isn't just to help a meeting impeded individual yet additionally to assist an outwardly weakened individual to connect with others. The framework upholds numerous spoken and communications through signing, including Czech, English, Turkish, and Russian, and the interpretation between these communicated in dialects is taken care of utilizing the Google Translate API. The acknowledgment of multi-lingual finger spelling and discourse was finished utilizing k-Nearest Neighbors Algorithm (k-NN) and HMMs, separately. In the fingerspelling synthesis model, a 3D enlivened symbol is utilized to communicate both manual and non-manual features of an offered hint. He Dicta-Sign[25] is a multi-lingual communication through signing research project that points to make Web 2.0 applications open for Deaf individuals so they can associate with each other. In their Sign Wiki model, the creators show how their system enables gesture-based communication clients to get data from the Web. Like Wikipedia, in which users are approached to enter text as a contribution from their console, gesture-based communication clients can search and alter any page they need, and collaborate with the framework by means of a Microsoft Kinect sensor in the Dicta-Sign Wiki. The Dicta-Sign is presently accessible in four languages: British Sign Language (BSL), German Sign Language (DGS), Greek Sign Language (GSL), and French Sign Language (LSF). In comparative way, Karpov et al.[26] present their multi-modal synthesizer system for Russian (RSL) and Czech (CSE) gesture-based communications that utilize a 3D vivified avatar for blend. VisualComm,[12,27] a Chinese Sign Language (CSL) acknowledgment and translation apparatus, means to assist the Deaf with speaking with hearing individuals utilizing the Microsoft Kinect sensor progressively. The framework can decipher a hard of hearing individual's sign phrases to text or discourse and a conference individual's content or discourse to gesture-based communication utilizing a 3D energized symbol. In view of 370 day by day states, VisualComm accomplishes a recognition rate of 94.2% and exhibits that 3D gesture-based communication acknowledgment should be possible in real-time by utilizing the modalities given by the Microsoft Kinect sensor.

8.4 DIGITAL HUMANITIES

Advanced humanities (DH) is a space of academic action at the crossing point of registering or computerized innovations and the disciplines of the humanities. It remembers the methodical utilization of computerized assets for the humanities, just as the examination of their application.[1,2] DH can be characterized as better approaches for doing grant that include shared, transdisciplinary and computationally drew in exploration, instructing, and publishing.[3] It carries advanced apparatuses and techniques to the investigation of the humanities with the acknowledgment that the printed word is as of now not the principle vehicle for information creation and circulation.

By creating and utilizing new applications and procedures, DH makes new sorts of instructing and exploration conceivable, while simultaneously contemplating and scrutinizing what these mean for social legacy and computerized culture.[2] Thus, a particular component of DH is its development of a two-way connection between the humanities and the advanced: the field both utilizes innovation chasing after humanities examination and subjects innovation to humanistic addressing and cross examination, frequently at the same time. The meaning of the advanced humanities is as a rule constantly figured by researchers and experts. Since the field is continually developing and changing, explicit definitions can immediately become obsolete or superfluously limit future potential. The second volume of debates in the digital humanities recognizes the trouble in characterizing the field: "Alongside the computerized documents, quantitative investigations, and apparatus building projects that once described the field, DH presently envelops a wide scope of techniques and practices: perceptions of huge picture sets, 3D displaying of recorded ancient rarities, "conceived advanced" theses, hash tag activism and the examination thereof, substitute reality games, portable maker spaces, and that's just the beginning.

In what has been called "enormous tent" DH, it can now and again be hard to decide with any particularity what, decisively, advanced humanities work involves. Generally, the computerized humanities created out of humanities processing and has gotten related with different fields, like humanistic figuring, social registering, and media examines. In substantial terms, the computerized humanities accept an assortment of subjects, from curating on the web assortments of essential sources (fundamentally printed) to the information mining of huge social informational indexes to theme demonstrating. Advanced humanities fuse both digitized (remediated) and conceived computerized materials and join the procedures from conventional humanities disciplines (like way of talking, history, reasoning, phonetics, writing, workmanship, paleontology, music, and social examinations) and sociologies, with apparatuses given by figuring (like hypertext, hypermedia, information representation, data recovery, information mining, insights, text mining, and computerized planning), and computerized distributing. Related subfields of computerized humanities have arisen like programming considers, stage examines, and basic code contemplates. Fields that equal the computerized humanities incorporate new media studies and data science just as media hypothesis of organization, game investigations, especially in regions identified with advanced humanities project plan and creation, and social examination.

In viable terms, a significant differentiation inside advanced humanities is the emphasis on the information being handled. For handling literary information, advanced humanities expands on a long and broad history of computerized version, computational semantics and regular language preparing and fostered a free and profoundly particular innovation stack (generally cumulating in the determinations of the text encoding initiative). This piece of the field is now and again hence put aside from digital humanities overall as "advanced philology" or "computational philology." For the creation and investigation of advanced versions of items or curios, computerized philologists approach computerized practices, techniques, and innovations, for example, optical person acknowledgment that are giving freedoms to adjust the field to the advanced age.

The possibility of interface and related ideas like plan and convenience, are the absolute generally vexed in contemporary processing. Meanings of interface normally summon the picture of a "surface" or a "limit" where at least two "frameworks," "gadgets," or "substances" come into "contact" or "connect." Though these terms support spatial translation, most interfaces likewise exemplify worldly, haptic, and psychological components. The directing wheel of a vehicle, the control board of a VCR, and the handle of an entryway are altogether instances of ordinary interfaces that show these measurements. With regards to PCs and processing, "interface" is regularly utilized conversely with "graphical UI," or GUI, most often experienced as a work area windows climate. The order line brief is maybe the most popular option in contrast to the GUI, however, there are a plenty of others, including screen peruses, movement trackers, unmistakable UIs (TUIs, stunningly delivered in the 2002 film Minority Report), and vivid or increased registering conditions. In the humanities, in the interim, it is progressively entirely expected to experience the possibility that a book or a page is a sort of interface, a reaction to the arrangement that the shows of composition and print culture are no less mechanically resolved than those of the computerized world. Something like one onlooker, Steven Johnson, has characterized our present recorded second as an "interface culture," a term he employs to accept not just the pervasiveness of PCs and electronic gadgets yet in addition the manner by which interface has come to work as a sort of saying or social getting sorted out standard—what the modern novel was to the nineteenth century or TV to the rural American 1950s are his models.

FIGURE 8.7 Role of AI and HCI in humanities.

However, much it is discussed; in any case, interface can now and again appear to be minimal cherished. Ease of use master Donald A. Norman states: "The genuine issue with interface is that it is an interface. Interfaces disrupt everything. I would prefer not to zero in my energies on interface. I need to zero in at work" (2002: 210). Nicholas Negroponte holds that the "secret" of interface configuration is to "make it disappear" (1995: 93). To additionally convolute issue, interface is frequently, practically speaking, an exceptionally recursive wonder. Take the experience of a client sitting at her PC and perusing the Web, maybe getting to content at a computerized humanities website. The website's inside plan forces one layer of interface between the client and the substance, and the internet browser—its catches and menus and edges—quickly force another. The client's work area climate and working framework then, at that point force a third layer of interface. The ergonomics of the circumstance (we'll accept our client is working with a console and mouse, taking a gander at a screen situated the suggested 18 inches away) make still another layer of interface, a layer which becomes clear when one considers choices, for example, getting to a similar substance with a PDA or a wearable gadget, or in a room-based computer-generated experience setting like a CAVE. Significantly, each of these "layers," as I have been calling them, shows the potential for collaboration with each other just as with the client. The work area climate administers the conduct of the program programming, whose highlights and capacities thus straightforwardly influence numerous parts of the client's cooperation with the sites' inside plan and content.

While all that I have quite recently been practicing is recognizable enough in software engineering circles, especially the area known as human–PC connection (HCI, additionally now and then distinguished as human–PC interface), parts of this account might appear to be tricky to perusers prepared in humanities disciplines. It would not be difficult to come by somebody willing to contend that my whole situation is the result of one more unacknowledged interface, a sort of normal social passage whose socially built belief systems administer our assumptions as to innovation, portrayal, and admittance to data. In addition, in the situation outlined over, my differentiation between various layers of interface and something I nonchalantly called "content" is one that contradicts many years of work in scholarly and social analysis, where structure and content are naturally perceived as inseparable from each other. Accordingly, the heaviness of set up shrewdness in a field like interface configuration lays on a basic disengage with the predominant scholarly presumptions of most humanists—that an "interface," regardless of whether the windows and symbols of a site or the position of a sonnet on a page, can some way or another be ontologically decoupled from whatever "content" it ends up epitomizing. This is exactly where Brenda Laurel starts her difference in computers as theater, her compelling study of standard interface hypothesis:

Typically, we ponder intuitive figuring as far as two things: application and interface. In the prevailing perspective, these two things are theoretically particular: an application gives explicit usefulness to explicit objectives, and an interface addresses that usefulness to individuals. The interface is what we speak with—the thing we "talk" to—what intervenes among us and the inward operations of the machine. The interface is regularly planned last, after the application is altogether considered and maybe even executed; it is joined to a previous heap of "usefulness" to fill in as its contact surface.

Shrub is writing to challenge this common perspective. However, in all actuality, according to a designer's point of view, the interface is frequently reasonably unmistakable, yet in addition computationally particular. As John M. Carroll calls attention to, it wasn't until the relatively late approach of dialects like visual basic that it even became pragmatic to program both UI and application's basic usefulness with a similar code (Carroll 2002: xxxii). Like the historical backdrop of hand-press printing, which instructs us that paper-production, type-setting, scratching or etching, and bookbinding came to envelop altogether different areas of work and specialized mastery (once in a while housed under a similar rooftop), asset improvement in the computerized world is regularly exceptionally divided and compartmentalized.

Interfaces are for the most part examined in unremarkable and utilitarian terms, however, figuring legend contains a lot of occasions where helpless interface configuration has had deadly and cataclysmic outcomes. One infamous scene included the Therac-25, a machine utilized for malignancy radiation treatment during the 1980s, whose unwieldy programming interface added to a few patients passing from overexposure. Moreover, the little plane mishap that killed vocalist lyricist John Denver has been credited to an inadequately planned cockpit interface, explicitly the situation of a fuel switch. While the stakes in computerized humanities research are (joyfully) other than life and passing, interface is regardless an essential segment of any task.

To be sure, interface presents various fascinating and remarkable issues for the advanced humanist. Justifiably determined by logical and utilitarian necessities, the interface is additionally where portrayal and its chaperon philosophies are generally prominent to our basic eyes. Apparently married to the ideal of ease of use, the interface is likewise where we send our most innovative highlights and creative twists. Time after time set up as the last period of an undertaking under a tight cut-off time and a much more tight spending plan, the interface turns into the first and in quite a while the select insight of the venture for its end clients. Apparently, the most innovative or instinctive phase of the advancement interaction, the interface is likewise conceivably the most observational, subject to the thorough quantitative ease of use testing spearheaded by HCI. This section makes neither endeavor to offer a thorough overview of the tremendous expert writing on interface and convenience, nor does it try to fill in as a plan groundwork or manual for best practices (Perusers intrigued by those themes ought to counsel the part's ideas for additional perusing.

8.5 BUSINESS INTELLIGENCE

Business intelligence is a bunch of techniques, cycles, models, and advances that change crude information into significant and helpful data used to empower more viable key, strategic, and functional experiences and dynamic.

This theory focuses on the business knowledge frameworks (BIS). BIS are intelligent frameworks planned to help chiefs use correspondence advances, information, reports, information, and insightful models to recognize and tackle issues; accordingly, a BI framework can be known as a choice emotionally supportive network (DSS).[9] We focus on online analytical processing (OLAP) apparatuses that empower clients to intuitively investigate multidimensional information according to various points of view. The center of the OLAP innovation is the information 3D square as shown in Figure 8.2, which is a multidimensional dataset model. The model comprises of measurements and measures. The actions are mathematical information like income, cost, deals, and spending plan. Those are arranged by measurements, which are utilized to bunch the information like the "bunch by" administrator in social datasets. Commonplace measurements are time, area, and item, and they are regularly coordinated in progressive systems. A progression is a construction that characterizes levels of granularity of a measurement and the connection between those levels. A period measurement can for instance have hours as the best granularity and higher up the progression can contain days, months, quarters, and a long time. Previously, business knowledge frameworks were utilized to answer OLAP inquiries in table-based reports that addressed various questions independently. The new age of BIS has two remarkable highlights such as coordination of various inquiry answers and perception of these answers (for example, utilizing perception dashboards). This work has been vigorously impacted by the modern climate on the BI act of SAP Research. SAP is one of the market chiefs in big business application programming, giving significant venture devices, for example, enterprise resource planning (ERP), customer relationship management (CRM), and BI frameworks like SAP crystal reports, SAP business objects dashboards, and SAP business objects explorer. The fundamental objective of information representation is its capacity to envision information, imparting data plainly and successfully. It doesn't imply that information representation needs to look exhausting to be practical or incredibly complex to look delightful. To pass on thoughts adequately, both tasteful structure and usefulness need to go connected at the hip, giving experiences into a fairly inadequate and complex data perception assumes a significant part in supporting information investigation and sum up the fundamental attributes of an informational index in a straightforward structure, frequently with visual designs or outlines. It allows the insightful thinking cycle to turn out to be quicker and more engaged.[12] Representation research looks to enhance human cognizance by utilizing human visual capacities to figure out theoretical data nevertheless as clarified by James et al.,[14] explicit sorts of visual portrayals and connection strategies permit clients to see, investigate, and see a lot of data on the double. In the space of multidimensional representation these include scatterplot frameworks, multidimensional glyphs (like StarGlyphs and Chernoff Faces), equal directions or starplots, multidimensional scaling, even perception procedures and composed numerous perspectives. This proposal focuses on facilitated and numerous perspectives like dashboards, portrayed straightaway, as they are the most famous inside the BI space. All things considered as clarified by James et al.,[14] explicit sorts of visual portrayals and connection methods permit clients to see, investigate, and see a lot of data immediately. In the space of multidimensional representation these include scatterplot networks,[15] multidimensional glyphs, (for example, StarGlyphs[16–18] and Chernoff Faces), equal directions[19–21] or starplots, multidimensional scaling,[22] plain perception procedures what's more, planned and numerous perspectives.[25] This theory focuses on facilitated and numerous perspectives like dashboards, portrayed straightaway, as they are the most well-known inside the BI space. To be handled and examined, like observing environment and climate, science, medication, and business. Our work center is business visual analytics that guides in examining enormous business informational indexes and allows taking right choices. By the by our emphasis is fundamentally on the communication and representation challenges, and less on the programmed investigation strategies.

The primary objective of HCI is to make PCs, all the more user-friendly by limiting the obstruction between the human's objective of what they need to achieve and the PC's comprehension of the client's undertaking. Perhaps the main HCI plan techniques is user-focused plan Figure 8.3, in which clients become the dominant focal point in the plan of any PC framework, and is the plan approach we followed to comprehend our clients' requirements and propose arrangements in this work.

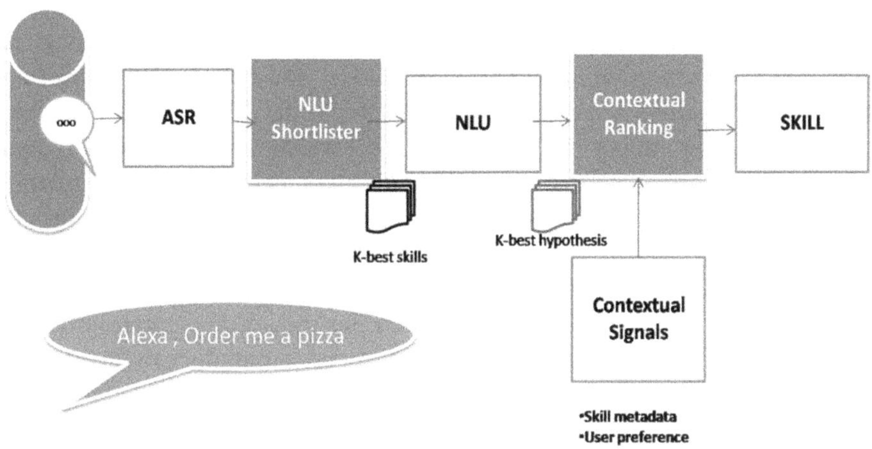

FIGURE 8.8 Role of HCI and AI in buisness intelligence.

Human-focused plan measure characterizes an overall interaction for including human-focused exercises all through an advancement life-cycle. The four exercises are: (1) Specify the setting of utilization: identify individuals who will utilize the item, what they will utilize it for, and under what conditions they will utilize it. (2) Specify necessities: identify any business prerequisites or client objectives that should be met for the item to be effective. (3) Create plan arrangements: this piece of the interaction might be done in stages, working from a harsh idea to a total plan. (4) Evaluate plans: the main piece of this cycle is that assessment—preferably through convenience testing with real clients—is, however, basic as quality testing may be to acceptable programming improvement. A focal idea in HCI is that of ease of use, that is, "The degree to which an item can be utilized by indicated clients to accomplish determined objectives with adequacy, effectiveness, and fulfilment in a predefined setting of utilization." Numerous techniques exist for considering convenience, however, the most essential and helpful is client trying, which has three segments as distinguished by Nielsen[34]: (1) Get hold of some delegate clients, for our situation business intelligence specialists or tenderfoots. (2) Ask the clients to perform agent undertakings with the plan. (3) Observe what the clients do, where they succeed, and where they experience issues with the UI. This is the strategy we have followed to work on the ease of use of the various models carried out during this theory. The measure of information in our reality has been detonating, and effectively breaking down enormous information gives an edge to the two organizations just as people. BI betters break down enormous informational collections, and one of the more significant apparatuses accessible to BI examiners are perception dashboards. By and by, these dashboards stay hard to make and modify by people with little preparing. Besides, they actually miss the mark in supporting nitty gritty investigation record keeping, and correspondence of examination results to other people. This postulation centers around improving client cooperation with BI dashboards, through (1) simple creation and customization of dashboard for a more extensive client range, (2) adding novel explanation supports for examination, and (3) utilization of visual narrating for imparting dashboard investigation results. The propensity for growing dashboard end-client populace with assorted necessities, requires clients, who are regularly tenderfoots in data perception, to get fast admittance to adaptable and simple to fabricate business intelligence data dashboards. This is reflected in progresses in effectively adjustable dashboard perception frameworks (for example Ref. [35,36]), it corresponds with progresses in effectively available data perception conditions for beginners, (for example, ManyEyes[37]), and is additionally reflected in the expanding focal point of the InfoVis research local area on the most proficient method to help amateur clients.[38,39] By and by, there are as yet numerous inquiries regarding how fledgling clients associate with genuine data representation frameworks and that's only the tip of the iceberg so with dashboards. In the BI investigation space, programming merchants center for the most part around adding more highlights into the following dashboard discharge, with no an ideal opportunity to do the exploration expected to find what really works, or even to step back and see how their items are truly being utilized. To discover the responses to such inquiries, we assembled experiences with respect to fledgling client rehearses on exploration views (EV), a framework that permits beginner representation clients to handily construct and tweak business intelligence data dashboards. EV gives a natural climate to progressively making, revamping, looking, and investigating numerous visual information portrayals from different information sources. These angles help clients to more readily recover, test, and acclimate themselves with their information. We give new rules that expand past work on planning for perception beginners, with regards to dashboards. Only assortments of visual portrayals don't represent themselves; to become significant they require understanding and clarification.

Because of the restrictions of human memory[40] recollecting every single thinking subtlety and achievements in the investigation interaction becomes testing. We report dashboard constraints in investigation record-keeping, information revelation, and social occasion experiences. We led top to bottom meetings with dashboard specialists and revealed the need of sorting out their representations by commenting on them with their discoveries. This prompted new explanation needs for multi-outline perception frameworks, on which we based the plan of a dashboard model that upholds information and setting mindful comments that can be shared across representations and datasets. We zeroed in especially on clever explanation perspectives, for example, multi-target comments, comment straight forwardness across graphs, just as joining comments to an information layer divided between various information sources and BI applications, permitting cross application comments. Albeit the work practices of BI experts' might vary, in the end they all develop stories that include: assortments of visual parts (dashboards, graphs, or tables) to help their accounts, verbal clarifications during presentation(s), or itemized literary explanations and portrayals as composed reports and outside assets, for example, wikis. This cycle stays awkward and requires incessant exchanging between various programming (dashboard investigation instruments, screen catches, content managers, show or drawing devices, and so on) BI investigators need a simpler instrument to share their examination results and recount stories with their information. Our work means to help BI investigators with their narrating needs, to foster high examination esteem. Our meetings show that their recorded experiences and information should be coordinated and moved to other people, just as stay accessible for future clients. To accomplish this, we need a planning and sharing instrument to work with the exchange of information and encounters. We accordingly gathered new necessities identified with BI visual narrating. In light of these necessities, we planned and carried out a narrating model apparatus that is coordinated in a dashboard investigation device and permits simple progress from examination to story creation and sharing. Section 5 gives subtleties on this work.

The measure of information in our reality has been detonating, and effectively investigating huge information gives an edge to the two organizations just as people. BI betters dissect huge informational collections, and one of the more significant apparatuses accessible to BI experts is representation dashboards. By and by, these dashboards stay hard to make and alter by people with little preparing. Besides, they actually miss the mark in supporting point by point examination record keeping, and correspondence of investigation results to other people. This postulation centers around the upgrading client cooperation with BI dashboards, through (1) simple creation and customization of dashboard for a more extensive client range, (2) adding novel explanation support for investigation, and (3) utilization of visual narrating for conveying dashboard examination results.

8.6 MARKET ANALYSIS

A market investigation is a quantitative and subjective appraisal of a market. It investigates the size of the market both in volume and in esteem, the different client portions and purchasing behaviors, the opposition, and the financial climate as far as boundaries to passage and guideline.

The "Hyper-Converged Infrastructure (HCI) Market" report offers a nitty-gritty outline of colossal development potential as far as income age and this development factors that will affecting the worldwide business scope. The report shows most recent market bits of knowledge, expanding development openings, business techniques, and plans received by significant players. Hyper-converged infrastructure (HCI) market report gives key measurements on market size, share, and changing serious scene by covering major topographical locales. It likewise offers an exhaustive examination of market elements, current market patterns, item portfolio with SWOT and store network investigation.

Hyper-merged foundation is a product characterized IT framework that virtualizes the entirety of the components of ordinary "equipment characterized" frameworks. HCI incorporates, at the very least, virtualized figuring, a virtualized SAN and virtualized organizing. HCI ordinarily runs on business off-the-rack workers.

The fundamental factors that are driving the market development incorporate expanding interest for information assurance and catastrophe recuperation, profoundly versatile arrangements, and expanded utilization for virtualization. In any case, disadvantage of packaged arrangements is the main consideration impeding the market. What's more, developing acknowledgment of hyper combination arrangements in coming years would be a huge development opportunity for the market. This exploration study looks at the current market patterns identified with the interest, supply, and deals, notwithstanding the new turns of events. Significant drivers, restrictions, and openings have been covered to give a comprehensive image of the market. The investigation presents inside and out data with respect to the turn of events, patterns, and industry arrangements and guidelines carried out in every one of the geological locale. Further, the by and large administrative structure of the market has been thoroughly covered to bring to the table partners a superior comprehension of the key variables influencing the general market climate.

The hyper-merged foundation market size was esteemed at \$3.84 billion of every 2018, and is projected to reach \$33.16 billion by 2026, developing at a CAGR of 30.7% from 2019 to 2026. Hyper-joined foundation (HCI) is a solitary framework system that incorporates capacity, registering, and network assets that assist organizations with effectively sending and make do with a solitary UI.

Hyper-intermingling framework has acquired high foothold from the new past among ventures to attract its capability to smooth out the sending of new jobs, ease foundation the executives, and enhance framework costs. This has additionally upheld the development of the hyper-combined foundation market during the estimate time frame. HCI engineering is intended for a pay-more only as costs arise cost model and can scale steadily which has emphatically affected the development of the market. Business-basic applications which are right now conveyed on three-level IT foundation will progress to hyper-converged framework as it offers incorporated stack frameworks, coordinated framework frameworks, and coordinated reference designs.

Among the business upward, the BFSI business overwhelmed the worldwide hyper-joined framework market as far as income in 2018 and is required to proceed with this pattern during the estimate time frame. As the monetary area is exceptionally compelling toward meeting their client needs. With rising digitization has provoked this area to convey incredible computerized encounters. Accordingly, expansion of strong advanced change systems currently incorporates hyper-combined foundation to build its framework readiness. Besides, monetary firms are slanting toward HCI to drive readiness and lower costs for the essential drives like dispersed it and far off workplaces/branch workplaces, virtual work area arrangement, business-basic applications, and calamity recuperation. This is a central point expected to drive the development of the hyper-met framework market among the BFSI business. The hyper-united foundation market in North America overwhelmed as far as income share in 2018. Ascend in shift toward inventive innovation that gives cloud-like financial aspects to existing server farms without the need to think twice about its presentation, accessibility, or dependability is a main consideration that drives the selection of HCI market in the district. This permit HCI adopters to observe worked on functional productivity, versatility, and sped up arrangement time.

The report centers around the hyper-joined foundation market development possibilities, limitations, and investigation. The investigation gives Porter's five powers examination of the hyper-combined framework industry to comprehend the effect of different factors like dealing force of providers, serious power of contenders, danger of new contestants, danger of substitutes, and bartering force of purchasers available.

The hyper-joined foundation market division incorporates part, application, industry vertical, and area. By segment, it is sorted into equipment and programming. Based on application, it is partitioned into far off office or branch office, virtualization work area framework (VDI), server farm solidification, reinforcement recuperation or fiasco recuperation, basic applications virtualization, and others. Contingent upon industry vertical, it is sorted into BFSI, IT, and broadcast communications, government, medical care, fabricating, energy and utilities, schooling, and others. In view of area, the market is broke down across North America, Europe, Asia-Pacific, and LAMEA.

The central members working in the worldwide hyper-combined foundation market gauge incorporate Cisco Systems, Inc., Dell Inc., Hewlett Packard Enterprise Company, Huawei Technologies Co., Ltd., Microsoft Corporation, NetApp, Inc., Nutanix, Inc., Pivot3, Scale Computing, and Vmware, Inc. Significant players working in this market have seen high development sought after for hyper-combined foundation, particularly due to on-going server farm modernization tasks or drives among server farm industry across the globe. This investigation incorporates hyper-united foundation market examination, patterns, and future assessments to decide the impending venture pockets. Developing requirement for information insurance and fiasco recuperation and decrease in capital expenditure (CAPEX) and operational expenditure (OPEX) are central point driving the development of the hyper-combined framework market patterns. These components are credited to the development of the market as significant foundation merchants are embracing half and half and multi-cloud organizations as either reinforcement systems or information calamity recuperation choices or as an option for on-premises framework. Nonetheless, seller lock-in is a central point expected to frustrate the market development somewhat. Moreover, ascend in interests in server farm foundations is an entrepreneurial factor of the market.

Hyper-merged foundation (HCI) is a solitary framework structure that incorporates capacity, processing, and organization assets that assist organizations with effectively sending and deal with a solitary UI. Significant drivers for the development of hyper-converged infrastructure market size incorporate benefits like diminished capital venture and working consumption and adaptability for recuperation from fiascos.

This investigation incorporates the market opportunity, examination, patterns, and future assessments to decide the fast approaching venture pockets. The hyper-converged infrastructure market report presents key drivers, restrictions, and openings for the hyper-joined foundation industry. The HCI market in North America overwhelmed as far as income share in 2018. Ascend in shift toward inventive innovation that gives cloud-like financial aspects to existing server farms without the need to think twice about its presentation, accessibility, or dependability is a central point that drives the reception of HCI market in the district. This permits the HCI adopters to observe worked on functional productivity, adaptability, and sped up sending time.

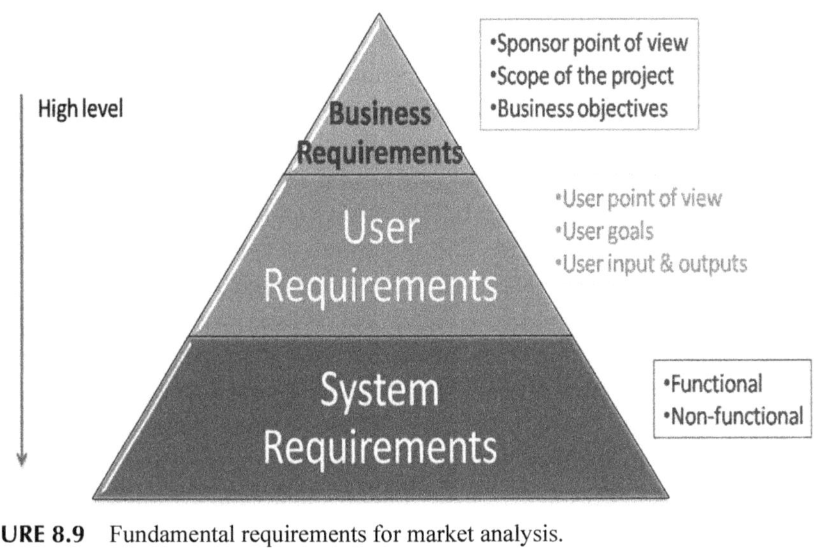

FIGURE 8.9 Fundamental requirements for market analysis.

8.7 CONCLUSION

Portraying the use of a PC as a semiotic interaction among people and PCs is both expressive and takes into account considerable generative force. Considering the semiotic viewpoint, it could bring about an assortment of undertakings on the crossing point of AI and HCI that have both more accessible techniques and a more profound hypothetical establishment. These could be utilized to acknowledge Douglas Engelbart's vision of "working on the compelling usage of the scholarly force of society's concern solvers."[1] This could appear as client focused, assessment zeroed in probes frameworks that use the method for man-made reasoning, design acknowledgment, and information science. Exploratory information investigation, as portrayed here Ref. [12], could turn into an appropriate strategy both as a technique and as a subject for collaboration plan. Particularly, since the quantity of accessible sensors and the delivered information are quickly expanding. There is now research that intends to empower and work with a mutual perspective of the computerized material.[18] This is a decent beginning stage, despite the fact that it is as yet restricted to materials. In a more all-encompassing manner, the explorative methodology could be reached out to entire themes and examination questions. An exploratory information investigation approach could likewise help solid ideas. Solid ideas are contestable, faultless, meaningful, and conceptual examples.[5] The information science toolset could help recognize and validate solid examples like social route, for example, by imagining multidimensional information, uncovering relationships, and empowering exploratory information investigation. All the more by and large, an association of AI and HCI could empower clients to all the more effectively and altogether investigate an assortment of data sources and information streams. A vital interdisciplinary examination project is "The Human Speechome Project Symbol Grounding and Beyond." The Speechome utilized varying media chronicles for a longitudinal report on language advancement of a solitary kid. The information was then examined to see more about language securing.[15] This can be utilized for research; however, it could likewise be applied to educate the lives regarding individuals, for example, concerning their own wellbeing and their accounts. Lieberman proposes that penmanship and voice acknowledgment will turn into a significant application area for both HCI and AI innovation.[7] Despite the fact that they acquired an awful standing as the original of executions didn't work, Moore's law will ultimately make factual language models sufficiently incredible to dependably settle penmanship and voice acknowledgment.[7]

Profound learning research previously showed that it is feasible to prepare a face indicator without marking pictures as containing a face or not and that solo learning is able to do consequently learning significant level ideas, for example, feline countenances and human bodies.[6] There have been different distributions with respect to intelligent AI.[2,17,19] Intuitive AI was proposed as "a characteristic method of incorporating foundation information into the demonstrating stage."[19] One of the objectives was to empower clients to "work connected at the hip with AI frameworks."[17] This was investigated utilizing a verbally process study,[17] a technique generally utilized in HCI. The collaborations of man-made consciousness and human–computer connection, particularly intelligent AI, could turn out to be extremely important in clinical and business applications. It could likewise turn into a pivotal instrument in acknowledging Seymour Papert's fantasy of "computational familiarity for everybody."[14] Resnick recommended that this will require another age of innovations, exercises, and instructive techniques.[14] Papert evoked the idea of a Mathland, where numerical reasoning is regular—like communicating in French is normal in France.[13] Papert composed that while kids might have issues learning French in French class in a US school, all youngsters in France will ultimately learn French. Such a Mathland could be acknowledged utilizing intelligent AI. The framework could both evaluate how much a client definitely thinks about programming (which equivalents to how enormous their jargon is) and what the person is possible attempting to accomplish (which equivalents to their goals and which is like how a human language educator poses the right inquiries to direct an understudy). Like in human interaction and correspondence, the best associations depend on a profound comprehension of one another and effectively gathering the inward state just as the aims of the other. Like an instructor, who can say for sure what an understudy comprehended and what the person didn't and who likewise realizes how to help that person advance (a Socratic thought of the educator as a birthing specialist). Measurable AI strategies could give clients the perfect measure of help and help and empower designers to make better client encounters. Proactive and strong innovation could become accommodating in close to home wellbeing following, brilliant home administration, individual accounting the executives just as interdisciplinary science. Be that as it may, it could likewise make programming simpler and significantly further develop training. Computer-based intelligence and HCI could commonly profit with a nearer joint effort. The AI people group could build their utilization of HCI techniques like the Thinking so anyone might hear convention and Wizard of Oz prototyping and consider social constructivist positions. HCI could additionally expand human capacities utilizing new advancements. The two disciplines can be hypothetically established by semiotic hypothesis. An interdisciplinary group of AI and HCI scientists could foster uncommon applications, for example, by further developing schooling and making homes more brilliant. Since the time PCs were conceived, there have been different connections among individuals and PCs to make PCs more receptive to the people's necessities. The consistent improvement in human interest and interest drives the advancement of HCI innovation and astute robot advances. A lot of exploration works has been completed to make HCI more regular and amicable and, simultaneously, make robots more wise and versatile. With the quick improvement of AI innovation as of late, it gives uncommon advancement freedoms to the examination of these two advances. This paper sums up the advancement status of HCI from the part of communication capacities and presented the connected advances of a keen robot. From there on, the difficulties for these two fields later on improvement and conceivable examination approaches are elucidated. The conversation introduced in this segment is wide and illustrative, however, not totally comprehensive. Some encouraging employments of AI methods to fabricate great arrangements were not unequivocally examined, and may be viewed as when the goal is to carry out some specific viewpoints in game openness. Some illustration of such AI strategies and their uses are: (1) Both artificial neural network and case-based reasoning should be thought of if the answer for an issue can be situated in past encounters. The previous is proper in non-emblematic or sub-emblematic settings, while the last ought to be utilized when using emblematic portrayal is possible; (2) Evolutionary computing is appropriate to take care of issues in which there are a bunch of boundaries to be optimized; (3) Mappings and forecast issues are very much addressed for certain particular artificial neural networks architectures; (4) Fuzzy approximated reasoning is appropriate to manage vulnerability—a typical component found in the human conduct presentation and investigation—and to perform fine settings in streamlining and planning process. Figure 8.1 sums up, through a connection lattice, the chances of utilizing AI to execute solutions to make client communication in games more open. Lines continue the issues—in regards to the HCI scope—proposed through the conversation in this segment; sections list the AI procedures early chose.

The space of assistive innovations advancement identified with games, particularly in regards to the utilization of AI to enhance the legitimate assets of client interaction with the different kinds of uncommon necessities, actually requires thorough investigations which can direct endeavors to innovative work of accessible games. One of the possible spaces of work is the guide ping of the genuine cooperation between gamers with extraordinary necessities and existing games, both for the exemplary games and for those which as of now have openness highlights. Instances of ongoing works are introduced by Zaharias and Papargyris (2009) and Kort and Ijsselsteijn (2008), which albeit not zeroed in on gamers with extraordinary requirements, fill in as motivation for research that intends to more readily comprehend the viewpoints identified with the accessibility world. The previous examinations the connections between gamers' way of life and games convenience; and the last talks about how games can impact the social associations of gamers in the genuine world. Specifically in the setting examined in this section, there are two regions which actually need of strongholds to arrive at worthy degrees of value concerning availability: ressions, body developments, or discourse construction (oral, composed, or gestured); (5) Translation/planning between various sorts of dialects and portrayals: oral language to gestural language; composed language to pictographic language; chart is portrayal to sound portrayal; complex discourse to basic discourse.

The primary goal of this part was to examine a few viewpoints identified with game availability, high-lighting how the HCI and AI regions can manage this inquiry. In this, contents were examined a few perspectives which conceivably address a hindrance for the adequate utilization of a game by a client with extraordinary requirements, and some potential answers for defeat such difficulties. In reality, the AI region as of now gives different instruments which could be applied to implement the uncommon highlights in games, in any case, the endeavors to utilize them viably for the accessible.

KEYWORDS

- **HCI**
- **AI**
- **framework**
- **frequency**

REFERENCES

1. Wang, T. M.; Tao, Y.; Liu, H. Current Researches and Future Development Trend of Intelligent Robot: A Review. *Int. J. Automat. Comput.* **2018,** *15* (5), 525–546.

2. Abdelaziz, A. H. Comparing Fusion Models for DNN-Based Audiovisual Continuous Speech Recognition. *IEEE/ACM Trans. Audio Speech Lang. Process.* **2018,** *26* (3), 475–484.
3. Ren, S.; He, K.; Girshick, R.; Zhang, X.; Sun, J. Object Detection Networks on Convolutional Feature Maps. *IEEE Trans. Pattern Anal. Mach. Intell.* **2017,** *39* (7), 1476–1481.
4. Young, T.; Hazarika, D.; Poria, S.; Cambria, E. Recent Trends in Deep Learning Based Natural Language Processing. *IEEE Computat. Intell. Magaz.* **2018,** *13* (3),55–75.
5. Tirri, K.; Nokelainen, P. Measuring Multiple Intelligences and Moral Sensitivities in Education; SensePublishers, 2011.
6. Kim, J.; El-Khamy, M.; Lee, J. Residual LSTM: Design of a Deep Recurrent Architecture for Distant Speech Recognition, 2017. arXiv preprint arXiv:1701.03360
7. Chen, Y. N.; Wang, W. Y.; Gershman, A.; Rudnicky, A. Matrix Factorization with Knowledge Graph Propagation for Unsupervised Spoken Language Understanding. *ACL-IJCNLP* **2015,** *1*, 483–494.
8. Yu, A. W.; Dohan, D.; Luong, M. T.; et al. QANet: Combining Local Convolution with Global Self-Attention for Reading Comprehension, 2018. arXiv preprint arXiv:1804.09541.
9. Liu, F.; Flanigan, J.; Thomson, S.; Sadeh, N.; Smith, N. A. Toward Abstractive Summarization Using Semantic Representations. arXiv preprint arXiv:1805.10399.
10. Carlisle, J. H. Evaluating the Impact of o±ce Automation on Top Management Communication, In Proceedings of National Computer Conference and Exposition, 1976; pp 611–616.
11. Card, S. K.; Moran, T. P.; Newell, A. The Keystroke-Level Model for User Performance Time with Interactive Systems. *Commun. ACM* **1980,** *23* (7), 396–410.
12. Hewett, T. T.; Baecker, R.; et al. ACM SIGCHI Curricula for Human-Computer Interaction. *ACM* **1992.** DOI: 10.1145/2594128.
13. D'Amico, G.; Bimbo, A. D.; Dini, F.; et al. Natural Human-Computer Interaction, In *Multimedia Interaction and Intelligent User Interfaces*; Springer: London, 2010; pp 85–106.
14. Wang, H. A.; Tian, F. Natural and e±cient Human Computer Interaction, In *10000 Selected Problems in Sciences*; Science Press: Beijing, 2011; pp 625–627.
15. Cao, X. What is the Natural Characteristics of Natural User Interface. *Commun. China Comput. Federat.* **2011,** *11*, 14–18.
16. Legg, S.; Hutter, M. A Collection of de¯nitions of Intelligence. *Front. Artif. Intell. Appl.* **2007,** *157*, 17–24.
17. Gottfredson, L. S. Mainstream Science on Intelligence: An Editorial with 52 Signatories, History, and Bibliography. *Intelligence* **1997,** *24* (1), 13–23.
18. Neisser, U.; Boodoo, G.; Bouchard, T. J.; et al. Intelligence: Knowns and Unknowns. *Am. Psychol.* **1996,** *51* (2), 77–101.
19. Ren, F.; Bao, Y. *Int. J. Info. Tech. Dec. Mak.* **2020,** *19*, 5–47. www.worldscientific.com
20. Matthews, G.; Zeidner, M.; Roberts, R. D. In *The Science of Emotional Intelligence: Knowns and Unknowns*; Oxford University Press: New York, NY, 2007.
21. Gardner, H. Frames of Mind: The Theory of Multiple Intelligences. *Quart. Rev. Biol.* **1985,** *4* (3), 19–35.
22. Legg, S.; Hutter, M. Universal Intelligence: A de¯nition of Machine Intelligence, In *Minds and Machines*; Kluwer Academic Publishers, 2007; vol. *17* (4), pp 391–444.

23. Zhou, H.; Huang, M.; Zhang, T.; Zhu, X.; Liu, B. Emotional Chatting Machine: Emotional Conversation Generation with Internal and External Memory, 2017. arXiv preprint arXiv:1704.01074.
24. Ren, F.; Huang, Z. Automatic Facial Expression Learning Method Based on Humanoid Robot XIN-REN. *IEEE Trans. Hum. Mach. Syst.* **2016,** *46* (6),
25. Minsky, M. L. In *The Society of Mind*; Simon & Schuster Press, 1988.
26. Picard, R. W. A®ective Computing; MIT Press, 1997.
27. Nagamachi, M. Kansei/A®ective Engineering; CRC Press, 2011.
28. Tu, X. Arti¯cial Emotion, The Paper Assembly of the 10th Annual CAAI; Guangzhou, China, 2000.
29. Wang, Z.; Xie, L. Arti¯cial Psychology-an Attainable Scienti¯c Research on the Human Brain; IPMM: Honolulu, 1999; vol 2, pp 1067–1072.
30. Wang, Z. Arti¯cial Psychology and Arti¯cial Emotion. *CAAI Trans. Intell. Syst.* **2006,** *1* (1), 38–43.
31. Ledoux, J. E. Emotion Circuits in the Brain. *Ann. Rev. Neurosci.* **1999,** *23* (23), 155–184.
32. Cardinal, R. N.; Parkinson, J. A.; Hall, J.; et al. Emotion and Motivation: The Role of the
33. Amygdala, Ventral Striatum, and Prefrontal Cortex. *Neurosci. Biobehav. Rev.* **2002,** *26* (3) 321–352.
34. Hossain, M. S.; Muhammad, G. Audio-visual Emotion Recognition Using Multidirectional Regression and Ridgelet Transform. *J. Multimodal User Interfaces* **2016,** *10* (4), 325–333.
35. Shaver, P.; Schwartz, J.; Kirson, D.; et al. Emotion Knowledge: Further Exploration of a Prototype Approach. *J. Pers. Soc. Psychol.* **1987,** *52* (6), 1061.
36. Nicolaou, M. A.; Zafeiriou, S.; Pantic, M. In *Correlated-Spaces Regression for Learning Continuous Emotion Dimensions*, ACM International Conference on Multimedia, 2013; pp 773–776.
37. Barrett, L. F. Discrete Emotions or Dimensions? The Role of Valence Focus and Arousal Focus. *Cogn. Emot.* **1998,** *12* (4), 579–599.
38. Kamarol, S. K. A.; Jaward, M. H.; Kälviäinen, H.; et al. Joint Facial Expression Recognition and Intensity Estimation Based on Weighted Votes of Image Sequences. *Pattern Recognit. Lett.* **2017,** *92* (C), 25–32.
39. Mehrabian, A. Pleasure-Arousal-Dominance: A General Framework for Describing and Measuring Individual Di®erences in Temperament. *Curr. Psychol.* **1996,** *14* (4), 261–292.
40. Breazeal, C. Function Meets Style: Insights from Emotion Theory Applied to HRI. *IEEE Trans. Syst. Man Cybern. C* **2004,** *34* (2), 187–194.
41. Ren, F.; Quan, C.; Matsumoto, K. Enriching Mental Engineering. *Int. J. Innov. Comput. Inf. Control* **2013,** *9* (8), 3271–3284.
42. Xiang, H.; Jiang, P.; Xiao, S.; et al. A Model of Mental State Transition Network. *IEEJ Trans. Electron. Inf. Syst.* **2007,** *127* (3), 434–442.
43. Mathavan, S. A.; et al. An Emotion Recognition Based Fitness Application for Fitness Blenders. *Euro. J. Mol. Clin. Med.* **2020,** *7* (2), 5280–5288.
44. Ren, F. A®ective Information Processing and Recognizing Human Emotion. *Electron. Notes Theor. Comput. Sci.* **2009,** *225*, 39–50.

CHAPTER 9

COMPUTER-AIDED AUTOMATIC DETECTION AND DIAGNOSIS OF CERVICAL CANCER BY USING FEATURE MARKERS

P. SUKUMAR[1], R MURUGASAMI[1], A. RAJAN[2], and S. SHARMILA[3]

[1]*Department of Computer Science and Engineering, Nandha Engineering College (Autonomous), Erode, Tamil Nadu, India*

[2]*Department of ECE, Sree Rama Engineering College, Thirupathi, Andhra Pradesh, India*

[3]*Department of Civil Engineering, Nandha Engineering College, Erode, Tamil Nadu, India*

ABSTRACT

A computer-aided automatic detection and diagnosis method for cervix cancer using Pap smear image is described in this chapter. Cervix disease is the foremost reason behind feminine genital cancers and also severe reason for feminine disease death around the world. A papanicolaou smear is a process performed by the therapist during which a trail of cells are taken from the cervix-uteri using a minor swab and examined for any abnormal microscopic appearances due to HPV infections. The proposed methodology constitutes the following stages: preprocessing, feature extraction, nuclei region segmentation, and classification. Morphological operations are used to segment the nuclei cell region. The features are extracted from the

Computational Imaging and Analytics in Biomedical Engineering: Algorithms and Applications.
T. R. Ganesh Babu, PhD, U. Saravanakumar, PhD & Balachandra Pattanaik, PhD (Eds.)

re-processed papnicolaou spot cell image. False error rate can be decreased by using the automated process.

9.1 INTRODUCTION

Cancer sickness is most common in women affecting the genital region characterized by cervical tumor, with an increasing incidence in the last periods. Cervix cancer deaths occur most normally among women aged between 16 and 43 years. The region of cervix, also called uterine cervix, is located in the lower part of the uterus (in women). It connects the body of the uterus, that is, endo-cervix with the birth canal by means of exo-cervix.

The cervix is covered with cells namely, squamous cells and the glandular cells. Cervical cancers originate in the core layer of the cervix, intersection of the vagina and at the uterus due to Human Papilloma Virus (HPV) mostly transmitted through sexual intercourse from male to female. Two common types of cervical cancers exist, namely, adenocarcinoma and squamous cell carcinoma, such that 80–90% of the cervical cancers occur due to the squamous cell carcinoma, which begins at the junction of the endo-cervix and the exo-cervix.

Women who have had more than three full-term pregnancies, and 17 years aged or younger women while they are at their first full-time pregnancies are subjected to a higher threat of having cervix cancers as per the Statistics of Indian Cancer Society. Even, passive smoking, which damages the DNA around the cervical cells, is considered a cofactor for getting the hazard of cervix cancer in women.

The symptoms of cervical cancer are abnormal vaginal bleeding, post-menopausal bleeding, pelvic pain, etc. In advanced stages of cervical cancer, woman may experience bone fractures around the pelvic area, vaginal discharge with blood, leakage of urine, etc.

Surgery is the primary treatment in the majority of women with endometrial cancer. Radiotherapy is also used when adjuvant therapy is indicated. However, indication for adjuvant therapy varies from country to country. Chemotherapy and hormonal treatment are only given in clinical trial contexts or as palliation in case of an advanced stage.

Cervical cancer is generally diagnosed using Pap smear test and women being screened for at least once in their lifetime at an age of 30–40 may reduce their risk of cancer by 25–36%. Papanicolaou smear is a screening test for cervix cancer in women used to detect nonvisible Human Papilloma Virus (HPV) infection.

A papanicolaou smear is a process performed by the therapist during which a trail of cells are taken from the cervix-uteri using a minor swab and examined for any abnormal microscopic appearances due to HPV infections.

Papanicolaou smear test is done either at the age of eighteen or at earlier. Papanicolaou smear test should be done 1–5 years depending on the women's risk factor. Smoking may make cervical cells more susceptible to infection and the development of cervical cancer because the carcinogens in cigarettes pass from the lungs into the blood stream and into the cervical mucus.

A woman's socioeconomic status is important because it is an indicator of her access to health care. Women from lower socioeconomic backgrounds tend to have less access to basic healthcare, which decreases the chance that abnormal cervical cell changes will be caught before they lead to cancer.

Women with first-degree relatives who have had breast cancer are more likely to perform breast self-examinations than are women without a family history of breast cancer. Daughters whose mothers have had breast cancer are also more likely to be involved in the medical setting and are more likely to seek out relevant information. Such women are more likely to have physical examinations and to have such examinations more frequently.

No doubt the daughters of these women are also less informed about Pap smears since mother–daughter communication surrounding anything perceived to be sexual is likely to be low. Generational acculturation may also influence mother–daughter communication; as a generation becomes more acculturated they are more likely to take on the morals/values of the host country. Open and candid discussions about sex or the body may become more frequent when a family has lived in this culture for two or more generations.

Studies show that women, in general, are severely undereducated about the Pap smear and about HPV. The purpose of the Pap smear is poorly understood. Some women think a Pap smear screens for Sexually Transmitted Infections (STIs) other than HPV, for HIV, or for pregnancy. Others think that the test screens for reproductive tract cancers other than cervical cancer such as ovarian, endometrial, and uterine cancer.

Both age and education level influence women's perceptions of the importance of Pap smears. As a woman ages, she is more likely to believe that Pap smears are important to her reproductive health.

This is even more obvious as women's education levels increase. Adolescents and older women with minimal education are more likely than more educated women to fear Pap smear testing and to have more misconceptions about the purpose of Pap smears.

The detection of cervical cancers from the papanicolaou smear images is a challenging task in medicinal image processing. This can be improved in two ways. One way is by selecting suitable well-defined exact features and the other is by selecting the best classifier.

Several spontaneous and semi-spontaneous methods have been suggested in various times to identify various periods of cervical cancer. These methods were not supported in attaining the purposes of providing dignified variables which could eradicate the clarification errors and interobserver discrepancy.

9.2 OBJECTIVES OF THE STUDY

- To develop an automated process detection process using medical images.
- To apply multi-resolution Gabor transform for cervical cancer detection.
- False error rate can be decreased by using automated process.
- To improve the performance evaluation parameters.
- Validate the results obtained.

9.3 LITERATURE SURVEY

9.3.1 GENERAL

There are lots of researches dealing with general diagnosis of cervical cancers. In this chapter, we have considered some papers which presented various diagnosing techniques for cervical cancers, including pap smear segmentation, texture-based detection, etc. In our literature survey, we go over some related works that account for high efficient cervical cancer diagnosis schemes as well as its performance in classification which are in proximity to our proposed concept.

9.3.2 TECHNIQUES FOR CERVICAL CANCER DETECTION

Carcinoma formation in the internal lining of cervix. Various stages of cervical cancer can be identified from the location of infection and the extent to which the region is affected.

Varghese et al. (2022) fixated on learning the association between DNA methylation and miRNA expression, which had provided new visions into the pathogenesis of cervical cancer. DNA-based methylation markers, along with usual screening technologies, may help in improving the exactness of the current selection techniques used for analysis and prognosis of the disease. Taken together, recognized DNA promoter methylation, foremost to its cancer causing agent, can be a conceivable mark for analysis and remedy. Revisions interrogating the translational impact of methylation in early detection of cervical cancer may help in a better prediction for the afflicted individuals, leading to longer survival for cervical cancer patients.

The suggested system consists of the resulting stages as preprocessing, Gabor transform, feature extraction, and classification. Gabor transform is used to convert the time domain cervical image into multi-resolution domain and then the watershed subdivision technique is used to categorize the abnormal regions from the cervical image. Structures are detached from the segmented region and classified using ANFIS. Morphological processes are identified to spot the cancer area in the image. The planned automation technique for cervical cancer detection is applied on the publicly available datasets. The planned system achieves exactness of 99.5% with respect to ground truth images.

Mariarputham and Stephen (2015) presented texture-based representation. They used texture characteristics and selected grading. Infectious cell is characterized based on parameter analysis. The set of tools was used to extract the information like local binary pattern and histogram analysis. The output of support vector machine provided accuracy rate of carcinoma 84.10% for moderate dysplasia

Edwin et al. (2015) used neural networks and SVM classifier to screen the disease using Pap smear cell images. Average precision 92.33% precision is obtained in this paper. Oscanoa et al. (2015) have proposed an efficient recognition method of ecto-cervical cell employed.

Jusman et al. (2014) have studied the techniques, their benefits and drawbacks. The alphanumeric data of the selection procedures were used as data for the processor rerun system as changed in the skilled study. Four phases of the processer system used in their study were enhancement, feature extractions, feature assortment, and categorization. The computer classification based on spectra data achieved better exactness data. The exactness in classification using neural networks was 78.7%. The overall shows that cytology structures and the features provided advanced exactness of about 90%.

Huagng et al. (2014) have established a procedure for segmenting nucleus and cytoplasm counters. Efficient system categorizes the papanicolaou smear cells into anyone of four dissimilar categorizations of classes using SVM. Two trials were conducted to authenticate the sorting concert which showed the best routine outputs. In the principal experiment, the outcomes presented that average truths of 97.16% were obtained individually, for separating dissimilar varieties of cells. In another trial, 70% of the cell pictures were used for training and 30% for investigation, attaining an exactness of 96.12%

An algorithm for Pap smear cell nuclei segmentation was proposed which uses an elastic segmentation algorithm for determining the exact shape of the nuclei. Then, noise is removed filters, and edges are extracted with a Canny edge detection algorithm.

Finally, a randomized Hough transform processed by a level set algorithm is applied to find the candidate nuclei. The segmentation results were higher than conventional methods. In Bamford et al. (1996), morphological watersheds were applied over gray scale images of low firmness and results in the identification of the locality of remote cells in each image, whereas cell nuclei were not detected.

9.4 METHODOLOGY

9.4.1 *GENERAL*

The proposed system constitutes reprocessing, feature succession, and nuclei partition. Internal structure processes are used to partition cell center area. The Gray equal, wavelet, and GLCM structures are mined from standard plus diseased cell center.

The mined structures are proficient and categorized using ANFIS categorization. The planned method attained 98.74% exactness for dysplasia cell subdivision.

9.4.2 *MATERIALS AND METHODS*

The DTU/Herlev Pap Smear Database (2005) contains papnicolaou-spot facts attained from images of fit and tumorous spot coming from the Herlev University. Images are remained by several investigators for their certain revisions.

9.4.3 METHODS

The planned papnicolaou spot cell subdivision process contains a reprocessing process primarily. The reprocessing is completed to eliminate the noises and then improve the better particulars of the image. Formerly, features are mined from the reprocessed image in the training method from regular normal and abnormal cell images.

These features are used to train the ANFIS categorization in the training mode as shown in Figure 9.1. Figure 9.2 labels the testing mode of categorization in which the papnicolaou spot image is categorized into normal or dysplasia based on its mined features.

The dysplasia image attained is exposed to morphological processes for nuclei subdivision and finally the routine of subdivision is assessed.

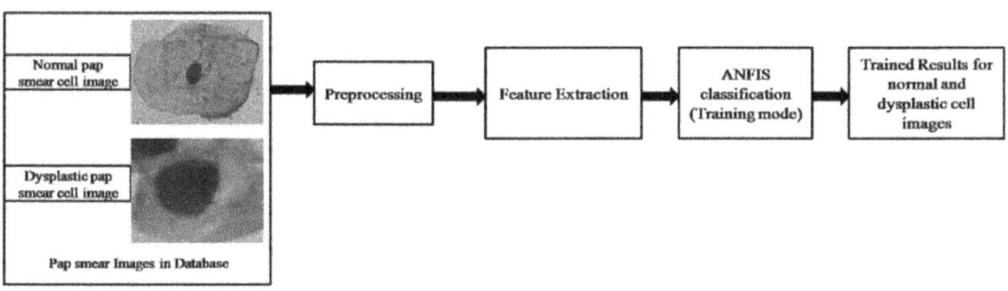

FIGURE 9.1 Dysplasis nucleus segmentation in training mode.

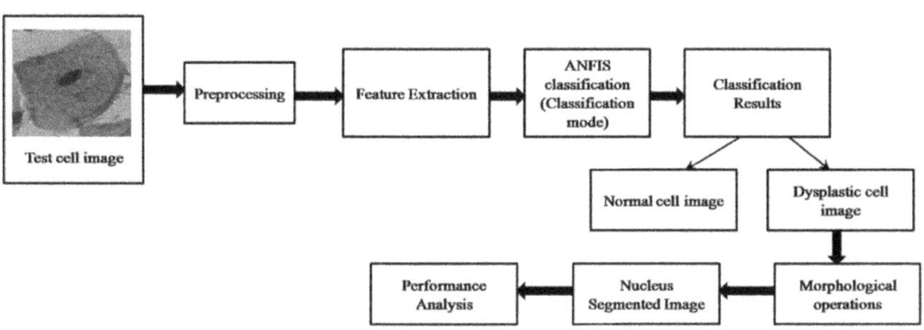

FIGURE 9.2 Dysplasis nucleus subdivision in categorization method.

9.5 FEATURE EXTRACTION

The features are extracted from the re-processed papnicolaou spot cell image. Features are used to distinguish the usual and irregular papnicolaou spot images for cervical cancer identification.

Local binary pattern, GLCM, wavelet, and Law's consistency structures are extracted, purpose of analysis of cervical cancer in papnicolaou spot images.

9.5.1 LOCAL BINARY PATTERN OPERATIVE

LBP is extremely skilled operative employed created on texture features. LBP label examines image element in an image by thresholding the adjacent pels element of each image element and substitutes the image element value with a binary numeral. It is added forcefully to mono gray-scale variations initiated by radiance differences. LBP operator has a computational simplicity; henceforth potential for examining the images in the present access.

9.5.1.1 ORIGIN OF LBP OPERATIVE

In difference to the LBP using 8 image elements in a 3 × 3 pels block, this general construction of the operative sets no boundaries to the size of the adjacent region or to the amount of sample points. Reflect a monotonous image f (x, y) and g_p represents the gray level of a subjective picture element. Furthermore, let g_p represent the rate of sample point which is equally set apart round adjacent sample facts in addition to that range R surface of (x, y)

$$sp = I\,(xy, yt), s = 0, \ldots\ldots s-2 \tag{9.1}$$

$$xy = x + R\sin(2\pi p \,/\, Y) \tag{9.2}$$

$$ys = y - s\cos(2\pi p \,/\, y) \tag{9.3}$$

$$p = I\,(xy, yt), p = 0, \ldots\ldots P-2 \tag{9.4}$$

$$xp = x + R\cos(2\pi p \,/\, Y) \tag{9.5}$$

The boundary operators are operated in the monotonous image to obtain the expected value.

Standard operative is resultant from dual circulation. It is obtained from threshold changes based on rules. The standard operative is

$$LBP_{P,R}(g_c, y_c) = \sum_{p=0}^{p=1} s(s_p - s_c)2^p \tag{9.6}$$

9.5.1.2 *LBP PROCESS OVER DIRECTIVE PIXEL AREA*

LBP operative describes a specific area roughness by 2 balancing procedures, namely, dull scale contrast. Unique LBP operative thresholds the 3 × 3 near image element around the center image element, thereby producing separate tags for each adjoining picture element in the image and changing the picture element with an expected numeric. Thus, the statistical principles based on $2^8 = 256$ sole markers are used as texture signifier.

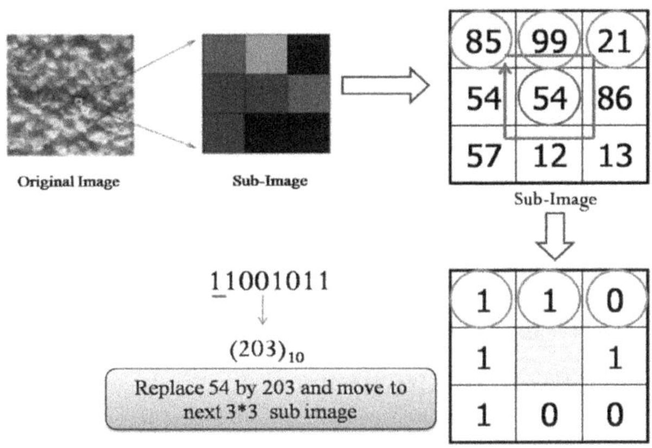

FIGURE 9.3 LBP reckoning operation.

9.5.1.3 PROCEDURE FOR LBP RECKONING OPERATION

Step 1: Original image is used for analysis in the order of three rows and three columns

Step 2: Location of sub-image based on medical image applications

Step 3: Conversion of sub-image into mathematical transformation

Step 4: Values attained based on the transformation results

Step 5: Transformation results attained based on mathematical formulation principle.

By altering the sum of picture element in the adjacency and its radius using a round adjacency and bilinear inserting values, the LBP might be applied for dissimilar sized adjacency area. The gray-scale differences in the local nearest might be used as the consistent difference portion. The picture element adjacency is based on adjacent points.

Figure 9.3 shows the operational procedure of Reckoning operation. Original image is subdivided into a number of segments as per the matrix computation. Values are considered based on intensity measures as per the operational procedure with 3 × 3 matrix.

The center pixel value is 54; it is compared with the other values in the matrix. The values greater than center pixel value should be considered as binary number high component as 1. The values lesser than center pixel value should be considered as binary number low component as 0.

Now the sub-image value is converted into binary value as 1 and 0. The intensity range of pixels varies based on the present value. The matrix block contains 1 as high resolution and the matrix block contains 0 as low resolution.

Sub-image is converted into binary values and it is converted into decimal value. Before the conversion of decimal operations, in clockwise order the values are considered 1,101,011 by converting into decimal value as 203.

$$H_s = \sum\nolimits_{x,y} I\{f_l(k,z) = b\}\, s = 0,1,\ldots.,s-1 \tag{9.7}$$

where, "n" is the numeral of LBP labels.

9.5.2 GLCM FEATURES

The co-occurrence structures can be mined from each co-occurrence medium. Co-occurrence features are energy, contrast, correlation, and homogeneity. GLCM Medium at 45 degree.

1	5	3	4
2	2	4	1
3	4	5	5
4	2	1	2

The data are modeled based on repeated occurrence from the original medium. It is calculated based on specific diagonal parameters. Contrast is computed using this technique. The comparisons between first order to first order, second order to second order, third order to third order, fourth order to fourth order, and finally fifth order to fifth order are framed in the matrix of computation and the expected values are attained based on the above principles.

GLCM	1	2	3	4	5
1	0	1	0	0	0
2	0	0	0	1	1
3	1	1	0	0	0
4	1	0	0	0	1
5	0	1	0	1	0

Contrast

- It measures the strength distinction among a picture element and its nearest.

Energy

- Yields the amount of shaped origins in the specific matrix.

Homogeneousness

- Yields a charge that actions the nearness of the circulation of basics in the array of element.

Recorrelation

- Yields a portion of how connected a picture element to nearest in comparison with the original values.

Law's Energy Texture Features

- Law's texture energy measures are derived from three simple vectors of length 3,
- L3 = (1,2,1), E3 = (–1,0,1), S3 = (1,2, –1).
- Convolving these features.

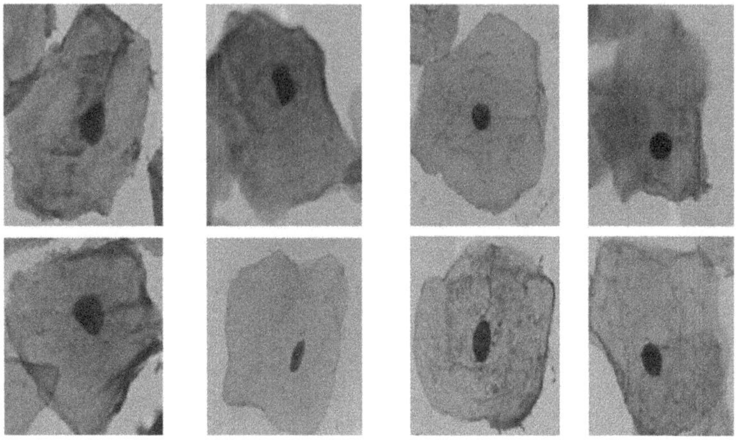

FIGURE 9.4 Trained pap smear cell images (normal case).

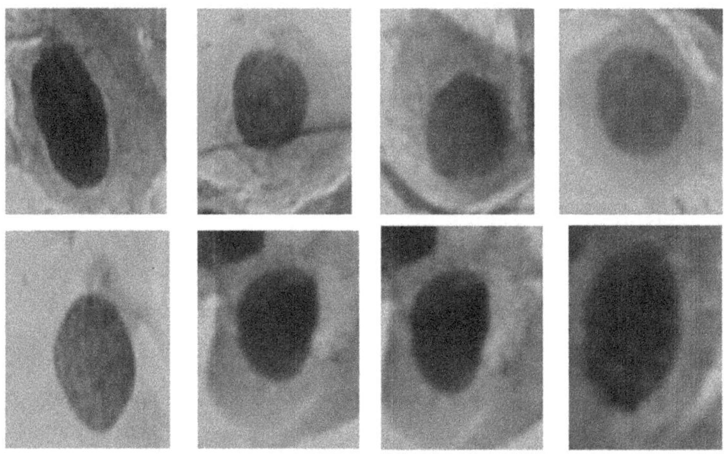

FIGURE 9.5 Trained pap smear cell images (abnormal-dysplastic cell).

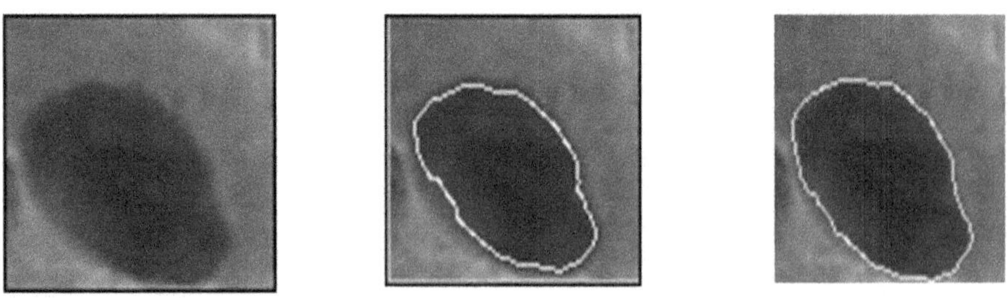

FIGURE 9.6 (a) Dysplasia Papanicolaou smear cell image. (b) Dysplasia cell subdivided image using planned technique. (c) Dysplasia cell physically subdivided image.

9.5.3 PERFORMANCE EVALUATION PARAMETER

- Sensitivities [Ses = T_{pos} / (T_{post} + F_{negt})]
- Specificities [Sps = T_{neg} / (T_{negt} + F_{post})]
- Positive predictive values [Ppvs = T_{post} / (T_{pots}+F_{post})]
- Negative predictive values [Npvs=T_{negt} / (T_{negt}+F_{negt})]
- Accuracies [Accs=(T_{post}+T_{negt}) / (T_{post}+F_{negt}+T_{negt}+F_{post})]

TABLE 9.1 Performance Evaluation for Dysplastic Cell Segmentation.

Performance evaluation parameters	Quantitative results (%)
Sensitivity	92.68
Specificity	99.65
Positive predictive values	51.0
Negative predictive values	93.01
Accuracy	98.74

TABLE 9.2 Mined GLCM Features for Regular and Dysplasia Cells.

GLCM features	Standard cells	Dysplasia cells
Energies	1.56×10^{-5}	1.623×10^{-5}
Entropies	1.1078	1.106
Autocorrelations	1.616×10^{4}	1.596×10^{4}
Contrasts	1.13×10^{4}	1.306×10^{4}

9.6 CONCLUSION

- Cervix disease is foremost reason for feminine genital cancers and also severe reason for feminine disease death around the world.
- Computer-aided automated detection technique for cervical cancer is proposed in this chapter.
- The mined structures are skilled and categorized with support of classifier.
- Planned system achieves 100% classification rate and further processes, detects the edges of the diseased region.
- Routine of the planned system is tested using specific images.

KEYWORDS

- **pap smear**
- **human papilloma virus**
- **cervical cancer**
- **medical imaging**
- **image analysis**
- **morphological operations**

REFERENCES

1. WHO Report. *Cervical Cancer Screening in Developing Countries*; Report of the World Health Organization, 2002. http://whqlibdoc.who.int/publications/002/9241545720.pdf
2. National Cancer Institute. *SEER Incidence and US Mortality Rates and Trends for the Top 15 Cancer Sites by Race/Ethnicity*; National Health Institute: Bethesda, 2005.
3. World Health Organization. *Global Burden of Disease Report: Causes of Death in 2004*; Global Burden of Disease Report, World Health Organization: Geneva, 2004.
4. Roland, K. B.; Benard, V. B.; Greek, A.; Hawkins, N. A.; Manninen, D.; Saraiya, M. Primary Care provider Practices and Beliefs Related to Cervical Cancer Screening with the HPV Test in Federally Qualified Health Centers. *Prev. Med.* **2013,** *57* (5), 419–425.
5. American Cancer Society (ACS). What is Cervical Cancer [Online] 2010. http://www.cancer.org/cancer/cervicalcancer/detailedguide/cervical-cancer-what-is-cervical-cancer.
6. American Cancer Society (ACS). Key Statistics About Cervical Cancer [Online] 2015. http://www.cancer.org/cancer/cervicalcancer/detailedguide/cervical-cancer-key-statistics.
7. Song, D.; Kim, E.; Huang, X.; Patruno, J.; Muñoz-Avila, H.; Heflin, J.; Long, R. L.; Antani, S. Multimodal Entity Coreference for Cervical Dysplasia Diagnosis. *IEEE Trans. Med. Imaging* **2015,** *34* (1), 229–235.
8. Gordon, S.; Zimmerman, G.; Greenspan, H. In *Image Segmentation of Uterine Cervix Images for Indexing in PACS*, Proceedings of 17th IEEE Symposium on Computer-Based Medical System, 2004; pp 298–303.
9. Ji, Q.; Engel, J.; Craine, E. Classifying Cervix Tissue Patterns with Texture Analysis. *Pattern Recognit.* **2000,** *33* (9), 1561–1574.
10. Park, S. Y.; Sargent, D.; Lieberman, R.; Gustafsson, U. Domain-Specific Image Analysis for Cervical Neoplasia Detection Based on Conditional Random Fields. *IEEE Trans. Med. Imaging* **2011,** *30* (3), 867–878.
11. Horng, J. T.; Hu, K. C.; Wu, L. C.; Huang, H. D.; Lin, F. M.; Huang, S. L.; Lai, H. C.; Chu, T. Y. Identifying the Combination of Genetic Factors that Determine Susceptibility to Cervical Cancer. *IEEE Trans. Inf. Technol. Biomed.* **2004,** *8* (1), 59–66.
12. Greenspan, H.; Gordon, S.; Zimmerman, G.; Lotenberg, S.; Jeronimo, J.; Antani, S.; Long, R. Automatic Detection of Anatomical Landmarks in Uterine Cervix Images. *IEEE Trans. Med. Imaging* **2009,** *28* (3), 454–468.

13. Alush, A.; Greenspan, H.; Goldberger, J. Automated and Interactive Lesion Detection and Segmentation in Uterine Cervix Images. *IEEE Trans. Med. Imaging* **2010,** *29* (2), 488–501.
14. Herrero, R.; Schiffman, M.; Bratti, C.; Hildesheim, A.; Balmaceda, I.; Sherman, M. Design and Methods of a Population-Based Natural History Study of Cervical Neoplasia in a Rural Province of Costa Rica: The Guanacaste Project. *Rev. Panam. Salud. Publica.* **1997,** *1*, 362–375.
15. Holland, J. H. In *Adaptation in Natural and Artificial Systems*, University of Michigan Press: Ann Arbor, MI, 1975.
16. Kim, E.; Huang, X. A Data Driven Approach to Cervigram Image Analysis and Classification; Color Medical Image Analysis, Series, In *Lecture Notes in Computational Vision and Biomechanics*; Celebi, M. E., Schaefer, G., Eds.; Springer: Amsterdam, The Netherlands, 2013; vol *6*, pp 1–13.
17. Chang, S.; Mirabal, Y.; Atkinson, E.; Cox, D.; Malpica, A.; Follen, M.; Richards-Kortum, R. Combined Reflectance and Fluorescence Spectroscopy for In Vivo Detection of Cervical Pre-cancer. *J. Lower Genital Tract Dis.* **2005,** *10* (2), 024031.
18. Davey, E.; Assuncao, J.; Irwig, L.; Macaskill, P.; Chan, S. F.; Richards, A.; Farnsworth, A. Accuracy of Reading Liquid Based Cytology Slides Using the ThinPrep Image Compared with Conventional Cytology: Prospective Study. *Br. Med. J.* **2007,** *335* (7609), 31.

CHAPTER 10

A STUDY ON SENTIMENT ANALYSIS

M. MOORTHY[1], YOHANNES BEKUMA BAKARE[2], and BALACHANDRA PATTANAIK[2]

[1]Muthayammal Engineering College, Rasipuram, India

[2]Department of Electrical and Computer Engineering, College of Engineering and Technology, Wollega University, Ethiopia, Africa

ABSTRACT

The collection of people opinion with the use of social media in terms of product selection is referred as sentiment analysis. Social media is the method of sharing data with a large number of people. It can be addressed as a medium of propagating information through an interface. In this paper, fundamental concept and different algorithms used in sentiment analysis were explained; also, it gives information about datasets for helping researchers to do research in sentiment analysis. Based on the analysis of responses in terms of views and feedbacks, three types of sentiments can be found—positive, negative, and neutral.

10.1 INTRODUCTION

The channels used as social media have stormy contact environments; it is imperative to relay sensitive knowledge regarding person's thinking on any kind of device, concept, or policy through these social network channels.[1]

To both customers and suppliers, this data is valuable. In general, persons view other people's opinions about the product especially during any kind

Computational Imaging and Analytics in Biomedical Engineering: Algorithms and Applications.
T. R. Ganesh Babu, PhD, U. Saravanakumar, PhD & Balachandra Pattanaik, PhD (Eds.)

of purchasing. Based on the customer's sentiment, the manufacturer can learn about its product benefits and drawbacks. For researchers, it is a very interesting area to examine and sum up the opinions conveyed in this broad opinion text content in the development of data science instead of human devices ready to analyze and solve the problems by finding the patterns in every data set on their own.

A machine-learning algorithm uses a specific type of data to reply to set of questions using the features extracted from the data. Besides, the advancement of data science has helped the creation of models that analyzes large volumes of complex data quickly and precisely. To obtain the highest value from big data, businesses need to know precisely how to fit the correct algorithm with a particular learning process or resources (Machine Learning and its Applications Outsource to India).[3]

Social media is the method of sharing data with a huge and vast audience. It can be addressed as a medium of propagating information through an interface. Social media in tandem with social networks helps individuals cater their content to a wider society and reach out to more people for sharing or promotion.[4]

Most utilize social destinations to express their feelings, convictions, or suppositions about things, spots, or characters.[5]

Sentiment analysis, as a technique, can be used in extracting meaningful information from any given set of documents.[6]

When it comes to human interactions and communication between individuals, emotions are of utmost importance. Identifying the underlying emotions from text is beneficial in deciding the axioms of human–computer interaction that governs communication and many additional key factors.[7–10]

Sentiment analysis is the procedure of categorizing the views expressed over a particular object. With the advent of varied technological tools, it has become an important measure to be aware of the mass view in business, products, or in matters of common like and dislike. Tracking down the emotion behind the posts on social media can help relate the context in which the user shall react and progress.

In this paper, we propose effective ways to perform sentiment analysis or opinion-mining using various resources. The next part of the paper is arranged as follows: Section 10.2 of the paper presents sentiment classification; Section 10.3 briefs about the approaches for sentiment analysis; Section 10.4 talks about the machine learning-based sentiment analysis methodology; Section 10.5 talks about the outline of machine learning techniques; Section 10.6 deals with the architecture of sentimental analysis for social media analytics; Section 10.7 discusses about open source sentiment analysis tools; and Section 10.8 presents the limitations of sentiment analysis. The paper is concluded in Section 10.9.

10.2 SENTIMENT CLASSIFICATION

Sentiment analysis is an automated method of determining whether a usage-produced text conveys a positive, negative, or common view of an object (i.e., the item, the individual, the subject, the case, etc.). Sentiment classification can be achieved at the four levels such as Document level, Sentence level, and Aspect or Feature level[2] as shown by the Figure 10.1.

A. Document level: In this level, the available record is classified into a positive or negative class as a simple information category. In this type of classification, the whole document can be represented as "positive," "negative," or "neutral.[11]
B. Sentence level: In this type of classification, all the sentences can be differentiated by subjective or objective, also each sentence can be represented as "positive," "negative," or "neutral." At this level, each sentence is named "positive," "negative," or "fair-minded.[11]
C. Aspect or Feature level: The Aspect or Feature-level classification involves the determination and finding the features from the available data. Also, in which the whole document or the sentences can be classified by considering some parts of sentences or record which is called as "perspective-level assessment grouping."[11]

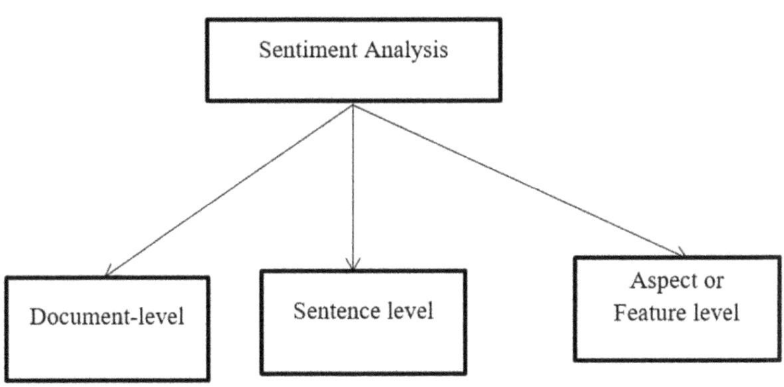

FIGURE 10.1 Sentiment classification.

10.3 APPROACHES FOR SENTIMENT ANALYSIS

There are several techniques available for sentiment analysis:

A. Lexicon-based approach:

The optimistic and pessimistic words available in a dictionary used by Lexicon are applied to assess the polarity of opinion. The count of optimistic and pessimistic words is discussed in the text. If the text is more positive, a positive score will be assigned to the text. The text is awarded a negative score if it has high amount of negative or pessimistic words. The score will be neutral if the text contains the same number of negative or pessimistic words. A lexicon of opinion (positive and negative opinions) is developed to finalize the word as positive or negative. There are numerous ways to build and compile a dictionary.[12]

- Dictionary-based approach: A small number of words of opinion with established guidelines are gathered manually.[12] In corpora like WordNet or thesaurus, synonyms, and opposite from these words are then searched and appended to the group until a new term reaches the process of assembling, but it is not fast. This method has the inconvenience of depending on the dictionary scale, the intensity of the sentiment classification. As the dictionary size increases, this approach is wrong.
- Corpus-based approach: This approach mainly depends on thinking patterns of enormous companies. The created words are context-specific and it needs a large dataset labeled.

B. Machine learning-based approach:

Machine learning techniques in the classification of sentiment depends on the use of well-known machine learning technology on text data as shown by Figure 10.2. The classification of the sentiment based on machine learning can be categorized primarily into supervised and unsupervised methods of learning.[13]

- Supervised learning: Supervised methods of learning rely on labeled training manuals. Supervised learning is an effective classification method and has been used with very promising results for classifying opinions. The regularly used supervised classification techniques in sentiment analysis are Support Vector Machine (SVM), Naïve Bayes (NB) Maximum Entropy (ME), and Artificial Neural Network (NN), and Decision Tree (DT) classifiers. Some less commonly used algorithms are Logistic Regression (LR), K-Nearest Neighbor (KNN), Random Forest (RF), and Bayesian Network (BN).[13]

- Unsupervised learning: This technique does not use pre-listed data to train the classifier, unlike supervised learning. The more common instance of unsupervised machine learning algorithms are Kmeans and Apriori Algorithms. Unsupervised machine learning may also be divided into clusters and associations.[14]

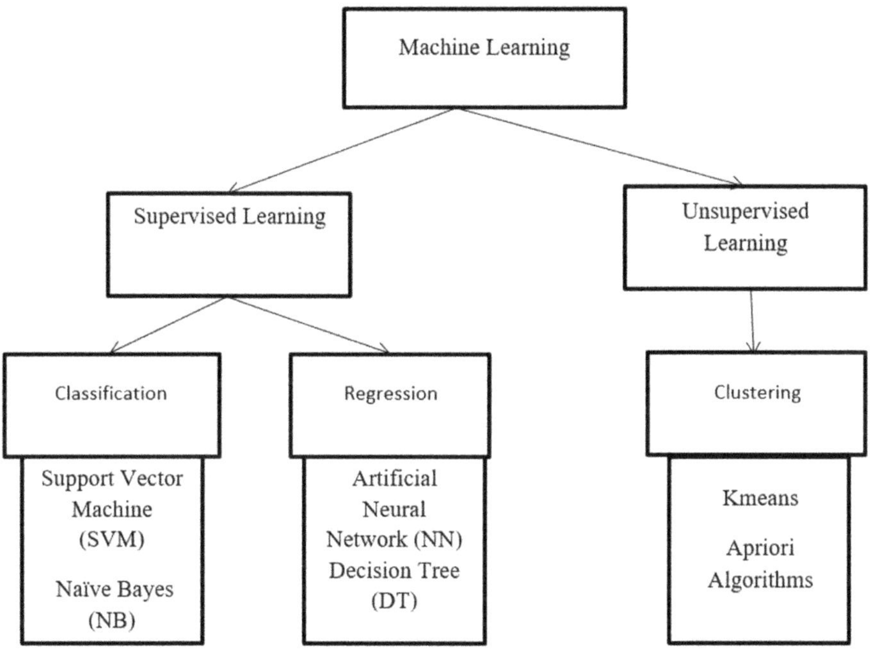

FIGURE 10.2 Machine learning-based approach.

C. Hybrid-based approach:

The hybrid-based approach uses both ML and lexicon-based classification approach. Few research techniques propose a mixture of lexicon-based and automated learning techniques to enhance the classification of sentiment. This hybrid approach is primarily advantageous as it can achieve the best of both. The combination of Lexicon and Learning has demonstrated increased accuracy.

10.4 MACHINE LEARNING-BASED SENTIMENT ANALYSIS METHODOLOGY

The polarity of an analysis data is calculated by various techniques. Machine learning basic sentiment analysis technique is the most popular and efficient. As discussed below, the polarity in analysis data and the most successful algorithm are calculated.[15] Data collection for any kind of text classification task-specific in size as to the number of words, data sets can be used. Such data sets were used after slight preprocessing for sentiment analysis such as case folding, word deletion, etc.

Data Preprocessing: This pre-processing phase seeks to prepare text data for further processing.

Feature Selection and Feature Vector Construction: A computer is not able to process text data straight away, which is an inherent problem. Text data must also be numerically interpreted. Terms are usually used as the characteristics of the text. This gives the text representation a high dimension. Features need to be filtered to reduce dimensions and remove noise to improve classification performance and processing efficiency.

Classification Algorithms for Sentiment Analysis: Several popular and commonly used classification algorithms such as the Multinomial Naïve Bayes Algorithm or the K-Nearest Neighbor Algorithm are commonly used to identify sentiment polarity of users' opinions based on given opinion data Support Vector Machines Algorithm. Evaluation Metrics Measuring every algorithm's output using parameters such as confusion matrix, efficiency, recall, and F-measurement.

10.5 OUTLINE OF MACHINE LEARNING TECHNIQUES

Mechanical learning is an Artificial Intelligence (AI) branch, which investigates machines for the establishment of new knowledge and skills and the identification of existing knowledge. In the area of data mining, computer vision, processing of natural languages, search engines, biometrics, medical diagnostics, credit card fraud detection, a market analysis of stocks, DNA sequence, speech and handwriting recognition, strategy games and robotics, machine learner has been widely used.[16] The popular machine learning algorithms are:

1. **Linear Regression**: A linear regression is defined as the value of the dependent or reliant variable is estimated using independent

variables for statistical techniques? A relationship consists of mapping a dependent and independent variable on a line and that line is known as regression line shown by Y = a*X + where Y is the Dependent variable, X is the Independent Variable is the Intercept and a is the slope.

2. **Logistic Regression**: This approach is used to define the discrete dependent variable from the set of separate variables. Logistic regression provides the coefficients to estimate a probability logistic transformation.
3. **Decision Tree**: For classification and regression, the decision tree may be used tree structure. In which the available dataset is divided into set of subsets and the decision tree is used to develop a training model to estimate the type of the final variable.
4. **Support vector machine:** A Binary Classifier (BC) is a Support Vector Machine (SVM). On the n-dimensional point, row data is drawn. In this, a hyper plane separating the data sets is drawn. This enhanced separation maximizes the training data margin.
5. **Naive-Bayes**: This method is based on the theorem of Bayes used by increasingly sophisticated classification methods. It is a classification technique. It learns how an entity with certain characteristics belonging to a certain category or class is possible.
6. **KNN:** This technique is used for classification and regression. This is one of the simple machine learning algorithms. It saves the cases and searches most k-neighbors it resembles for new data. It saves the cases. With a testing dataset, KNN makes clear predictions.
7. **K-means Clustering**: It is an unsupervised algorithm for learning to reach the cap. The initial partition is achieved by Euclidean distance for grouping the datasets into clusters.
8. **Random Forest**: It is the category of the supervised algorithm. The random forest algorithm involves the generation of several decision trees. The regression as well classification process used these decision trees together. Based on the targets and features of the training dataset this algorithm contain rules.
9. **Dimensionality Reduction Algorithms**: This implies that the number of random variables is decreased by acquiring those key variables. Function extraction and selection of features are methods for reducing dimensionality. The main component analysis can be done by Principal Component Analysis (PCA), which is a method of removing primary variables from a wide range of variables.

10. **Gradient boosting and Ada Boost Algorithms**: The algorithm of gradient boosts is a classification and regression algorithm. AdaBoost selects only those features that enhance model prediction. It operates by selecting a base algorithm such as decision trees and refining it iteratively by taking into account the wrong examples in the training data set.

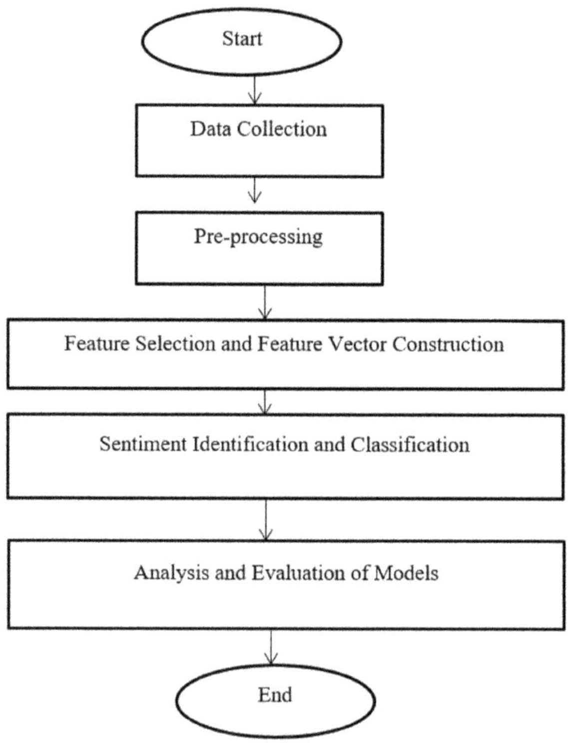

FIGURE 10.3 Flow chart of machine learning-based sentiment analysis technique.

10.6 ARCHITECTURE OF SENTIMENTAL ANALYSIS FOR SOCIAL MEDIA ANALYTICS

Feeling assessment, opinion of customers, feedback, writing emotions, and attitudes involves in the process of investigation of human feelings. The review of the product as many is negative and positive is a decision-making process. The key method of the rejection norm and the classification of negative and positive feelings earned by the users or customers in the social community.[17] Data Pre-processing: The preprocessing technique is more useful to identify and remove meaningless, noisy, and incompatible data. Eliminating URLs: URLs will not help to inspect the emotion in the non-formal text. Questions: The terms question will not help to reduce the ambiguity of polarity such as when, when, who, how, etc. Removing Special Characters: Special characteristics like, and, (), [] {}/' are separated to eliminate inconsistencies by the function of polarity. Removal of Retweets: The re-tweeting process doubles the tweet of another user and redistributes it to peers. This also happens whenever a user tries to tweet another user. Retweets are normally reduced.

10.7 OPEN SOURCE SENTIMENT ANALYSIS TOOLS

The following open source tools are all free and available for building and maintaining your own sentiment analysis infrastructures and other NLP systems. However, do keep in mind that in order to make use of the tools below, you or someone on your team will need the necessary programming and development skills to handle the coding and ML integration.

- **NLTK**: This includes lexical analysis, named entity recognition, tokenization, PoS tagging, and sentiment analysis. It also offers some great starter resources.
- **Spark NLP**: Considered by many as one of the most widely used NLP libraries, Spark NLP is 100% open source, scalable, and includes full support for Python, Scala, and Java. You'll find a whole host of NLP features, pre-trained models and pipelines in multiple languages. There's also an active Slack community for discussion and troubleshooting.
- **TextBlob:** Built on the shoulders of NLTK, TextBlob is like an extension that simplifies many of NLTK's functions. It offers an easy to understand interface for tasks including sentiment analysis, PoS tagging, and noun phrase extraction. TextBlob is a recommended natural language processing tool for beginners.
- **Doccano**: This open source text annotation tool has been designed specifically for text annotation. It allows for the creation of labeled data for sentiment analysis, named entity recognition, and text summarization. This is a good option to look at for smaller datasets and building initial proof of concept projects.

10.8 LIMITATION OF SENTIMENT ANALYSIS

Customer's sentiment about the product or brand may be influenced by many factors. He might have a bad day and it may directly influence his remark negatively. Also, sentiment can change over a period based on his mood. So, it is advisable to go with large sample of data. It will be difficult for an algorithm to understand the sarcasm and ironic language while interpret the sentiment in isolation. So, there is a need to train the model rigorously.

10.9 CONCLUSIONS

Analyzing the amount of data that is being generated through internet media and micro blogging by users of facilities available like Twitter, Reddit,Facebook, etc., and understanding the behavior of people using internet is useful; it can generate both informative and revenue for the business. Using sentiment features rather than conventional text classification gives high accuracy. This system is used to rank satisfactory classifiers for sentiments and helps business organizations for making their future business plans related to the product. Science and technology becomes fruitful only when it is put to solve real problems. In this context, sentiment analysis can serve as the solution to several business problems. It is the nexus of several technologies and accumulated knowledge. It is safe to say that sentiment analysis is a multi-disciplinary domain. To accurately infer sentiments and emotions, we draw knowledge from computer science, psychology, linguistics, machine learning, data mining, deep learning, and statistics.[20] Sentiment analysis is in an evolving phase and shows great promise. In terms of technological evolution, sentiment analysis has lost its passive state of rule-based implementations and entered into an assertive state augmented by advanced data mining and machine learning techniques.

KEYWORDS

- **sentiment analysis**
- **approaches**
- **open source sentiment tools**
- **sentiment analysis classification**
- **social network channels**

REFERENCES

1. Yi, S.; Liu, X. Machine Learning Based Customer Sentiment Analysis for Recommending Shoppers, Shops Based on Customers' Review. *Complex Intell. Syst.* **2020,** *1* (1). DOI: https://doi.org/10.1007/s40747-020-00155-2
2. Vohra, S.; Teraiya, J. A Comparative Study of Sentiment Analysis Techniques. *Int. J. Inf. Knowl. Res. Comput. Eng.* **2013,** *2* (2), 313–317.
3. Machine Learning & its Applications Outsource to India. [Online] May 18, 2020. https://www.outsource2india.com/software/articles/machine-learning-applications-how-it-works-whouses-it.asp.
4. Leskovec, J. In *Social Media Analytics: Tracking, Modeling and Predicting the Flow of Information Through Networks*, Proceedings of 20th International Conference Companion World Wide Web, 2011; pp 277–278.
5. Hasan, A.; Moin, S.; Karim, A.; Shamshirband, S. Machine Learning-Based Sentiment Analysis for Twitter Accounts. *Math. Computat. Applicat.* **2016,** *21* (1), ISSN: 2297–8747.
6. Padmaja, S.; et al. Opinion Mining and Sentiment Analysis – An Assessment of Peoples' Belief: A Survey. *Int. J. Ad Hoc Sens. Ubiq. Comput.* **2013,** *4* (1).
7. Sahu, T. P.; Ahuja, S. In Sentiment Analysis of Movie Reviews: A Study on Feature Selection & Classification Algorithms, 2016 International Conference on Microelectronics, Computing and Communications (MicroCom), 2016.
8. Akhtar, Md. S.; Kumar, A.; Ekbal, A.; Bhattacharyya, P. In *A Hybrid Deep Learning Architecture for Sentiment Analysis*, International Conference on Computational Linguistics: Technical Papers, 2016; pp 482–493.
9. Anil Kumar, K. M.; Rajasimha, N.; Reddy, M.; Rajanarayana, A.; Nadgir, K. In *Analysis of Users' Sentiments from Kannada Web Documents*, International Conference on Communication Networks, 2015; vol *54*, pp 247–256.
10. Mittal, N.; Aggarwal, B.; Chouhan, G.; Bania, N.; Pareek, P. In *Sentiment Analysis of Hindi Review Based on Negation and Discourse Relation*, International Joint Conference on Natural Language Processing, 2013; pp 45–50.
11. Wagh, R.; Punde, P. In *Survey on Sentiment Analysis using Twitter Dataset*, 2nd International Conference on Electronics, Communication and Aerospace Technology (ICECA 2018) IEEE Conference, 2018; ISBN: 978-1-5386-0965-1.
12. Medhat, W.; Hassan, A.; Korashy, H. Sentiment Analysis Algorithms and Applications: A Survey. *Ain Shams Eng. J.* **2014,** *5* (4), 1093–1113. DOI: https://doi.org/10.1016/j.asej.2014.04.011
13. Aydogan, E.; Akcayol, M. A. In *A Comprehensive Survey for Sentiment Analysis Tasks Using Machine Learning Techniques*, International Symposium on Innovations in Intelligent Systems and Applications, 2016; vol *1* (1), pp 1–7. DOI: https://doi.org/10.1109/INISTA.2016.7571856
14. Ahmad, M.; Aftab, S.; Muhammad, S. S.; Ahmad, S. Machine Learning Techniques for Sentiment Analysis: A Review. *Int. J Multidiscip. Sci. Eng.* **2017,** *8* (3), 27–35.
15. Yogi, T. N.; Paudel, N. Comparative Analysis of Machine Learning Based Classification Algorithms for Sentiment Analysis. *Int. J. Innov. Sci. Eng. Technol.* **2020,** *7* (6), 1–9.
16. Patel, A. Machine Learning Algorithm Overview. Medium [Online] May 18, 2020. https://medium.com/ml-research-lab/machine-learning-algorithm-overview-5816a2e6303.

17. Mahendran, N.; Mekala, T. A Survey: Sentiment Analysis Using Machine Learning Techniques for Social Media Analytics. *Int. J. Pure Appl. Math.* **2018,** *118* (8), 419–422.
18. Abdul-Mageed, M.; Diab, M. T.; Korayem, M. In *Subjectivity and Sentiment Analysis of Modern Standard Arabic*, Proceedings of the 49th Annual Meeting of the Association for Computational Linguistics: Human Language Technologies: Short papers, 2011; vol *2*.
19. Nakov, P.; Ritter, A.; Rosenthal, S.; Sebastiani|, F.; Stoyanov, V. In *SemEval-2016 Task 4: Sentiment Analysis in Twitter*, Proceedings of SemEval2016; Association for Computational Linguistics, 2016.
20. Xie, H.; Wong, T.; Wang, F. L.; et al. Editorial: Affective and Sentimental Computing. *Int. J. Mach. Learn. Cybern.* **2019,** *10*, 2043–2044.

CHAPTER 11

APPLICATIONS OF MAGNETIC RESONANCE IMAGING TECHNIQUES AND ITS ADVANCEMENTS

V. RAMESH BABU,[1] S. MARY CYNTHIA,[2] K. SAVIMA,[3] and G. LAKSHMI VARA PRASAD[4]

[1]*Department of CSE, Sri Venkateswara College of Engineering, Sriperumbudur, India*

[2]*Department of ECE, Jeppiaar Institute of Technology, Chennai, India*

[3]*Department of Computer Science, S.T.E.T. Women's College, Mannargudi, Tamil Nadu, India*

[4]*Department of IT, QIS College of Engineering and Technology, Andhra Pradesh, India*

ABSTRACT

Magnetic resonance imaging (MRI) is widely used in biomedical research and clinical applications. The emergence of new techniques increases the applications of MR imaging remarkably. MRI is established with the use of magnetization properties and RF signal for generating images of organs to diagnose diseases. In this paper, the types of MRI and their several applications were explained and it gives the significance of different MRI images in medical field. In general, the value of MRI is affected by some factors like the availability of experts, number of scanners, and cost. The MRI scanning process is comparatively more secure than other methods, like X-ray CT and PET, in terms of ionization.

Computational Imaging and Analytics in Biomedical Engineering: Algorithms and Applications.
T. R. Ganesh Babu, PhD, U. Saravanakumar, PhD & Balachandra Pattanaik, PhD (Eds.)

11.1 INTRODUCTION

Magnetic resonance imaging (MRI) is widely used in biomedical research and clinical applications. The emergence of new techniques increases the applications of MR imaging remarkably. MRI is established with the use of magnetization properties. To capture MRI, a strong magnetic field is applied to randomly oriented protons of human body available in water molecules to make changes in the alignment. Then this magnetization is disordered by applying external radio frequency wave. This absorbed RF energy is then emitted with the help of several relaxation processes and the protons realigned. Subsequently, the emitted signals are determined.

The frequency information present in the emitted signal for each location is then converted into corresponding intensity levels. By changing the cycle of RF pulses enforced and acquired different MRI images can be constructed. The value of time required between two applied pulses is called repetition time (TR) and the value of time between the distribution of RF pulse and the collection of the echo signal is called time to echo (TE).

11.2 TYPES OF MRI

11.2.1 T1-WEIGHTED IMAGES

T1 images indicate the time required for the protons to realign with the direction of applied magnetic field which leads to short TE and repetition time. Generally, fat takes short time for realignment compared with water so fat appears bright while water emerges dark. The T1-weighted image should contain short TR otherwise all protons appear with the same intensity value. If the selected TR time is shorter than tissue's recovery time then only contrast images can be obtained.

11.2.2 T2-WEIGHTED IMAGES

The longer value of TE and TR times are used in the development of T2-weighted images. The transverse relaxation time (T2) is defined as the value of time required to give up phase adherence of protons.

11.2.2.1 APPLICATIONS

11.2.2.1.1 Identify Hematologic Marrow Diseases

Initially, from the T1- and T2-weighted MRI images, the bone marrow was segmented and lot of features were extracted; among them the important characteristics were selected based on principle component analysis (PCA) and least absolute shrinkage and selection operator (LASSO).

Then, at last, classification models random forest (RF) and logistic regression (LR) were used to classify bone chondrosarcoma, metastatic diseases, and osteoporosis.[9]

11.2.2.1.2 Prediction and Classification of Gliomas

Gliomas is a type of brain tumor. The conventional MRI images were used to classify Gliomas in any one of the four grades given by World Health Organization.[10] This classification plays vital role in decision making and planning of medical treatment.

11.2.2.1.3 Achilles Tendon Ruptures (ATRs) Treatment

T2-weighted images were used for the diagnosis and treatment of Achilles tendon ruptures (ATR). The experimental results emphasized that this method provides better positive correlation and higher specificity compared with manual contour tracing (MCT).[11]

11.2.2.1.4 Diagnosis of Colorectal Cancer

The T2-weighted imaging method provides better results compared with other methods to discriminate colorectal cancer. This scanned image can give required information about medical treatment. The features of the collected data, after extraction, were applied to deep learning algorithms for classification process.[12]

11.2.2.1.5 Prediction of Out-of-Field Recurrence (OFR) of Cervical Cancer

The selection of the method of medical treatment for cancer is based on the stage of the disease. Also, the level of recurrence for the same variety of cancer may change from person to person. The conventional T1-, T2-weighted images used to estimate Out-of-field recurrence (OFR) subsequences based on this the treatment method is chosen from surgery, radiotherapy, and chemotherapy methods.[13]

11.2.2.1.6 Early Detection of Lung Cancer

T2-weighted MR images along with convolutional neural network is used to identify the lung nodule region. A false positive (FP) degradation method is used to protect the true nodule. [14]

11.2.2.1.7 Automated Stroke Lesion Segmentation

The fully automated quantitative segmentation of stroke lesion with the help of T2-weighted MRI images as input is a very efficient method compared with manual segmentation.[15]

11.2.3 FLUID ATTENUATED INVERSION RECOVERY (FLAIR)

The length a T2-weighted image depends on the value of TE and TR times selected. So the CSF fluid is completely repressed and abnormalities appears bright which yields easy determination abnormalities. In FLAIR images, grey matter occurs brighter than white matter as like T2-weighted images but cerebra spinal fluid (CSF) appears dark.

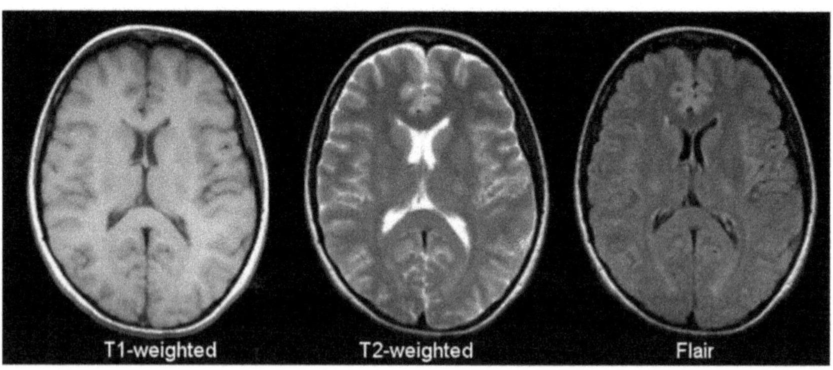

FIGURE 11.1 T1-, T2-weighted, and FLAIR magnetic resonance images.

11.2.4 SPIN ECHO MAGNETIC RESONANCE IMAGING

The SEM image captured with the help of spin-echo pulse sequence consists of 90° excitation pulse and a 180° inversion or refocusing pulse. These pulses are applied to the tissues present in the region of interest.

11.2.5 GRE (GRADIENT ECHO IMAGING)

The GRE image is obtained with the use of gradient-echo sequences which are having flip-angle changing over a range of 10–80°. If the flip angle value is large then it gives more T1-weighting to the image and if the value is small then it provides T2-weighting to the image.

11.2.6 DIFFUSION TENSOR IMAGING (DTI)

Diffusion tensor imaging (DTI) is a type of MRI which is based on the flow of water molecules present in the white matter of central nervous system (CNS). Since DTI provides the information about the structural connectivity of the brain white matter, its demand has increased over the last two decades. Because of limited resolution and contrast, the conventional MRI techniques were not able give information about axonal organization. But it is possible using DTI because it primarily depends on the diffusion of water molecules; its value is high in the axonal bundles compared to the normal direction so the axonal direction can be easily determined.

11.2.6.1 MAPPING CEREBRAL CONNECTIVITY CHANGES AFTER MTBI

Since TBI is a very diverse neurological form, WM structures are affected differently by injury. The identification of WM degradation for mTBI patients is possible by combining (1) automatic; atlas informed labeling of WM streamline clusters, (2) streamline prototyping, and (3) Riemannian matching of elastic curves to quantitate within-subject WM changes, focusing on the arcuate fasciculus. The mTBI commonly results in traumatic axonal injury (TAI) but the information about manifestations of disease is not clearly available in T1- and T2-weighted MR images; it creates curiosity in mapping the issues of mTBI onto the white matter functionality of the brain over time. It is possible with the use of DTI.[1]

The fractional anisotropy (FA) of water in the brain is quantified if its value is less—it indicates that the occurrence of TAI. In this work, small white matter bundles shorter than 4 cm length were neglected because clustering algorithms used in DTI is not suitable for processing short streamlines.

11.2.6.2 AUTISM SPECTRUM DISORDER DIAGNOSIS

Different types of MRI are used to diagnose the autism spectrum disorder (ASD); whereas structural MRI (sMRI) can be used to study physiological characteristics and brain functions can be studied by using functional MRI (fMRI); also DTI is involved in the diagnosis of ASD disorder by studying brain connectivity.

In general, ASD, called as autism, which contain variety of symptoms like struggling in social interaction, interpersonal skills, and restricted and repetitive behaviors.

In which water molecules direction is measured in minimum of six directions, by using this diffusion of water molecules in any other direction can be determined. The mathematical representation of these directions can be represented by diffusion tensor which is a 3×3 matrix and it is graphically represented by an ellipsoid. A lot of characteristics may be taken out from this diffusion tensor matrix, especially FA, axial and radial diffusivity, and mean diffusivity (MD) which gives information about connectivity and microstructure of white matter. Also the parameters calculated from these important features such as trace, skewness, rotational invariance, and others can be used to diagnosis ASD effectively.[2] Image fusing will improve the accuracy of diagnosis. This process contain three important steps. The first step is preprocessing step which eliminates image artifacts result of improper operation of imager and non-brain tissues. Feature extraction is the second step for that any efficient atlas-based segmentation technique can be used to calculate, extract, and select features. The final step is the classification step the linear SVM classifier is used to classify ASD and TD (typically developed) subjects. In this work, six different output features FA, mean diffusivity (MD), axial diffusivity (AD), radial diffusivities in the directions of two minor axis of diffusion ellipsoid, and skewness were used for the determination of anisotropy.

11.2.6.3 DETECTION OF FOCUSED ULTRASOUND-INDUCED BLOOD-BRAIN BARRIER OPENING

Also, DTI was used to detect blood–brain barrier (BBB) opening at the without the use of MRI contrast agent. In this method, diffusion-weighted images are captured over several directions minimum six directions in conjunction with an image captured with the absence of weights in order to populate the diffusion tensor which is a three-by-three, symmetric, positive definite matrix.[3] In this, the word diffusion means that movement of particles of a body move with the same velocity along parallel paths propelled by the thermal energy of particles. While movement, these molecules explore the neighboring tissues at a small scale. The consequences of this displacement corresponds to tissue structure is acquired by diffusion-weighted MRI images. So it is used in many applications like diagnosis of stroke, edema formation, subarachnoid hemorrhage, and multiple sclerosis.

FA mapping can identify structural changes in axons ensuing traumatic brain injury.

11.2.6.4 CARDIOVASCULAR DISEASE DIAGNOSIS

To improve the study of in vivo cardiac DTI features a dense-encoder and decoder trained in an unsupervised training method. Then the important characteristics obtained from diffusion-weighted images were fused based on maximum intensity principle.[4] This method improved the quality of image and calculated diffusion metrics.

11.2.7 FUNCTIONAL MAGNETIC RESONANCE IMAGING (FMRI)

The fMRI technique is used to determine the minute changes in blood flow that happened during brain functions. It can predict anomaly brain functions which cannot be determined with other imaging modalities.

11.2.7.1 DETERMINATION OF BRAIN FUNCTIONAL ACTIVITIES

In this work, resting state fMRI (rs-fMRI) is used to determine brain functional activities based on blood oxygen level-dependent (BOLD) signals then the ASD was diagnosed with the help of the combined framework of convolutional neural network with a prototype learning (CNNPL).[5] The ASD affected persons face problems in social interactions and communications.

11.2.7.2 DETECTION OF POST-TRAUMATIC STRESS DISORDER (PTSD)

In this work, post-traumatic stress disorder (PTSD) is detected with the use of rs-fMRI data and the highly affected brain region is detected with the help of artificial neural network (ANN). The resting-state fMRI is very helpful to provide functional relationship between the areas of the brain. The ANN is used to provide dominance level of classification among the affected brain regions left and right regions of the hippocampus, medial prefrontal cortex, and amygdala. The experimental results show that the left hippocampus is the highly influenced brain area in PTSD individuals.[6]

11.2.7.3 EARLY DETECTION OF ASD USING TASK-BASED FMRI

ASD is a neuro-developmental disorder which is mostly defined by impaired social interaction and communication. The severity of autism can be reduced with the help of early detection and initial treatment. This is possible for 12–14 months old persons and the use of structural MRI (sMRI), fMRI, and DTI. In this work, a computer-aided grading framework in infants and toddlers (between 12 and 40 months) dependent on the analysis of brain activation in response to a speech experiment.[7]

Task-based fMRI is used to determine blood oxygen level-dependent (BOLD) signals in all brain areas and their homogeneity in response to experiments in different domains. The conclusion of this work is task-based fMRI with machine learning technique is an effective tool for the early detection of autism disorder. In which task-based fMRI were captured during the stimulation of three categories of audio record they were played as simple forward speech, complex forward speech, and backward speech. This speech experiment is repeated every 6 min and 20 s and separated by rest blocks with the duration of 20s.

Task fMRI images were captured with the help of echo-planar imaging techniques. fMRI session scan composed of 154 volumes. Each volume is combines slices acquired with the alternating in the plus direction slice acquisition pattern. Initially pre-processing is done to make the fMRI images ready for the first level analysis using general linear model (GLM). The classification process is based on the features determined from the statistical maps created by GLM as well as group analysis. Each group is classified against remaining groups in terms of the parameters accuracy, sensitivity, and specificity. The best accuracy obtained for classifying between mild, moderate and severe against the remaining two groups are 73%, 84%, and 83%, respectively, by random forest classifier compared with SVM and multi-layer perceptron.

11.2.7.4 DETECTION OF EMOTIONAL CHANGES

In this work, emotional changes induced by social support can be detected with the use of fMRI images. This experiment consists of three stages high support stage, medium support stage, and low support stage.[8]

11.2.8 DIFFUSION KURTOSIS IMAGING (DKI)

In DTI images, diffusion are based on Gaussian distribution only. If diffusion behavior is non-Gaussian, then DKI are used so that they are captured at multiple b values with multiple gradient orientations. The generally used performance metrics for DKI are mean kurtosis and axial and radial kurtosis.

11.2.9 HIGH ANGULAR RESOLUTION DIFFUSION IMAGING (HARDI)

In white matter the diffusion is anisotropic and WM injury detection is challenging one because of crossing and kissing fibers. HARDI data having ability to resolve cross fibers and it is typically very useful in the case of complex tract arrangement of brain. But the main disadvantage of DKI and HARDI method is scanning time is large.

11.2.10 SUSCEPTIBILITY-WEIGHTED IMAGING (SWI)

Susceptibility-weighted imaging is a newly developed neuro imaging technique which uses magnitude and phase images. It is used to detect components having paramagnetic, diamagnetic, and ferromagnetic properties like blood products, iron, and calcium. These components couldn't find using conventional MRI. Also, SWI have the ability to discriminate calcium from hemorrhage. It is very essential for traumatic brain injury patients to classify their severity.

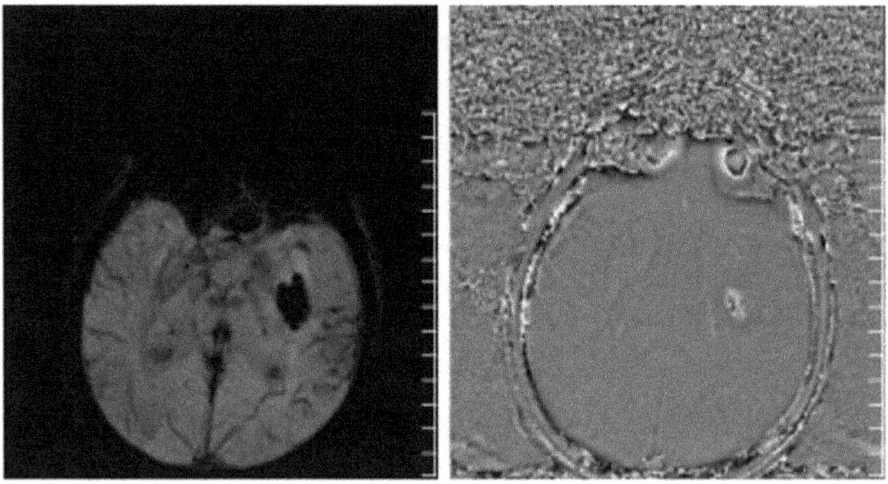

FIGURE 11.2 SWI magnitude and phase image.

KEYWORDS

- **T1 and T2-weighted MRI image**
- **FLAIR**
- **fMRI**
- **DTI**
- **DKI**
- **SWI**

REFERENCES

1. Irimia, A.; Fan, D.; Chaudhari, N. N.; Ngo, V.; Zhang, F.; Joshi, S. H.; O'Donnell, L. J. In *Mapping Cerebral Connectivity Changes After Mild Traumatic Brain Injury in Older Adults Using Diffusion Tensor Imaging and Riemannian Matching of Elastic Curves*, IEEE-17th International Symposium on Biomedical Imaging (ISBI), 2020; pp 1690–1693.

2. Elnakieb, Y. A.; Ali, Md. T.; Soliman, A.; Mahmoud, A. H.; Shalaby, A. M. Computer Aided Autism Diagnosis Using Diffusion Tensor Imaging. *IEEE* **2020,** 191298–191308.
3. Karakatsani, M. E.; Pouliopoulos, A. N.; Liu, M.; Jambawalikar, S. R.; Konofagou, E. E. Contrast-Free Detection of Focused Ultrasound-Induced Blood-Brain Barrier Opening Using Diffusion Tensor Imaging. *IEEE Trans. Biomed. Eng.* **2021,** *68* (8), 2499–2508.
4. Deng, Z.; Wang, L.; Wu, Q.; Chen, Q.; Cao, Y.; Wang, L.; Cheng, X.; Zhang, J.; Zhu, Y. Investigation of In Vivo Human Cardiac Diffusion Tensor Imaging Using Unsupervised Dense Encoder-Fusion-Decoder Network. *IEEE Access*, **2020,** *8*, 220140–220151.
5. Liang, Y.; Liu, B.; Zhang, H. A Convolutional Neural Network Combined With Prototype Learning Framework for Brain Functional Network Classification of Autism Spectrum Disorder. *IEEE* **2020,** *8*, 2193–2202.
6. Shahzad, M. N.; Ali, H.; Saba, T.; Rehman, A.; Kolivand, H.; Bahaj, S. A. Identifying Patients With PTSD Utilizing Resting-State fMRI Data and Neural Network Approach. *IEEE Access* **2021,** *9*, 107941–107954.
7. Haweel, R.; Shalaby, A.; Mahmoud, A. H.; Ghazal, Md.; Seada, N.; Ghoniemy, S.; Casanova, M. A Novel Grading System for Autism Severity Level Using Task-Based Functional MRI: A Response to Speech Study. *IEEE Access* **2021,** *9*, 100570–100582.
8. Candemir, C.; Gonul, A. S.; Selver, A. M. Automatic Detection of Emotional Changes Induced by Social Support Loss using fMRI. *IEEE Trans. Affect. Comput.* **2021,** 1–12.
9. Hwang, E. J.; Kim, S.; Jung, J. Y. Bone Marrow Radiomics of T1-Weighted Lumber Spinal MRI to Identify Diffuse Hematologic Marrow Diseases: Comparison With Human Readings. *IEEE Access* **2020,** *8*, 133321–133329.
10. Ge, C.; Gu, I. Y. H.; Jakola, A. S.; Yang, J. Enlarged Training Dataset by Pairwise GANs for Molecular-Based Brain Tumor Classification. *IEEE Access* **2020,** *8*, 22560–22570.
11. Regulsk, P. A.; Zielinski, J. Multi-Step Segmentation Algorithm for Quantitative Magnetic Resonance Imaging T2 Mapping of Ruptured Achilles Tendons. *IEEE* **2020,** *8*, 199995–200004.
12. Yang, T.; Liang, N.; Li, J.; Yang, Y. Intelligent Imaging Technology in Diagnosis of Colorectal Cancer Using Deep Learning. *IEEE Access* **2019,** *7*, 178839–178847.
13. Ikushima, H.; Haga, A.; Ando, K.; Kato, S.; Yuko, K.; Uno, T. Prediction of Out-of-Field Recurrence After Chemo Radiotherapy for Cervical Cancer Using a Combination Model of Clinical Parameters and Magnetic Resonance Imaging Radiomics: A Multi-institutional Study of the Japanese Radiation Oncology Study Group. *J. Radiat. Res.* **2022,** *63* (1), 98–106.
14. Li, Y.; Zhang, L.; Chen, H.; Yang, N. Lung Nodule Detection With Deep Learning in 3D Thoracic MR Images. *IEEE Access* **2019,** *7*, 37822–37832.
15. Liu, Z.; Cao, C.; Ding, S. Towards Clinical Diagnosis: Automated Stroke Lesion Segmentation on Multi-Spectral MR Image Using Convolutional Neural Network. *IEEE* **2018,** *6*, 57006–57016.

CHAPTER 12

A HYBRID CLUSTERING APPROACH FOR MEDICAL IMAGE SEGMENTATION

M. MALATHI[1], K. SEKAR[2], MAHENDRAKAN K[3], and P. SINTHIA[4]

[1]*Department of Electronics and Communication Engineering, Rajalakshmi Institute of Technology, Chennai, Tamil Nadu, India*

[2]*Department of Electrical and Electronics Engineering, Hindusthan College of Engineering and Technology, Coimbatore, Tamil Nadu, India*

[3]*Department of Electronics and Communication Engineering, Hindusthan Institute of Technology, Coimbatore, Tamil Nadu, India*

[4]*Department of Biomedical Engineering, Saveetha Engineering College, Chennai, Tamil Nadu, India*

ABSTRACT

Different medical imaging techniques were used to various different kinds of medical images. It provides the thorough interior composition of the body organs. Radiologist detects the abnormality of the body parts from this image. X-ray, CT, MRI, other tomographic modalities (SPECT, PET, or ultrasound) are the various medical imaging modalities. After image capturing segmentation is the next important process. The segmentation process helps to determine the region of interest partially or automatically. Manual segmentation refers to the partitioning and naming of an image by hand by a human operator or physician. On a three-dimensional volumetric image, segmentation is done slice by slice. Partitioning of medical images is an easy or difficult process based on the presence of artifacts.

Computational Imaging and Analytics in Biomedical Engineering: Algorithms and Applications.
T. R. Ganesh Babu, PhD, U. Saravanakumar, PhD & Balachandra Pattanaik, PhD (Eds.)

But manual segmentation is a time-consuming process. The experienced physicians have to completely check the images, slice by slice in order to remove the region of interest from the target image. The accuracy of the manual segmentation is based on the physician's experience. When the segmentation is performed manually it is possible to perform a large number of segmentation operations for images without any mistakes. Current trends in the diagnosis of diseases employ CT and MRI imaging techniques. Radiologists require computer-aided design to obtain useful clinical information from the images. The computer-aided design of the imaging modalities also helps in segmenting ample amount of data with the same accuracy. It means that the result is not affected by fatigue, missing of any manual steps. This automatic segmentation is performed by various segmentation algorithms. But the type of algorithm used depends on the type of brain imaging techniques, what kind of body part to be studied. Currently, a number of novel techniques have been proposed for the hybrid clustering algorithm. Hybrid clustering implements the combination of two different algorithms to perform the segmentation process. As a result, hybrid clustering techniques evolve the advantages of the two algorithms employed to perform the clustering process. The predominant algorithms used in hybrid clustering techniques are K-means, adaptive K-means, and spatial fuzzy c-means algorithm. By combining the K means algorithm along with a spatial fuzzy c-means algorithm, the hybrid clustering segmentation yields higher accuracy level for tumor detection and also minimizes the processing time required for the segmentation process.

12.1 USE OF SEGMENTATION IN MRI BRAIN IMAGE

To partition tissue and body parts, the segmentation technique is utilized. From Ref. [1], based on applications, the segmentation is used to detect the border in angiograms of coronary, surgical planning, tumor detection and segmentation, brain growth study, mass detection in mammograms, image registration, and heart segmentation in cardiac images. The region boundaries[1] were signified established on the segmentation of each homogeneous or a similar region. After image segmentation, it classifies the brain image[2] into the tissue class like white matter, gray matter, and cerebro-spinal fluid. Basically, segmentation is classified into two types

- Manual segmentation.
- Automatic segmentation.

12.1.1 MANUAL SEGMENTATION

Manual segmentation is a tedious process where the human operator or physician executes segmentation and labeling of an image by hands-on method. The 3-D volumetric image obtained as slice-by-slice pixels is the outcome of the segmentation process. The complexity of the technique depends upon the nature of the artifacts present in the images. But the entire task completion process for manual segmentation is time consuming. The experienced physicians have to completely check the images, slice by slice in order to remove the region of interest from the target image. The accuracy of the manual segmentation is based on the physician's experience. Hence, the above difficulty in the manual segmentation has to be overcome by automatic segmentation. The ground truth value for the quantitative analysis of automatic segmentation is also obtained by manual segmentation.

12.1.2 NEED FOR AUTOMATIC SEGMENTATION

When the segmentation is performed manually it is possible to perform a large number of segmentation operations for images without any mistakes. Current trends in the diagnosis of diseases employ CT and MRI imaging techniques. Radiologists require a computer-aided design to obtain useful clinical information from the images. The computer-aided design of the imaging modalities also helps in segmenting ample amount of data with the same accuracy. It means that the result is not affected by fatigue, missing of any manual steps. This automatic segmentation is performed by various segmentation algorithms. But the type of algorithm used depends on the type of brain imaging techniques, what kind of body part to be studied. During imaging techniques, the brain image is affected by many factors like an artifact, noise, etc. The authors in Ref. [7] state that the brain imaging is mostly affected by the partial volume effect, but in the thorax the motion artifact occurs. Certain problems are common to both CT, MRI medical images.[3] These can be listed as follows:

- Partial volume effect.
- Motion artifacts and ring artifacts.
- The occurrence of noise caused by the sensors and the associated electronic components.

There is no common algorithm for all the medical data, because every imaging system has its personal specifications and boundaries. To speed up the automatic segmentation process very high-speed computers are now available at moderate cost.

12.2 MRI IMAGING TECHNIQUE

An MRI technique uses a magnet, radio waves, and a computer to produce the internal structure of body parts. The patient is located on the bed that is connected to the magnet. The hydrogen atoms of the human body are aligned with the help of the strong magnetic field produced by the magnet. Further, this atom can be exposed to a beam of radio waves, and it rotates the photons of a human body and produces a weak signal. It has to be detected by the receiver of the MRI. Next, the signal from the receiver is sent to the computer in order to produce the image. Finally, the MRI method relatively produces complete organization of body parts.

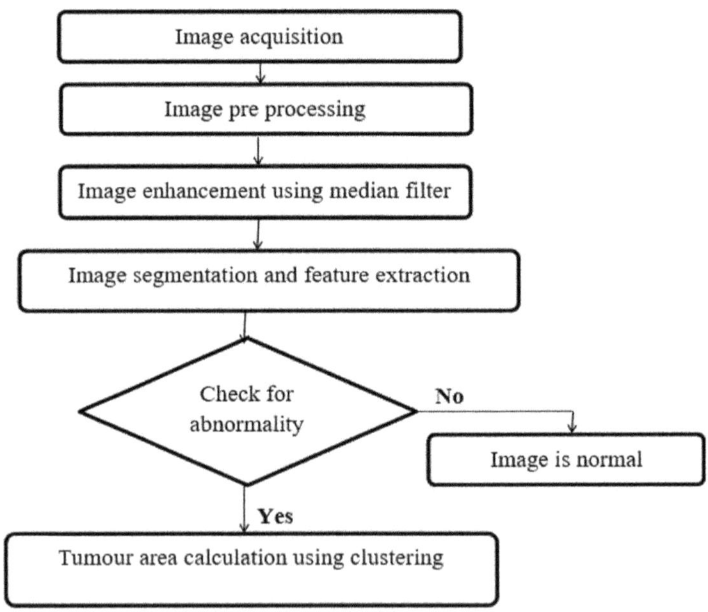

FIGURE 12.1 Flow Diagram of Image Segmentation.

Neurosurgeons or doctors mostly prefer MRI imaging technique,[4] because of its high accuracy in disease detection for the entire body. Neurosurgeons use the MRI technique to diagnose the abnormalities in the head like stroke, brain tumors, and inflammation of the spine. Neurosurgeon uses the MRI not only for the study of brain anatomy, but for the reliability of the spinal cord after trauma. MRI scanners generate 1500 images/second. With the help of MRI imaging, it generates high-contrast images for studying soft tissue anatomy. After doing many kinds of literature on brain tumor segmentation, MRI images are mostly used by researchers.

12.2.1 CLASSIFICATION OF SEGMENTATION

This topic briefly gives the various segmentation methods. And it also provides how the segmentation is performed on the MRI brain image. Andac Hamamci,Nadir Kucuk, Kutlay Karaman, Kayihan Engin& Gozde Unal 2012 state that MRI brain image segmentation is a challenging job since the captured image is affected by magnetic noise and other image artifacts. Hence, many segmentation methods are implemented for processing MRI images. But there is no one method which is appropriate for every image.[5] Every method will be suitable for certain specific images. For instance, the spatial information will be obtained from the texture features associated with an image. But the intensity-based approach basically depends on the gray level histogram, which does not provide spatial information. But the segmentation based on the theory of graph cuts is applied to any type of images like gray or binary images. An unsupervised fuzzy clustering finds many applications like sensing of remote areas, geology, biomedical, molecular or biological imaging.

12.2.2 IMAGE SEGMENTATION STRATEGIES

Generally, the entire segmentation algorithm is based on the two fundamental properties of intensity level values.

- Discontinuity.
- Similarity.

Discontinuity-based segmentation method partition of an image depends on the sudden changes in intensity. Similarity-based segmentation partition of an image is based on some predefined condition, that is, the similarity between the two regions. Figure 12.2 shows the classification of image processing techniques.

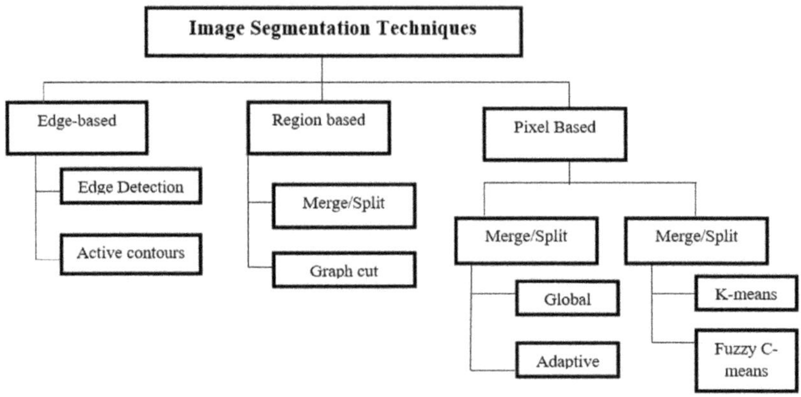

FIGURE 12.2 Various image segmentation techniques.

12.3 CLUSTERING METHODS

The following topics give the detailed discussion on clustering and its types. Shijuan He et al. (2001) told clustering is one of the simple unsupervised learning algorithms. It is defined as a grouping of pixels with similar intensities without using any training images. This classification operation on pixels is performed without the knowledge of prior information, The clustering algorithm trains by its own, using the available data.

12.3.1 CLASSIFICATION OF CLUSTERING

Basically, the clustering is classified into two types

- Hard computing clustering.
- Soft computing clustering.

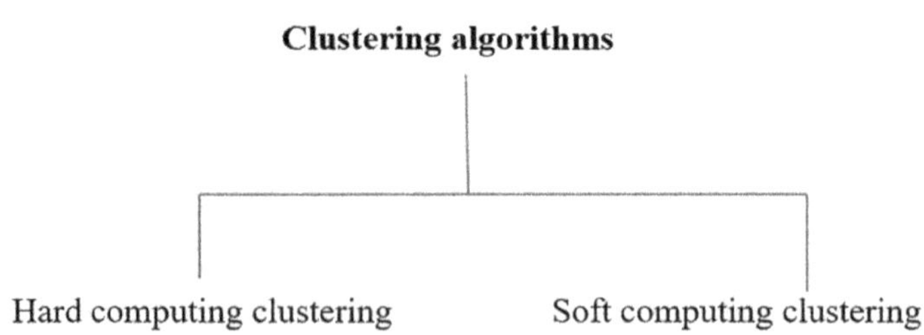

FIGURE 12.3 Clustering algorithms classification.

12.3.2 HARD COMPUTING CLUSTERING

Clustering is a technique which partitions the input image into different clusters by repeatedly calculating the centroid, and the pixel is forced to move to the nearest cluster center. This is called hard clustering,[9] because it pushes each pixel into particular cluster center through the continuous iteration. The authors in Ref. [9] state that there are three common types of hard clustering.

- K-means clustering.
- Fuzzy c-means clustering.
- Expectation and maximization (EM algorithm).

12.3.3 SOFT COMPUTING CLUSTERING

Conventional clustering creates a pattern by grouping the pixel belongs to one and only cluster. In hard clustering, the clusters are disconnected. But in soft computing, the pattern is related to every clustering by arranging the membership function, namely every cluster is a fuzzy set of all patterns. Fuzzy clustering is one of the best examples for the soft computing clustering.

12.3.3.1 K-MEANS CLUSTERING

Macqueen proposed the algorithm in the year 1997. This comes under the category of unsupervised algorithms. The algorithm is initiated by assigning random values of the number of cluster K. Next, centroid is computed from the cluster center. Each pixel value is estimated in contrast with the centroid. Then the pixel is located to the particular cluster having the shortest path among all. The same process[7] is repeated by reestimating the centroid for the next pixel. This process is repeated till convergence of the center. The algorithm steps are explained as follows:

Step 1: Choose random values for the C cluster center.

Step 2: Euclidean distance has been evaluated among each pixel to cluster center.

Step 3: Every pixel is assigned to the specific cluster, which has shortest distance.

Step 4: The chief objective of the algorithm is to reduce the squared error

$X_i - V_i$ is the Euclidean distance between $X_{i,}$ V_i

C is the number of clusters

C_i is the number of data points in the ith cluster. Next, calculate the cluster center by using the following formula:

$$C = \frac{1}{C_i}\sum_{j=1}^{C_i} X_i \quad (12.1)$$

The advantages are

- An easy implementation is possible.
- The algorithm is easy to understand.

The disadvantages are

- The choice for choosing a K-value is essential to accomplish careful segmentation.
- It is more sensitive to disturbances and artifacts.

Clustering finds many concerns in a variety of areas like image processing, data mining, Image retrieval, pattern recognition, image segmentation, etc.

12.3.3.2 FUZZY C-MEANS ALGORITHM

It is one of the best-unsupervised algorithms used for medical image segmentation. It is developed by DQMM[12] and modified by Bezdek. Fuzzy clustering finds the application in pattern recognition. This is an iterative type clustering method. It is one of a more suitable method for segmentation when the K-value is predefined. A Fuzzy algorithm is capable of processing the overlapped dataset. The technique is able to give good segmentation results for noise-free images. The algorithms are described as follows:

Step 1: After assigning the K-value, assign the membership value of each data point based on the cluster center and the data point. The main motive of the algorithm is to minimize the following function:

$$J = \sum_{i=1}^{N}\sum_{j=1}^{C}\left\|x_j - c_j\right\|^2 \quad (12.2)$$

where N is the number of data points

C refers to the required number of clusters.

δ_{ij} defines the value of membership for ith data point x_i in the cluster

$J = \left\|x_j - c_j\right\|$ represents the measure of the closeness of the datapoint x_i to the center vector c_j of the cluster j

The above formula estimates the distance between the data point and cluster center.

Step 2: Next, the data points near to the particular cluster center have the largest membership value of that specific center. The membership value is calculated by using the following formula:

Let x_j be a data point, and let its degree of membership to a particular cluster j be calculated as follows:

$$\delta_{ij} = \frac{1}{\sum_{k=1}^{C} \frac{x_i - C_i}{x_i - C_k}^{\frac{2}{m-1}}} \tag{12.3}$$

where m is the fuzziness coefficient and c_j is calculated as follows:

$$C_j = \frac{\sum_{i=1}^{N} \left(\delta_{ij}^{m} x_j\right)}{\sum_{i=1}^{N} \delta_{ij}^{m}} \tag{12.4}$$

δ_{ij}^{m} is degree of membership value.

The membership value lies between 0 and 1. It means that $\sum \delta_{ij} = 1$

Step 3: The fuzziness coefficient values lie between $1 < m < \infty$. The values define the amount of cluster that can overlap with each other. Number of iterations of the segmentation process depends on the accuracy of degrees of membership values. This accuracy of degrees of membership is measured using the amount of membership value varied from one iteration to the next.

$$\in = \Delta_i^N \Delta_j^c \left|\delta_{ij}^{k+1} - \delta_{ij}^{k}\right| \tag{12.5}$$

δ_{ij}^{k+1}, δ_{ij}^{k} are membership values for K, K+1 iterations, respectively.

Step 4: The algorithm ends when the cluster center is stabilized.

The disadvantages of FCM are

- All the membership values for a data point in the entire cluster are one, but the outlier points have the value more than one. It is difficult to process this kind of process.
- The algorithm is more suitable for noise-free images.

12.3.3.3 SPATIAL FUZZY C-MEANS ALGORITHM

The conventional algorithm uses the single membership value to characterize the desired pattern. It is not sufficient to perform the segmentation exactly. For the detection of brain tumors in MRI images, using pixel intensity as the sole parameter is not sufficient to classify the brain tissue. When any dissimilar structure appears, the conventional FCM[14] is not sufficient for segmentation. This can be avoided by adding the spatial information of neighboring pixels which is considered to define the probability function of each pixel. This spatial information helps to find new membership values for each pixel. It leads to reducing the problem due to noise and intensity inhomogeneity and increases the accuracy of the result.

Consider that X= $\{x_1, x_2, x_3..... x_n\}$ is the set of data points

C=$c_1,c_2,c_3,.....c_n\}$ is the set of centers.

The following two equations are used to calculate membership and the cluster center is updated for each iteration.

$$\mu_{ij} = \frac{1}{\sum_{k=1}^{c}\left[\frac{d_{ij}}{d_{ik}}\right]^{\frac{2}{m}-1}} \tag{12.6}$$

$$C_j = \sum_{i=1}^{n}\left(\frac{\left(\mu_{ij}{}^{m}\right)x_i}{\mu_{ij}{}^{m}}\right) \tag{12.7}$$

d_{ij}—refers the distance between the ith data and the jth cluster

C—represents the number of clusters

m—fuzziness index

μ_{ij}—membership of the ith data to the jth cluster data

n—number of data points

C_j—represents the jth cluster center

12.4 HYBRID SEGMENTATION

Determination of the correct choice of the segmentation algorithm for the given applications is a challenging task in medical image segmentation (Zhang Xiang et al. 2002). A novel method is introduced which combines any two segmentation methods to acquire the accurate lesion segments from the MRI brain image. The cognitive process of uniting any two segmentation algorithms is named as hybrid segmentation. The main purpose of combining the various algorithms is to take away the disadvantages of two different methods and to better the quality of segmentation. The EM segmentation[17] is combined with the active contour models, binary mathematical morphology to implement the segmentation of adult 2D images. The cerebral blood is segmented by combining model-based region growing with morphological segmentation. The authors in Ref. [9] combine fuzzy c-means with K-means which is called a KIFCM technique to perform accurate tumor detection from an MRI image.

The advantage of hybrid segmentation is

- The combination of two algorithms improves overall segmentation accuracy.

The disadvantages are listed as follows:

- It is difficult to implement the new model for segmentation.
- It reduces the computation time, but a large number of parameters should be tuned to perform segmentation for certain applications.
- Care should be taken while designing the modeling to provide the best quality of segmentation.

In this context, the tissue characteristics and all physical properties are closest to the in vivo properties. This kind of phantom images is produced by the MRI scanner. It provides the image which is more genuine than the images produced by software simulators. But the method is not flexible, like software simulators; it is more costly and needs more labor.

12.4.1 IMPLEMENTATION OF CLUSTERING TECHNIQUES

The following topic describes the brief discussion[18] about results for various segmentation algorithms. The algorithms are like K-means clustering, adaptive k-means clustering, spatial fuzzy, c-means clustering used for segmentation of brain image. The performance analysis between the various stages of results of K-means, adaptive k-means algorithms is compared in terms of accuracy, time, PSNR, and area. For the segmentation process the sample brain images are acquired from hospital. The next topic discusses the various results of the K means algorithm.

12.4.2 ABOUT MATLAB

The brain tumor segmentation algorithm is implemented using Mat Lab software. This is one of the commonly used computational tools in various fields like science and engineering, which includes the fields of physics, chemistry, math, and all engineering streams. It is used in a range of applications including

- Signal processing and communications.
- Image and video processing.
- Image segmentation.
- Control systems.
- Computational finance.
- Computational biology.

12.5 CONCLUSION

Three different types of algorithm are implemented in this research work. The hybrid segmentation which is proposed in this work is entirely based on providing the best results and the accurate calculations for estimating the area and time when compared to the other two algorithms. Because of the integration of the two algorithms, that is, K-means segmentation and spatial fuzzy C-means algorithm, the distinct features of the two methods are also combined.

It can be concluded that the compounding of two algorithms yields better results. Based on this motivation, the research work combines the two different algorithms, namely SFCM and K-means algorithm and it is called KISFCM. K-means algorithms are used to detect the tumor as faster than FCM, but it gives a good result for only smaller values of K. The next algorithm FCM is utilized to discover the tumor cells that are not adjacent by K-means. FCM is also not considering the spatial characteristics of brain images because of the fact that this kind of characteristics is very essential to sort out the complex structures. But it is a time-consuming segmentation process. Hence, the research work integrates the K-means with SFCM titled as KISFCM to acquire faithful area detection of various algorithms which are represented in terms of area and time.

KEYWORDS

- **segmentation**
- **clustering**
- **computer tomography (CT)**
- **magnetic resonance imaging (MRI)**
- **K-means**
- **fuzzy c-means**

REFERENCES

1. Banerjee, A.; Maji, P. Rough Sets and Stomped Normal Distribution for Simultaneous Segmentation and Bias Field Correction in Brain MR Images. *IEEE Trans. Image Process.* **2015,** *24* (12), 5764–5776.
2. Gooya, A.; Biros, G.; Davatzikos, C. Deformable Registration of Glioma Images Using EM Algorithm and Diffusion Reaction Modeling. *IEEE Trans. Med. Imaging* **2011,** *30* (2), 375–389.
3. Ism, A.; Direkoglu, C.; Sah, M. In *Review of MRI Based Brain Tumor Image Segmentation Using Deep Learning Methods*, Proceedings of 12th International Conference on Application of Fuzzy Systems and Soft Computing; Vienna, Austria, Aug 29–30, 2016.
4. Roniotis, A.; Manikis, G. C.; Sakkalis, V.; Zervakis, M. E.; Karatzanis, I.; Marias, K. High Grade Glioma Diffusive Modeling Using Statistical Tissue Information and Diffusion Tensors Extracted from Atlases. *IEEE Trans. Inf. Technol. Biomed.* **2012**, 16 (2), 255–263
5. Asanambigai, V.; Sasikala, J. Adaptive Chemical Reaction Based Spatial Fuzzy Clustering for Level Set Segmentation of Medical Images. *Ain Shams Eng. J.* **2016,** *9* (3), 459–467.
6. Islam, A.; Syed, M. S.; Khan, M. I. Multifractal Texture Estimation for Detection and Segmentation of Brain Tumors. *IEEE Trans. Biomed. Eng.* **2013,** 60 (11), 3204–3215.
7. Arizmendi, C.; Daniel, A. S.; Alfredo, V.; Enrique, R. Automated Classification of Brain Tumours from Short Echo Time In Vivo MRS Data Using Gaussian Decomposition and Bayesian Neural Networks. *Expert Syst. Appl.* **2014,** *41* (11), 5296–5307.
8. Chen, L.; Weng, Z.; Yoong, L.; Garland, M. An Efficient Algorithm for Automatic Phase Correction of NMR Spectra Based on Entropy Minimization. *J. Magn. Reson.* **2002,** *158* (1), 164–168.
9. Eman, A. M.; Mohammed, E.; Rashid, A. L. Brain Tumor Segmentation Based on a Hybrid Clustering Technique. *Egypt. Inform. J.* **2015,** *16* (1), 71–81.
10. Xing, F.; Xie, Y.; Yang, L. Automatic Learning-Based Framework for Robust Nucleus Segmentation. *IEEE Trans. Med. Imaging* **2016,** *35* (2), 550–566.
11. Hai, S.; Xing, F.; Yang, L. Robust Cell Detection of Histopathological Brain Tumor Images Using Sparse Reconstruction and Adaptive Dictionary Selection. *IEEE Trans. Med. Imaging* **2016,** *35* (6), 1575–1586.
12. Kalbkhani, H.; Mahrokh, G. S.; Behrooz, Z. V. Robust Algorithm for Brain Magnetic Resonance Image (MRI) Classification Based on GARCH Variances Series. *Biomed. Signal Process. Control* **2013,** *8* (6), 909–919.
13. Yao, J.; Chen, J.; Chow, C. Breast Tumor Analysis in Dynamic Contrast Enhanced MRI Using Texture Features and Wavelet Transform. *IEEE J. Select. Top. Signal Process.* **2009,** *3* (1), 94–100
14. Jainy, S.; Kumarb, V.; Gupta, I.; Khandelwalc, N.; Kamal, C. A Package SFERCB Segmentation, Feature Extraction, Reduction and Classification Analysis by Both SVM and ANN for Brain Tumors. *Appl. Soft Comput.* **2016,** *47*, 151–167.
15. Jothi, G.; Inbarani, H. H. Hybrid Tolerance Rough Set Firefly Based Supervised Featureselection for MRI Brain Tumor Image Classification. *Appl. Soft Comput.* **2016,** *46*, 639–651.

16. Sallemi, L.; Njeh, I.; Lehericy, S. Towards a Computer Aided Prognosis for Brain Glioblastomas Tumor Growth Estimation. *IEEE Trans. Nanobiosci.* **2015,** *14* (7), 727–733.
17. Valarmathy, G.; Sekar, K.; Balaji, V. An Automated Framework to Segment and Classify Gliomas Using Efficient Segmentation and Classification. *Int. J. Adv. Sci. Technol.* **2020,** *29* (10S), 7539–754.
18. Valarmathy, G.; Sekar, K.; Balaji, V. An Automated Framework to Segment and Classify Gliomas Using Efficient Shuffled Complex Evolution Convolutional Neural Network. *J. Med. Imag. Health Inf.* **2021,** *11*, 2765–2770.
19. Malathi, M.; Sujatha, K.; Sinthia, P. Brain Tumour Segmentation Using Clustering And EM Segmentation. *Int. J. Appl. Eng. Res.* **2015,** *10* (11), 29105–29119.
20. Malathi, M.; Sujatha, K.; Sinthia, P. Detection and Classification of Brain Tumour using Back Propagation Algorithm. *Int. J. Control Theory Appl.* **2016,** *9* (24), 299–306.

CHAPTER 13

APPROACHES FOR ANALYZING DENTAL IMAGES WITH MEDICAL IMAGE PROCESSING WITH ITS STATISTICS

G. BALANAGIREDDY[1], P. SUBRAMANIAN[2], J. MARTIN SAHAYARAJ[3], and S. MUTHUKUMAR[4]

[1]*Department of Electronics and Communication Engineering, Rajiv Gandhi University of Knowledge Technologies – Ongole Campus, Ongole, Andhra Pradesh, India*

[2]*Department of Computer Science Engineering, Mohamed Sathak A. J. College of Engineering, Chennai, India*

[3]*Department of Electronics and Communication Engineering, Sri Indu College of Engineering and Technology, Hyderabad, Telangana, India*

[4]*Department of Computer Science Engineering, B.S. Abdur Rahman Crescent Institute of Science and Technology, Chennai, Tamil Nadu, India*

ABSTRACT

In dental morphological images, there exhibits a lot of inconsistency due to noise, teeth visualized per image. So, providing insightful experience would nurture the needs for stomatological purposes. This chapter deals with such

Computational Imaging and Analytics in Biomedical Engineering: Algorithms and Applications.
T. R. Ganesh Babu, PhD, U. Saravanakumar, PhD & Balachandra Pattanaik, PhD (Eds.)

aspects considering radiographical images. Measuring the distance between the mandible inferior border and the superior border of the "alveolar" for the image is done manually. Thresholding via various methods such as "Haung," "Otsu" and their corresponding values are provided for analysis. Both thresholding algorithm used in the subset of the image dataset shows unbiased results of mandible region.

13.1 INTRODUCTION

The X-ray image acquired for dental purposes can be classified as intraoral and extraoral. The former states images are acquired within the mouth and the latter states that images are acquired outside the mouth. The majority of the survey has been carried out using "intraoral segmentation" via threshold-based segmentation. The extraoral dataset had been discussed in Ref. [1], with a review of varying segmentation algorithms. The oral cavity has been analyzed with malignant and benign lesions using deep learning techniques specifically for dentigerous cysts.[2]

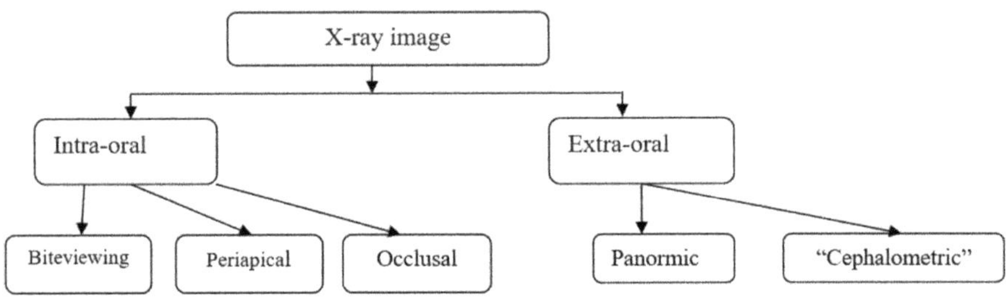

FIGURE 13.1 Two-dimensional classification of X-ray images.

Based on their geometrical projection, two-dimensional radiographic images are shown in Figure 13.1.[4] "Cephalometric" analysis using Artificial Intelligence has been done in Ref. [5], for localization of landmarks for classification and segmentation purposes. The impact of mandibular region segmentation in the image and the dose given for radiological perspective has been analyzed with scanners in phantoms.[12]. In Ref. [11], segmenting the "mandibular bone" from the teeth is difficult with grayscale images. Hence, preprocessing with a "Gaussian filter" and postprocessing algorithms have been carried out with active contours. Images were taken from the imaging repository as in Ref. [13], and the ImageJ tool has been used for processing.

Otsu level of thresholding divides the object into segments of two groups based on variance levels.[18] Haung level of thresholding incorporates object attributes for segmenting an image.[19] Both thresholding algorithms used in the image dataset show unbiased results of the mandible region in sample 1.

Section 13.2 deals with a literature survey of dental imaging, and segmentation. Section 13.3 deals with the steps involved in the proposed system development. Section 13.4 deals with statistical and medical imaging algorithms. Section 13.5 concludes the overall work with future scope.

13.2 LITERATURE SURVEY

Interpretation of X-rays for three-dimensional image constructions has evolved in various means like subtraction radiography, computer-aided learning for the acquired image.[10] Early, discussion of multislice computed tomography (CT) states that it provides better accuracy in terms of volume in vivo and in vitro studies.[6] In terms of detection accuracy, the images of bitewing and periapical were better than the panoramic images considering the posterior part of the teeth.[3] The discussions in Ref. [7] state the need for morphological operation differences between dentistry annotated data and automated segmented data. The states finding states the source of errors, and similarity indices obtained via Jaccard's coefficient. Local ternary pattern filter had been discussed in Ref. [8]; for image segmentation, it calculates three values specified for the threshold pixel. The first "1" value signifies the neighbor pixel has a value greater than the threshold constant plus the center of the pixel. The second "−1" value specifies the neighbor pixel has a value lesser than the center of the pixel minus the threshold constant. The third value "0" denotes that lies in a range greater than the threshold constant plus the center of the pixel to the threshold constant minus the center of the pixel.[8] Thus a connected component analysis and easy as well as automatic segmentation had been provided. Deep learning model had been used in Ref. [9], for instance, tooth segmentation and its region of interest. Bounding boxes had been used for analysis as a comparative index between the missing region and its ground truth. In Ref. [13], the image registration technique followed by extracting the mandible contour from the existing mandible atlas has been done. An inferior border has been extracted with a canny edge detector where vertical variance obtained using a Gaussian filter is 2 times lesser than horizontal variance. A semiautomatic model for incorporating manual segmentation has been associated with the body axis plane.[14] The acquired images and their age had been more than 20 years in Ref. [13] to avoid deciduous teeth. The demographic detail of age less than 20 years has been considered in Ref. [14]. Asymmetric mandibles exist wherein it is difficult to analyze the three-dimensional shape via conventional techniques. Further, the study states the influence of landmarks and angular measurements in accessing the morphology.[15]

An extensive review of dental X-rays is given in Ref. [16] as it states the evolution of image processing, deep learning, and machine learning practices. Variation of intensity in an irregular manner in some cases might degrade the image quality. In Ref. [17], the discussed works state strangeness in acquired medical images has to be visualized via image processing for extracting its features. A canny edge detector has been used in DICOM images for pinpointing the exactness of boundaries without loss of its features. Examining patients based on verbal exchange has been enhanced using deep learning techniques.[21] The support vector technique state it provides more accuracy for diagnostic and prognostic features.

13.3 METHODOLOGY

13.3.1 PROCESSING OF MANDIBLE REGION IN SAMPLE 1

Panoramic Images were obtained from the imaging archive[13] with a segmented mandible region. Length measurement across the mandible region has been performed by the mandible inferior border and the superior border of the “alveolar” for the image.

Then, Otsu thresholding has been done using the ImageJ plugin. It is a single intensity value that separates the pixels into its classes. It signifies there is minimal interclass variance that exhibits in an image.

Subsequently, Haung thresholding is done based on the function of Shannon entropy.

13.3.2 PROCESSING OF MANDIBLE REGION IN SAMPLE 2

Panoramic Images were obtained from the imaging archive[13] with a segmented mandible region.

Edge detection is done to process the edges. It is followed by similar multi-thresholding via the Haung model and its maximum threshold values.

13.4 RESULTS AND DISCUSSION

A complete image followed by two dentists segmented with the mandible region segmentation has been taken for analysis with ImageJ software. The ImageJ software has been used in Ref. [20] since it gives accuracy in results used for volumetric analysis in oral bone.

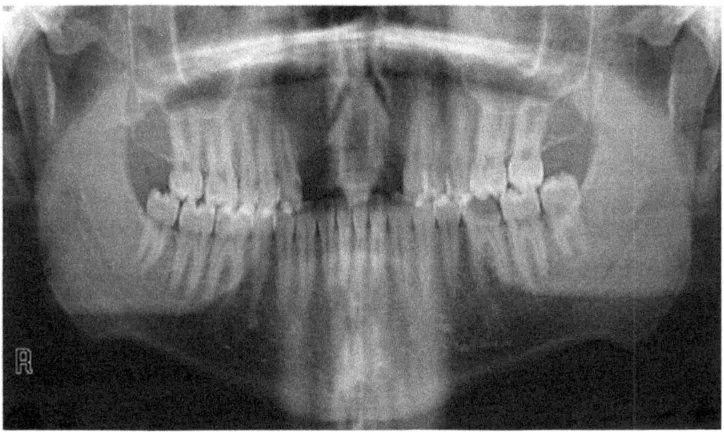

FIGURE 13.2 Original grayscale image-1 taken without grayscale segmentation.

Figure 13.2 shows full image with the mandible region included. Figure 13.3 is a segmented image available in the dataset by dentist 1. Figure 13.4 is a segmented image available in the dataset by dentist 2.

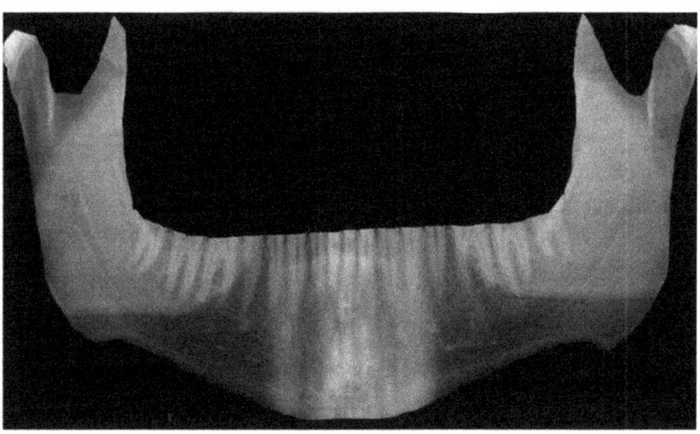

FIGURE 13.3 Segmented mandible region of interest by dentist 1.

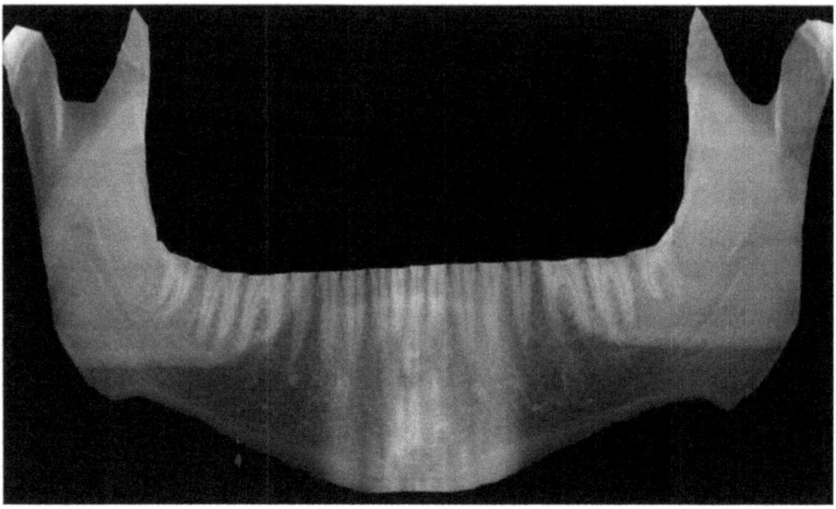

FIGURE 13.4 Segmented mandible region of interest by dentist 2.

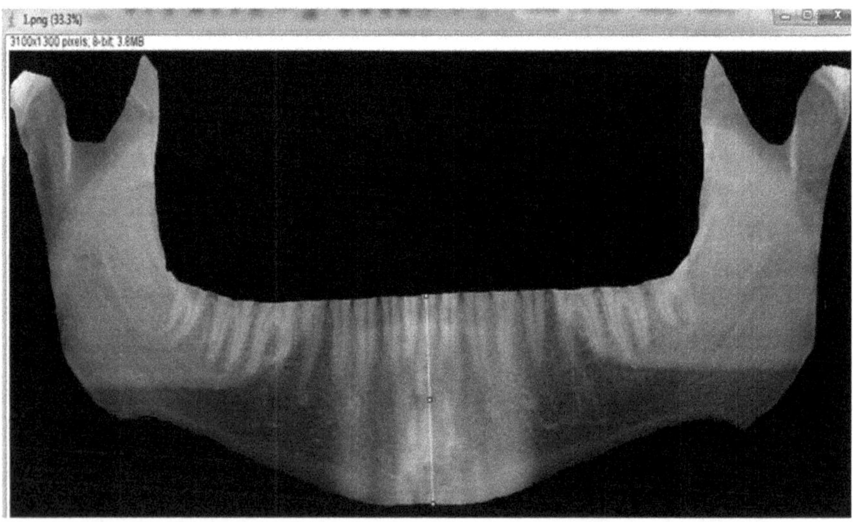

FIGURE 13.5 Segmented mandible length measured manually indicated via yellow lines.

In Figure 13.5, a length of 582.5 has been measured between mandible inferior border and the superior border of "alveolar" for image obtained from dentist 1.

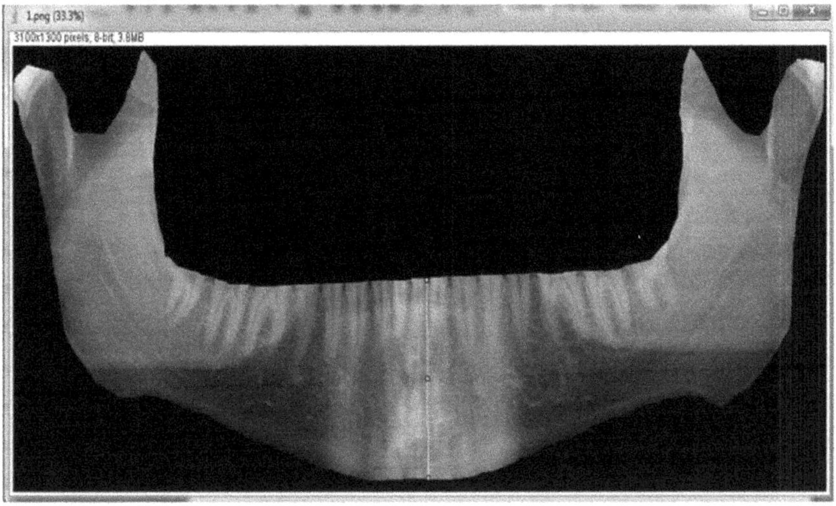

FIGURE 13.6 Segmented mandible length measured manually indicated via yellow lines.

In Figure 13.6, a length of 579 has been measured between the mandible inferior border and the superior border of the "alveolar" for the image obtained from dentist 2.

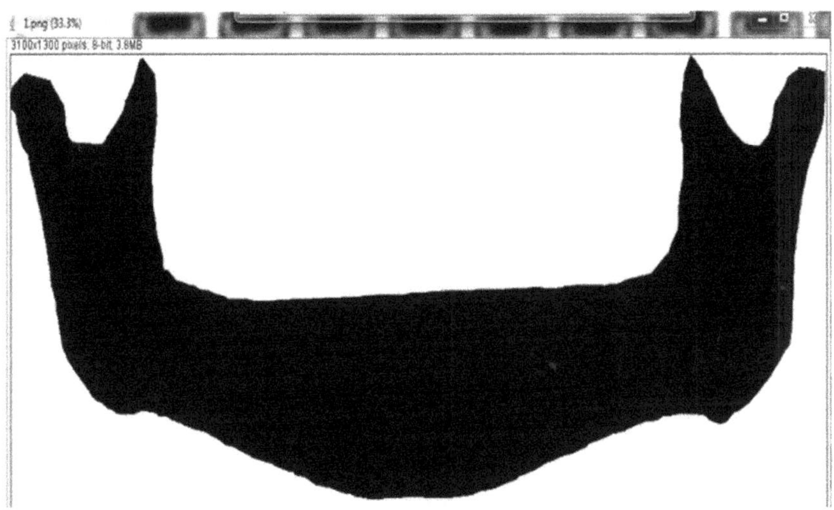

FIGURE 13.7 The threshold value of the image 2.

In Figure 13.7, a threshold value of 20 is chosen automatically with the Huang model. The minimum and maximum value used in the Haung model is shown in Figure 13.8.

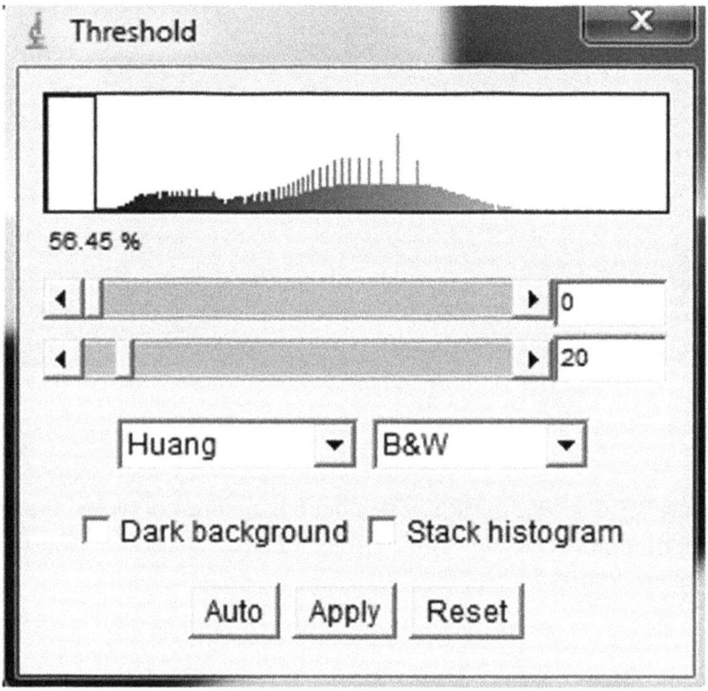

FIGURE 13.8 The threshold values used in Haung.

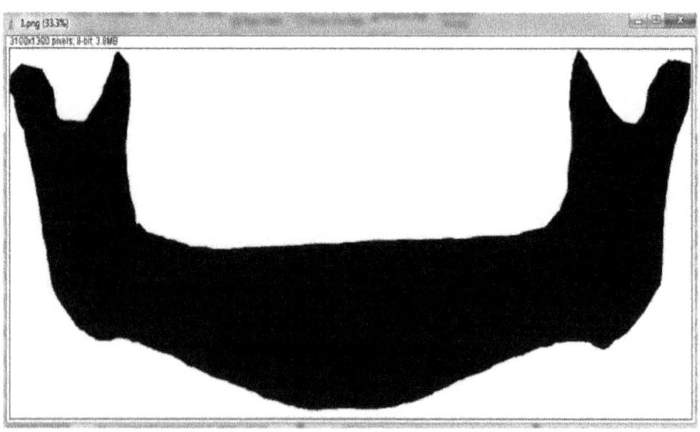

FIGURE 13.9 The threshold value of the image 2 via Otsu thresholding.

In Figure 13.9, a threshold value of 21 is chosen automatically with Otsu thresholding. The minimum and maximum value used in is shown in Figure 13.10.

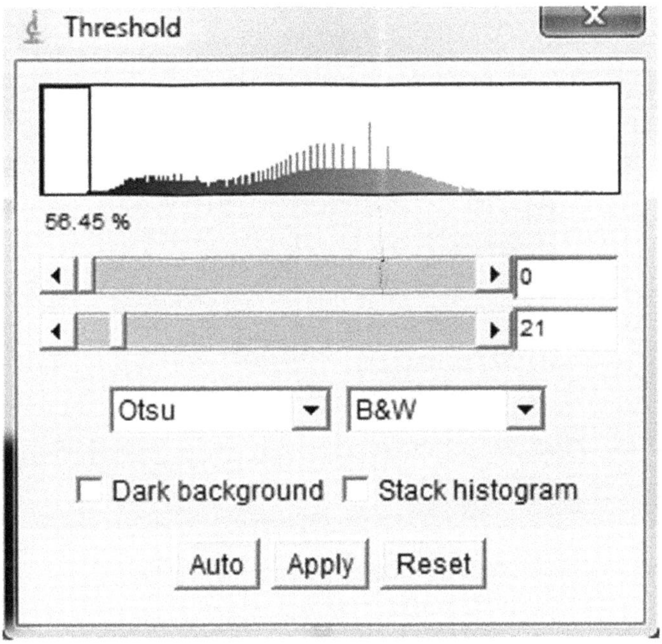

FIGURE 13.10 The threshold values used in "Otsu".

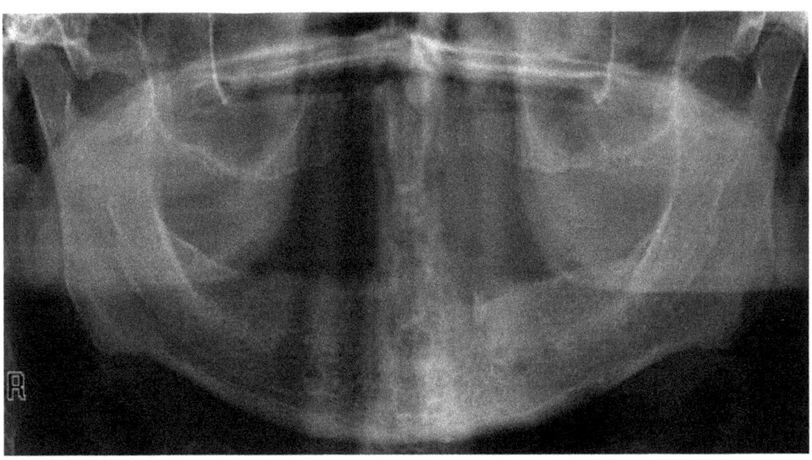

FIGURE 13.11 Original grayscale image 2 taken without grayscale segmentation.

Figure 13.11 shows whole image with the mandible region included. Figure 13.12 shows a segmented image available in the dataset by dentist 1 for an original image shown in Fig. 13.11. Figure 13.13 shows segmented image available in the dataset by dentist 2 for an original image shown in Fig. 13.12.

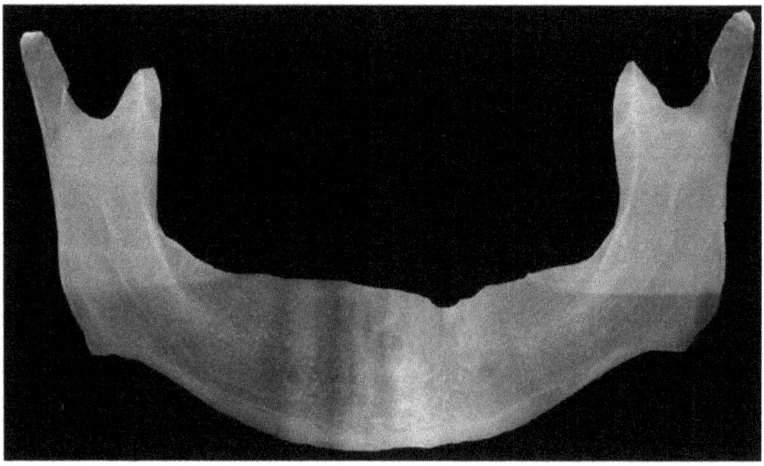

FIGURE 13.12 Segmented mandible region of interest by dentist 1.

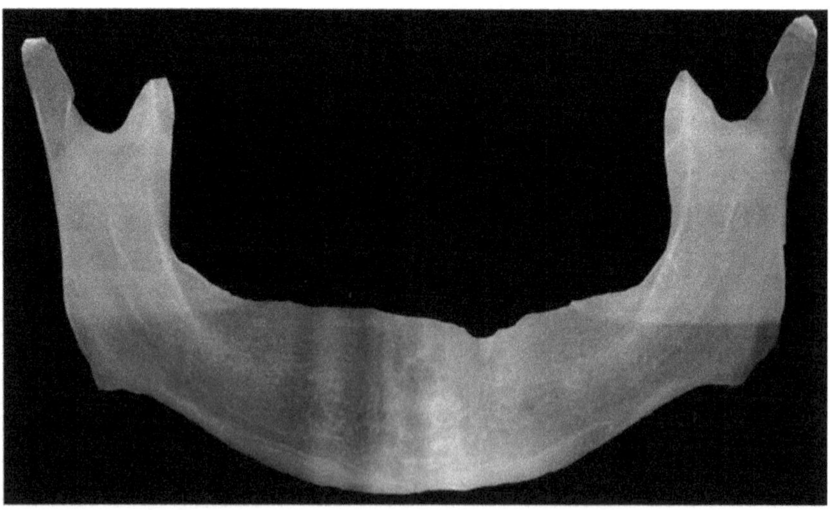

FIGURE 13.13 Segmented mandible region of interest by dentist 2.

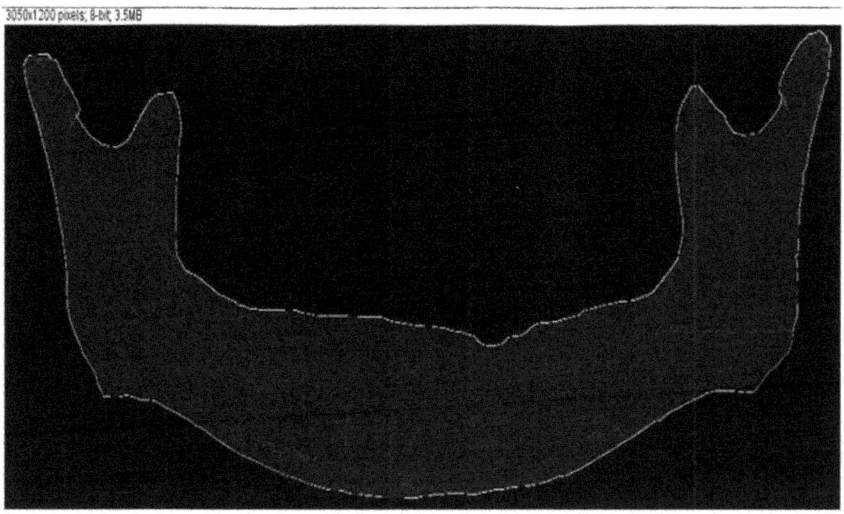

FIGURE 13.14 Edge detection technique for mandible region.

Figure 13.14 shows the edge detection region of interest for dentist 1 image shown in Figure 13.12. Figure 13.15 used the Haung threshold maximum value lies at 6 for interpretation.

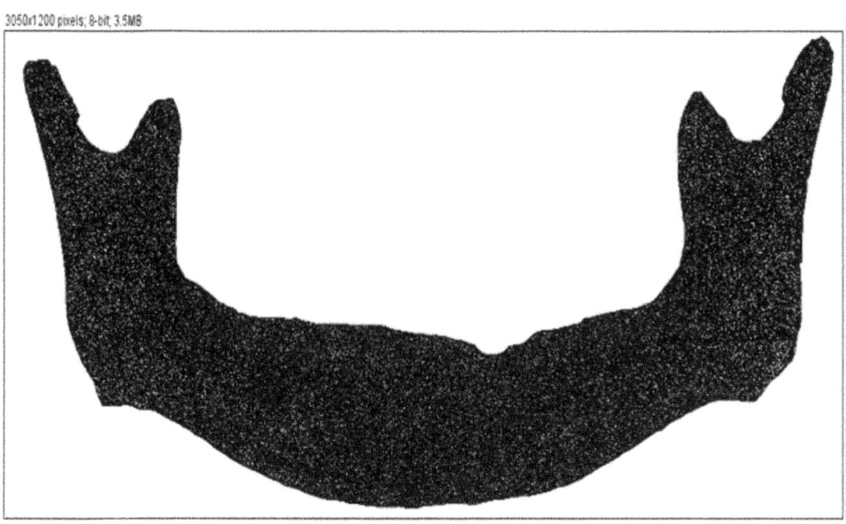

FIGURE 13.15 Haung threshold method for edge detection image for mandible region.

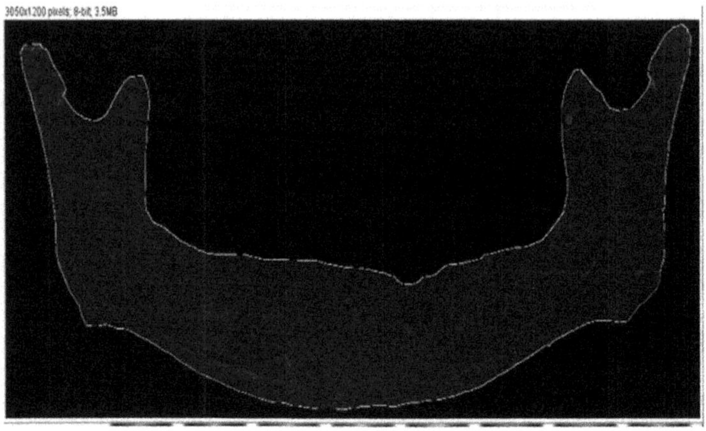

FIGURE 13.16 Edge detection technique for dentist image 2.

FIGURE 13.17 Haung thresholding for dentist image 2.

Figure 13.17 shows the Haung threshold for dentist image 2 where the maximum value lies at 6.

13.5 CONCLUSIONS

Two samples of oral panoramic X-ray images were taken for analysis followed by processing them via segmented mandible regions from experts.

In the first case, the image and its measurement between the inferior border and the superior border of the mandible region were performed. The work was further analyzed by thresholding the region via Haung and Otsu-based analysis. In sample 2, the image obtained was processed via edge detection and analyzed with subsequent Haung-based thresholding. The future work will incorporate the impact of age and related dosage associated with volume metrics analysis in the mandible region.

KEYWORDS

- **thresholding**
- **edge detection**
- **intraoral**
- **extraoral**
- **segmentation algorithms**

REFERENCES

1. Silva, G.; Oliveira, L.; Pithon, M. Automatic Segmenting Teeth in X-ray Images: Trends, a Novel Data Set, Benchmarking and Future Perspectives. *Expert Syst. App.* **2018,** *107*, 15–31.
2. Yang, H.; Jo, E.; Kim, H. J.; Cha, I. H.; Jung, Y. S.; Nam, W.; Kim, D. Deep Learning for Automated Detection of Cyst and Tumors of the Jaw in Panoramic Radiographs. *J. Clin. Med.* **2020,** *9* (6), 1839.
3. Akarslan, Z. Z.; Akdevelioglu, M.; Gungor, K.; Erten, H. A Comparison of the Diagnostic Accuracy of Bitewing, Periapical, Unfiltered and Filtered Digital Panoramic Images for Approximal Caries Detection in Posterior Teeth. *Dentomaxillofacial Radiol.* **2008,** *37* (8), 458–463.
4. Akarslan, Z. Z.; Peker, I. *Advances in Radiographic Techniques Used in Dentistry*; IntechOpen, 2015; Chapter 34.
5. Hung, K.; Montalvao, C.; Tanaka, R.; Kawai, T.; Bornstein, M. M. The Use and Performance of Artificial Intelligence Applications in Dental and Maxillofacial Radiology: A Systematic Review. *Dentomaxillofacial Radiol.* **2020,** *49* (1), 20190107.
6. Cavalcanti, M. D. G. P.; Ruprecht, A.; Vannier, M. W. 3D Volume Rendering Using Multislice CT for Dental Implants. *Dentomaxillofacial Radiol.* **2002,** *31* (4), 218–223.
7. Kanuri, N.; Abdelkarim, A. Z.; Rathore, S. A. Trainable WEKA (Waikato Environment for Knowledge Analysis) Segmentation Tool: Machine-Learning-Enabled Segmentation on Features of Panoramic Radiographs. *Cureus* **2022,** *14* (1).

8. Salih, O.; Duffy, K. J. The Local Ternary Pattern Encoder–Decoder Neural Network for Dental Image Segmentation. *IET Image Process.* **2022,** 1–11. https://doi.org/10.1049/ipr2.12416.
9. Park, J.; Lee, J.; Moon, S.; Lee, K. Deep Learning Based Detection of Missing Tooth Regions for Dental Implant Planning in Panoramic Radiographic Images. *Appl. Sci.* **2022,** *12* (3), 1595.
10. van der Stelt, P. F. From Pixel to Image Analysis. *Dentomaxillofacial Radiol.* **2021,** *50* (2), 20200305.
11. Nafi'iyah, N.; Fatichah, C.; Astuti, E. R.; Herumurti, D. The Use of Pre and Post Processing to Enhance Mandible Segmentation using Active Contours on Dental Panoramic Radiography Images. In *2020 3rd International Seminar on Research of Information Technology and Intelligent Systems (ISRITI)*; IEEE, Dec 2020; pp 661–666.
12. Loubele, M.; Jacobs, R.; Maes, F.; Denis, K.; White, S.; Coudyzer, W.; ... & Suetens, P. Image Quality vs Radiation Dose of Four Cone Beam Computed Tomography Scanners. *Dentomaxillofacial Radiol.* **2008,** *37* (6), 309–319.
13. Abdi, A. H.; Kasaei, S.; Mehdizadeh, M. Automatic Segmentation of Mandible in Panoramic X-ray. *J. Med. Imag.* **2015,** *2* (4), 044003.
14. Chuang, Y. J.; Doherty, B. M.; Adluru, N.; Chung, M. K.; Vorperian, H. K. A Novel Registration-Based Semi-Automatic Mandible Segmentation Pipeline Using Computed Tomography Images to Study Mandibular Development. *J. Comput. Assist. Tomogr.* **2018,** *42* (2), 306.
15. Inoue, K.; Nakano, H.; Sumida, T.; Yamada, T.; Otawa, N.; Fukuda, N.; Mori, Y. A Novel Measurement Method for the Morphology of the Mandibular Ramus Using Homologous Modelling. *Dentomaxillofacial Radiol.* **2015,** *44* (8), 20150062.
16. Kumar, A.; Bhadauria, H. S.; Singh, A. Descriptive Analysis of Dental X-ray Images Using Various Practical Methods: A Review. *PeerJ Comput. Sci.* **2021,** *7*, e620.
17. Chikmurge, D.; Harnale, S. Feature Extraction of DICOM Images Using Canny Edge Detection Algorithm. In *International Conference on Intelligent Computing and Applications*; Springer: Singapore, 2018; pp 185–196.
18. Otsu, N. A Threshold Selection Method from Gray-Level Histograms. *IEEE Trans. Syst. Man Cybern.* **1979,** *9* (1), 62–66.
19. Huang, L. K.; Wang, M. J. J. Image Thresholding by Minimizing the Measures of Fuzziness. *Pattern Recogn.* **1995,** *28* (1), 41–51.
20. Fadili, A.; Alehyane, N.; Halimi, A.; Zaoui, F. An Alternative Approach to Assessing Volume-of-Interest Accuracy Using CBCT and ImageJ Software: In Vitro Study. *Adv. Radiol.* **2015**.
21. Menon, N. G.; Shrivastava, A.; Bhavana, N. D.; Simon, J. Deep Learning Based Transcribing and Summarizing Clinical Conversations. In *2021 Fifth International Conference on I-SMAC (IoT in Social, Mobile, Analytics and Cloud) (I-SMAC)*; IEEE, 2021; pp 358–365.

CHAPTER 14

AN INVESTIGATION ON DIABETES USING MULTILAYER PERCEPTRON

J. SURENDHARAN[1], A. KARUNAMURTHY[2], R. PRAVEENA[3], and K. SHEBAGADEVI[3]

[1]HKBK College of Engineering, Bengaluru, Karnataka, India

[2]BWDA Arts and Science College, Villupuram, India

[3]Department of ECE, Muthayammal Engineering College, Namakkal, India

ABSTRACT

The breakthroughs in public healthcare infrastructure have resulted in a large influx of highly sensitive and critical healthcare information. The application of sophisticated data analysis techniques can aid in the early detection and prevention of a variety of fatal diseases. Diabetes can cause heart disease, renal disease, and nerve damage, all of which are life-threatening complications of the disease. The goal of this work is to identify, detect, and forecast the emergence of diabetes in its earliest stages by employing machine learning (ML) techniques and algorithms. When it comes to diabetes classification, a multilayer perceptron (MLP) is used. The experimental evaluation was carried out using the Pima Indian diabetes dataset. According to the study findings, MLP outperforms the competition in terms of accuracy, with an accuracy rate of 86.08%. Following this, a comparison of the suggested technique with the existing state of the art is carried out, proving the flexibility of the proposed approach to a wide range of public healthcare applications.

Computational Imaging and Analytics in Biomedical Engineering: Algorithms and Applications.
T. R. Ganesh Babu, PhD, U. Saravanakumar, PhD & Balachandra Pattanaik, PhD (Eds.)

14.1 INTRODUCTION

Diabetes is referred to as diabetes mellitus (DM) and hypoglycemia, often known as low blood sugar, in which the body is unable to synthesize glucose in the bloodstream. Type 1 diabetes is a prevalent type, while type 2 diabetes is the second most common.[1] Type 1 is more likely to develop in children than in adults. The immune system in a person attacks the pancreas with type 1, causing it to generate antibodies that cause the organ to malfunction and cease generating insulin. It is possible to develop substantial issues and the nervous system, despite the fact that it is less severe than type 1 diabetes.[2]

Diabetes mellitus is a key topic of medical study because of the disease's enormous social impact and the massive amount of data that is generated as a result of this impact.[3] Machine learning (ML) approaches in DM diagnosis, management, and other clinical administration elements are certainly of major concern. An ensemble strategy for diabetes classification utilizing ML has been proposed in this article, which builds on a range of methodologies that have been explored previously.[4–8]

Obstacles such as gestational DM and obesity occur during pregnancy and have a long-term influence on the health of both the mother and the child. In nonpregnant women, because of the relationship between the microvascular block risk and the level of glucose in the blood, several criteria for recognizing diabetes have been developed. As a result of the improvement in living standards, diabetes has become increasingly prevalent. The only way to avoid its problems is to discover and diagnose them early.[9]

There has been a significant amount of research on disease prediction, including diagnosis, prediction, classification, and treatment. Various ML[10–15] methods have been developed and applied to the identification and prediction of diseases throughout the last few years. The implementation of ML with conventional approaches has resulted in significant improvements. Machine learning has proved its ability to deal with enormous numbers of variables in an efficient and effective manner while simultaneously developing robust prediction models. Supervised ML methods are used to investigate the independent terms and variables of the dependent term.

Biomedical datasets are transformed into usable information through this process, which allows for top-tier clinical research while also enhancing patient care. As previously noted, there have been numerous improvements in ML approaches as a result of the requirement to classify diabetic patients. The Pima Indian diabetes dataset is available to the public. There is a total of 769 data points, 500 of which are free of diabetes and 268 of which are positive for the presence of diabetes.

According to the research history, a variety of ML algorithms is used on this dataset for the aim of disease classification, none of which has reached an accuracy of more than 76%, according to the research history. As a result, we came up with the idea of improvising as a group rather than individually. The subject of this research is ML models, and it investigates their performance, theory, and attributes in greater depth.

The classification approach has been adopted by scientists in place of the regression strategy for making disease predictions. Its performance has been assessed using the accuracy, precision, recall, and F1-score of the aforementioned algorithm as measures of its effectiveness.

14.2 RELATED WORKS

Healthcare professionals rely on correct diagnoses to perform their duties effectively. Diagnosing a patient's type of diabetes necessitates a variety of different tests, making it one of the most difficult duties faced by medical practitioners. Considering too many variables at the same time when diagnosing has the potential to produce erroneous conclusions. As a result, determining the kind of diabetes that someone has is incredibly difficult to accomplish. Recent years have seen a significant impact on the healthcare industry, particularly as a result of ML techniques. Numerous investigations have been conducted to determine the classification of diabetes.

Qawqzeh et al.[15] came up with a logistic regression for the classification of type 2 diabetes. It was built using data from 459 training, and 128 testing. When their proposed system was used to categorize 552 people as nondiabetic, they were able to attain an accuracy rate of 92%. The proposed method, on the other hand, has not been evaluated in comparison to existing methods.

Pethunachiyar[16] developed a technique for categorizing patients with DM by employing a support vector machine (SVM). According to him, linear SVM is more efficient than any of the other models that have been discussed thus far. Despite this, there is no comparison of current state-of-the-art technology, and there are no specifics about parameter selection.

Diabetes was classified using Naive Bayes (NB) classifiers and SVMs classification. These conclusions were drawn from the Pima Indian diabetes study, which used data from the Pima Indian diabetes study. The current as well as the level of accuracy obtained, are notably absent from the document.

Choubey et al.[18] conducted a study in which they compared several diabetes classification techniques. The datasets utilized were a local diabetes dataset and Pima Indian datasets. Feature engineering was carried out using principal component analysis (PCA) and linear discriminant analysis (LDA), both of which were shown to be beneficial in boosting the accuracy of the classification method and removing undesired features from the dataset.

An ML paradigm that they developed was used to identify and predict diabetes by Maniruzzaman et al.[19] They employed four ML techniques for the classification of diabetes: Naive Bayes, decision trees, AdaBoost, and random forests, among others. Additionally, they used three alternative partition techniques in addition to the 20 trials they conducted. The researchers used data from the National Health and Nutrition Survey (NHNS) for both diabetic and nondiabetic patients to put their innovative technique through its paces.

Ahuja et al.[20] conducted an examination of ML algorithms, including neural networks, deep learning, and multilayer perceptrons (MLPs), on the Pima dataset for diabetic classification. When compared to the data, MLP was determined to be superior to other classifiers. According to the authors, fine-tuning and efficient feature engineering can help to increase the performance of MLP algorithms. Recent research by Mohapatra et al.[21] has demonstrated the use of MLP to classify diabetes.

Singh and Singh[22] proposed an ensemble method for the prediction of type 2 diabetes to improve accuracy. The Pima dataset from the University of California, Irvine Machine Learning Repository was used in this work. The bootstrap approach with cross-validation was used to train the four base learners of the stacking ensemble, which were used to train the stacking ensemble four base learners. However, neither variable selection nor a comparison of the current state of the art is mentioned, though.

Kumari et al.[23] created a soft computing-based diabetes prediction system based on an ensemble of three commonly used supervised ML algorithms. It was discovered that they had used Pima datasets in their investigation. When they compared their system performance to that of state-of-the-art individual and ensemble approaches, they discovered that it outperformed them by 79%.

To forecast diabetes in its early or onset stage, Islam et al.[24] used a combination of techniques. Training methods included cross-validation and percentage splits, which were both employed in this study. They collected data from 529 Bangladeshi patients, both diabetic and nondiabetic, using questionnaires administered at a hospital in the nation. The experimental results reveal that the random forest algorithm outperforms them all by a significant margin. Although there is no comparison to the present state of the art, there is no clear reporting of the accuracy that was attained in this study.

The use of ML approaches to predict early and ongoing DM in females has been demonstrated in several studies.[25] They employed typical ML methods to construct a framework for predicting diabetes to better understand the disease.

Using ML models for diabetes prediction that were published between 2010 and 2019, Hussain and Naaz[26] did a comprehensive evaluation of the literature on this topic. They evaluated the algorithms based on the Matthews correlation coefficient and discovered that Naive Bayes and random forests outperformed the other algorithms in terms of overall performance.

14.3 PROPOSED METHOD

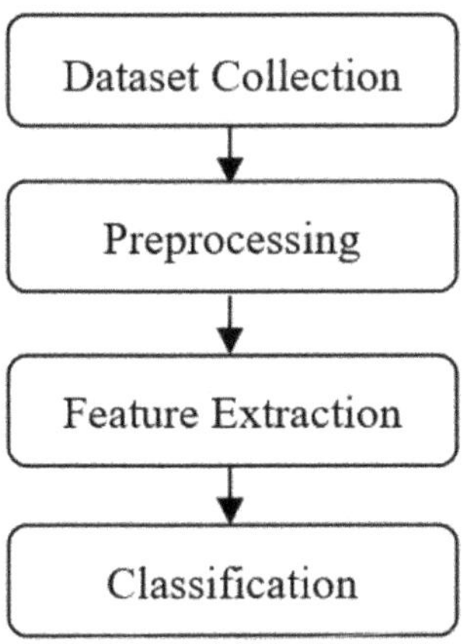

FIGURE 14.1 Proposed architecture for diabetes classification.

A variety of ML techniques have been applied to make decisions in the proposed diabetes classification system. For classification, we used and fine-tuned MLP for the first time, owing to its good performance in healthcare, notably in the prediction of diabetic complications, and the illustration is given in Figure 14.1.

14.3.1 CLASSIFICATION

The study makes improvements to three commonly used cutting-edge strategies for identifying diabetes that were previously published. For the most part, the proposed ways of putting a person into one of the two categories of diabetes are compared and contrasted against one another. The following are the various diabetic treatment options that have been offered.

MLP: Our experimental setup incorporates a well-tuned MLP for diabetes classification that was developed by our team of researchers. As indicated in Figure 14.2, a classification method is represented as a network of numerous layers that are connected to one another. Perceptrons are the fundamental building blocks of our paradigm, and they are linear combinations of input and weights. When developing Algorithm 14.1, we employed a sigmoid unit as the activation function.

The suggested method is comprised of three major phases, as follows: The weights are initially established, then the output layer (δk) output is calculated by applying the sigmoid activation function to the weights and layer outputs. Second, the inaccuracy of all concealed units is determined at the hidden layer level (δh).

Finally, to reduce network errors, all network weights (wi, j) are updated in a backward manner. A step-by-step summary of the entire diabetes classification procedure is provided by algorithm 1 for diabetes classification.

Algorithm 14.1 Classification using MLP.

	Input: Total patients, skin thickness, BP, age, BMI, HB weight
	Output: A Trained Model
	Method: Initialize weights with a random number
	while (Weights≤Threshold) **do**
	for Training samples, **do**
	Input samples is set as input to the network Compute the required output;
	end for
	for output *k*, **do**
	$\delta k = Ok(1 - Ok)(tk - Ok)$
	end for

	for hidden layer h **do**
	$\delta_h = O_h(1-O_h)\ \Sigma_{k\in output} w_{h,k}\delta_k$
	end for
	Update weight of each network $w_{i,j}$;
	$wi,j = wi,j + \Delta wi,j;\ \Delta wi,j = \eta\delta jxi,j$
	End

Because there are eight different variables in the MLP classification model, eight neurons of the MLP classification model. When computing the weights and inputs in the middle layer, the sigmoid unit will be employed as a calculator. At the end of the process, the output layer will calculate the results.

The hidden layers are used to compute the output data, starting with the values and weights of the input layer as a starting point for computation. Sigmoid activation functions are applied to each unit in the hidden layer to reduce the vast amount of data to a more manageable range between 0 and 1. This computation can be used by every middle layer. It is also treated with the same technique as the input layer, yielding results that can be used to create predictions regarding the development of diabetes.

14.4 RESULTS AND DISCUSSION

To put the proposed diabetes classification system to the test, the Pima Indian diabetes dataset is employed. A comparison study is also carried out, utilizing the most up-to-date computational techniques, which are also included in the package. On the basis of the experimental results, the suggested method outperforms the currently available algorithms in terms of performance. Included in this section are sections devoted to defining the dataset, performance indicators, and conducting a comparison study.

14.4.1 DATASET

This study made use of the Pima Indian diabetes dataset. Create an intelligent model for predicting if a person has diabetes based on some of the metrics contained in this dataset using data from this dataset. When it comes to the classification of diabetes, it is a binary classification problem to be solved. The variables are shown in Table 14.1.

TABLE 14.1 Description of Variables in the Dataset.

Attributes	Range	Mean	Standard deviation
Pregnancies	0–17	3.946	3.454
Glucose	0–199	124.025	32.800
BP	0–122	70.828	19.783
Skin thickness	0–99	21.013	16.298
Insulin	0–846	81.795	117.875
BMI	0–67	32.800	8.077
Diabetes pedigree function	0.078–2.4	0.482	0.338
Age	21–81	34.030	12.095
Outcome	Y/N	0.359	0.492

As shown in Figure 14.4, the dataset contains 768 records of female patients over the age of 21 who are either healthy or diabetic, as determined by their blood glucose levels. The distribution of feature values is depicted in Figure 14.5. For the outcome of the target variable, there are only two potential values: 0 and 1. This dataset was used to get an accurate diagnosis of type 2 diabetes. Pima Indian diabetes dataset determines whether or not a user is at risk of acquiring diabetes in the next four years based on their behavior.

To compare the proposed classification system with existing methodologies, this research employs the same experimental design as the Pima Indian dataset from the previous paper. The performance measures that were employed and the outcomes that were obtained for classifying or predicting are explained further below. It is also possible to make a comparison with previous investigations.

The performance is evaluated using three extensively used performance measures: recall, precision, and accuracy. Recall, precision, and accuracy are three frequently used, measures to evaluate the performance of the presented techniques.

Individuals who are not diabetic are appropriately identified as such in true positive (TP), whereas diabetic patients are correctly identified as such in true negative (TN). The patient has diabetes, as revealed by the false negative (FN), but a favorable result is indicated. The patient's false positive (FP) also suggests that he or she is a healthy individual who has been classified as having diabetes, which is a positive sign. The classification models were trained and evaluated using a 10-fold cross-validation approach, which was developed by the authors.

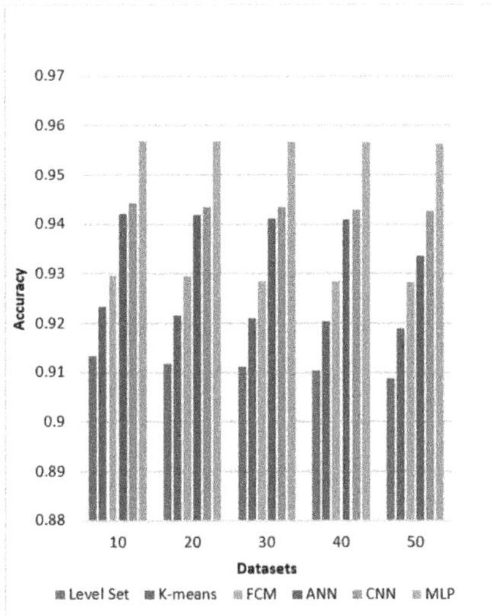

FIGURE 14.2 Accuracy.

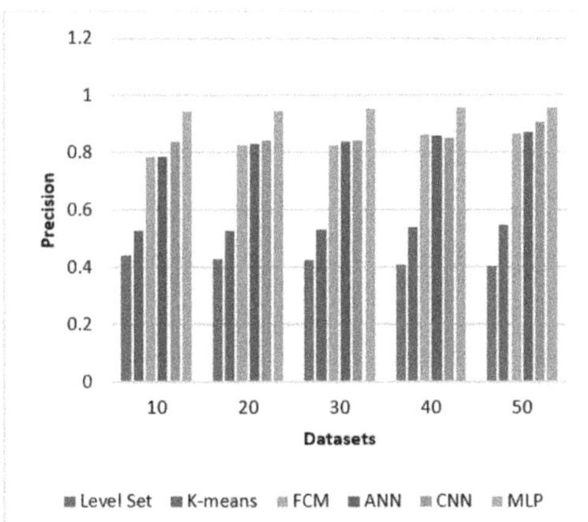

FIGURE 14.3 Precision.

The Pima dataset is used to assess three state-of-the-art diabetic classification classifiers that have been developed recently. When comparing the accuracy of the fine-tuned MLP algorithm in Figure 14.2 to that of present systems, the highest accuracy of 86.083% can be observed. The precision rate has been increased than the existing method which is shown in Figure 14.3. The comparison of the recall rate with the existing method is shown in Figure 14.4.

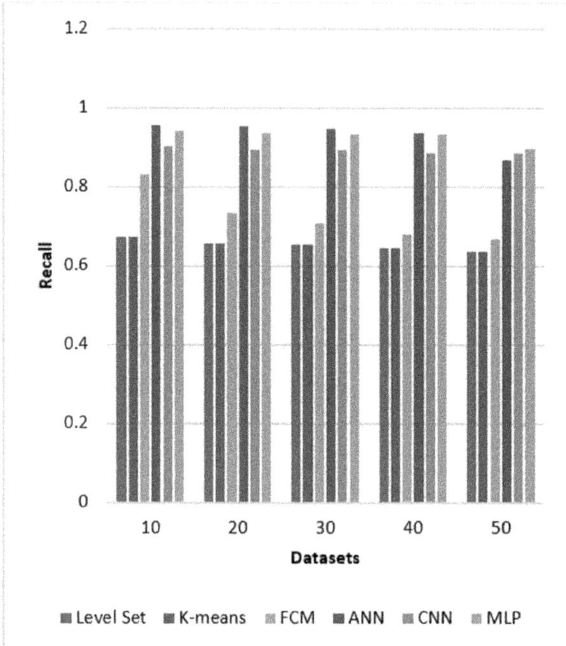

FIGURE 14.4 Recall.

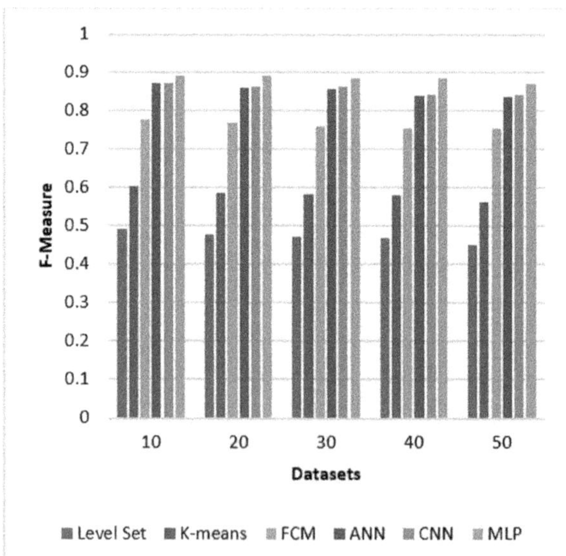

FIGURE 14.5 F-Measure.

In light of the findings, the calibrated MLP model might be utilized to accurately classify diabetes, as demonstrated. We believe that the classification approach we have described here will be beneficial to our hypothetical system in the future.

According to Figure 14.5, a proposed MLP algorithm outperforms with an accuracy of 86.083% when compared to the current state of the art (86.6% precision and 85.1% recall). These discoveries will have a significant impact on the planned hypothetical method for evaluating whether a patient has type 1 or type 2 diabetes.

14.5 CONCLUSIONS

In this paper, the authors proposed a model for supporting the healthcare business. The study developed an algorithm for the classification of diabetes that was based on MLPs. The primary purpose of the proposed system is to aid users in keeping track of their vital signs through the use of their mobile phones and other mobile devices. Users will be able to recognize their elevated risk of diabetes at an earlier stage as a result of the model projections about future blood glucose levels, which is an extra benefit. Diabetic patients are classified and predicted using MLP. The proposed methodologies are tested on the Pima Indian diabetes dataset, which is available online. In terms of accuracy, the two approaches outperform existing best practices by 86.083 and 87.26%, respectively, when compared to current best practices.

KEYWORDS

- **fatal diseases**
- **diabetes mellitus**
- **prediction**
- **classification**

REFERENCES

1. Miranda, M. A.; Macias-Velasco, J. F.; Lawson, H. A. Pancreatic β-Cell Heterogeneity in Health and Diabetes: Classes, Sources, and Subtypes. *Am. J. Physiol.-Endocrinol. Metabol.* **2021,** *320* (4), E716–E731.

2. Babalola, O. M.; Garcia, T. J.; Sefcik, E. F.; Peck, J. L. Improving Diabetes Education in Mexican American Older Adults. *J. Transcult. Nurs.* **2021,** *32* (6), 799–809.
3. Zaccardi, F.; Kloecker, D. E.; Buse, J. B.; Mathieu, C.; Khunti, K.; Davies, M. J. Use of Metformin and Cardiovascular Effects of New Classes of Glucose-Lowering Agents: A Meta-Analysis of Cardiovascular Outcome Trials in Type 2 Diabetes. *Diab. Care* **2021,** *44* (2), e32–e34.
4. Kannan, S.; Idrees, B. A.; Arulprakash, P.; Ranganathan, V.; Udayakumar, E.; Dhinakar, P. Analysis of Convolutional Recurrent Neural Network Classifier for COVID-19 Symptoms Over Computerised Tomography Images. *Int. J. Comput. App. Technol.* **2021,** *66* (3–4), 427–432.
5. Saravanan, D.; Surendiran, J. A New Framework for Video Data Retrieval Using Hierarchical Clustering Technique. *Int. J. Eng. Adv. Technol. (IJEAT)* ISSN: 2249-8958 **2019,** *8* (6S3).
6. Ojugo, A. A.; Ekurume, E. Predictive Intelligent Decision Support Model in Forecasting of the Diabetes Pandemic Using a Reinforcement Deep Learning Approach. *Int. J. Educ. Manag. Eng* **2021,** *11* (2), 40–48.
7. Khunti, K.; Knighton, P.; Zaccardi, F.; Bakhai, C.; Barron, E.; Holman, N. et al. Prescription of Glucose-Lowering Therapies and Risk of COVID-19 Mortality in People with Type 2 Diabetes: A Nationwide Observational Study in England. *Lancet Diab. Endocrinol.* **2021,** *9* (5), 293–303.
8. Syed, S. A.; Sheela Sobana Rani, K.; Mohammad, G. B.; Chennam, K. K.; Jaikumar, R.; Natarajan, Y. et al. Design of Resources Allocation in 6G Cybertwin Technology Using the Fuzzy Neuro Model in Healthcare Systems. *J. Healthcare Eng.* **2022**.
9. Greiver, M.; Havard, A.; Bowles, J. K.; Kalia, S.; Chen, T.; Aliarzadeh, B. et al. Trends in Diabetes Medication use in Australia, Canada, England, and Scotland: A Repeated Cross-Sectional Analysis in Primary Care. *Br. J. Gen. Pract.* **2021,** *71* (704), e209–e218.
10. Vidhya, R. G.; Batri, K. Segmentation Classification of Breast Cancer Using an Krill Herd Oprimization. *Med. Imaging Health Inf.* **2020,** *10* (6), 1294–1300.
11. Anandaraj, S. P.; Kirubakaran, N.; Ramesh, S.; Surendiran, J. Efficient Way to Detect Bone Cancer Using Image Segmentation. *International Journal of Engineering and Advanced Technology (IJEAT)* ISSN: 2249-8958, **2019,** *8* (6S3).
12. Ferrari, M.; Speight, J.; Beath, A.; Browne, J. L.; Mosely, K. The Information-Motivation-Behavioral Skills Model Explains Physical Activity Levels for Adults with Type 2 Diabetes Across all Weight Classes. *Psychol. Health Med.* **2021,** *26* (3), 381–394.
13. Shah, N.; Karguppikar, M.; Bhor, S.; Ladkat, D.; Khadilkar, V.; Khadilkar, A. Impact of Lockdown for COVID-19 Pandemic in Indian Children and Youth with Type 1 Diabetes from Different Socio-Economic Classes. *J. Pediatric Endocrinol. Metabol.* **2021,** *34* (2), 217–223.
14. Mansi, I. A.; Chansard, M.; Lingvay, I.; Zhang, S.; Halm, E. A.; Alvarez, C. A. Association of Statin Therapy Initiation with Diabetes Progression: A Retrospective Matched-Cohort Study. *JAMA Intern. Med.* **2021,** *181* (12), 1562–1574.
15. Bouyahya, A.; El Omari, N.; Elmenyiy, N.; Guaouguaou, F. E.; Balahbib, A.; Belmehdi, O. et al. Moroccan Antidiabetic Medicinal Plants: Ethnobotanical Studies, Phytochemical Bioactive Compounds, Preclinical Investigations, Toxicological Validations and Clinical Evidences; Challenges, Guidance and Perspectives for Future Management of Diabetes Worldwide. *Trends Food Sci. Technol.* **2021,** *115*, 147–254.

16. Saritha, G.; Saravanan, T.; Anbumani, K.; Surendiran, J. Digital Elevation Model and Terrain Mapping Using LiDAR. *Mater. Today: Proc.* **2021,** *46* (9), 3979–3983. ISSN 2214-7853.
17. Vidhya, R. G.; Saravanan, G.; Rajalakshmi, K. Mitosis Detection for Breast Cancer Grading. *Int. J. Adv. Sci. Technol.* **2020,** *29* (3), 4478–4485; Gupta, S.; Verma, H. K.; Bhardwaj, D. Classification of Diabetes Using Naive Bayes and Support Vector Machine as a Technique. In *Operations Management and Systems Engineering*; Springer: Singapore, 2021; pp 365–376.
18. Choubey, D. K.; Kumar, M.; Shukla, V.; Tripathi, S.; Dhandhania, V. K. Comparative Analysis of Classification Methods with PCA and LDA for Diabetes. *Curr. Diab. Rev.* **2020,** *16* (8), 833–850.
19. Satheeshwaran, U.; Sreekanth, N.; Surendiran, J. X-ray CT Reconstruction by Using Spatially Non Homogeneous ICD Optimization. *Int. J. Eng. Adv. Technol. (IJEAT)*, ISSN: 2249-8958, **2019,** *8* (6S3).
20. Ahuja, R.; Sharma, S. C.; Ali, M. A Diabetic Disease Prediction Model Based on Classification Algorithms. In *Annals of Emerging Technologies in Computing (AETiC), Print ISSN*, 2019; pp 2516–0281.
21. Mohapatra, S. K.; Swain, J. K.; Mohanty, M. N. Detection of Diabetes Using Multilayer Perceptron. In *International Conference on Intelligent Computing and Applications*; Springer: Singapore, 2019; pp 109–116.
22. Singh, N.; Singh, P. Stacking-Based Multi-Objective Evolutionary Ensemble Framework for Prediction of Diabetes Mellitus. *Biocybern. Biomed. Eng.* **2020,** *40* (1), 1–22.
23. Surendiran, J.; Saravanan, S. V.; Elizabeth Catherine, F. Glaucoma Detection Using Fuzzy C- Mean (FCM). *IJPT* **2016,** *8* (3), 16149–16163.
24. Islam, M. M.; Ferdousi, R.; Rahman, S.; Bushra, H. Y. Likelihood Prediction of Diabetes at Early Stage Using Data Mining Techniques. In *Computer Vision and Machine Intelligence in Medical Image Analysis*; Springer: Singapore, 2020; pp 113–125.
25. Malik, S.; Harous, S.; El-Sayed, H. Comparative Analysis of Machine Learning Algorithms for Early Prediction of Diabetes Mellitus in Women. In *International Symposium on Modelling and Implementation of Complex Systems*; Springer: Cham, 2020; pp 95–106.
26. Hussain, A.; Naaz, S. Prediction of Diabetes Mellitus: Comparative Study of Various Machine Learning Models. In *International Conference on Innovative Computing and Communications*; Springer: Singapore, 2021; pp 103–115.

CHAPTER 15

DERMOSCOPIC IMPLEMENTATION AND CLASSIFICATION ON MELANOMA DISEASE USING GRADIENT BOOST CLASSIFIER

B. BALAKUMAR[1], K. SAKTHI MURUGAN[2], N. SURESHKUMAR[3], A. PURUSHOTHAMAN[4]

[1]*Centre for Information Technology and Engineering, Manonmaniam Sundaranar University, Tirunelveli, India*

[2]*Department of ECE, PSN College of Engineering and Technology, Tirunelveli, India*

[3]*Department of ECE, Muthayammal College of Engineering, Rasipuram, India*

[4]*Department of ECE, Hindhusthan Institute of Technology, Coimbatore, India*

ABSTRACT

Melanoma is a form of skin cancer that grows when melanocyte become out of balance (the cells that render the skin tan or brown). Cancer starts with cells developing out of balance in the body. Cells may become cancer in virtually every part of the body and then spread to other parts of the body. The melanoma of certain other skin cancers is considerably less common. However, melanoma is harmful as it travels to certain parts of the body even faster than detected and handled. In this study, we propose a deep learning

Computational Imaging and Analytics in Biomedical Engineering: Algorithms and Applications.
T. R. Ganesh Babu, PhD, U. Saravanakumar, PhD & Balachandra Pattanaik, PhD (Eds.)

techniques for classify a deadly disease. Dermoscopic images are used in this study as a dataset directory. We use the classifier of gradient boost and feature extraction by NSCT to predict accuracy and specificity value of melanoma disease.

15.1 INTRODUCTION

Melanoma is one of the skin cancer's deadliest types of disease. Late diagnosis is curable, although only professionally qualified specialists can reliably diagnose the disease. Despite insufficient availability of resources, computer devices that can classify pathogens can save lives, decrease unwanted biopsies and rising prices. In order to do this, we propose a framework that combines recent advances in profound learning with existing method of machine learning, creates sets of methods which segment skin lesions and analyzes the detected region and the surrounding tissue to detect melanoma. In people's daily life, skin diseases are very normal.[1]

Millions of individuals in the United States suffer various forms of skin conditions annually. Diagnosing skin disorders also requires a strong degree of knowledge in the different aspects of their vision. Because human judgment is always arbitrary and hard to replicate, a computer-aided diagnostic device would be considered to obtain a more accurate and effective diagnosis. In this[2] article, we investigate the viability in the development of a deep convolutionary neural network (CNN), a standardized diagnostic method of skin diseases.

One out of five Americans was treated in their lifespan with cutaneous malignancy. While melanomas constitute less than 5% of all skin cancers in the United States, nearly 75% of all deaths due to skin cancer and more than 10,000 deaths are recorded in the United States alone per year. Early detection is important, as the average 5-year melanoma survival rate decreases from 99% in the earliest stages to around 14% in the latest level. We also[3] established a statistical approach that can proactively map skin lesions and diagnose cancer sooner by physicians and patients.

Second, many AL training functions depend on model instability, but such model vulnerability is not always reflected by profound learning methods. In this article, we incorporate in a realistic way recent developments in Bayesian[5] profound learning in the active learning sense. With very limited current literature, we are designing an active learning system for higher dimensional data that have been incredibly challenging to date.

We demonstrate our active learning strategies with images, using special models such as Bayesian CNNs, to greatly enhance current active learning approaches.

Dermoscopy is a technology for capturing skin images, and these images are important for the study of different forms of skin diseases. Malignant melanoma is a form of skin cancer that often causes death to its extent. Previous melanoma diagnosis[6] delays mortality and doctors should treat patients in order to improve their likelihood of survival. There are several machine learning methods to diagnose the melanoma by its characteristics. Melanoma is a type of skin cancer that is most active and increasing. There is a development in work exploring the usage of photographs taken by automated cells to computerize-reported lesions of skin for malignancy. The analysis of these images[7] is usually difficult due to disturbing factors like light variations and light reflections on the surface of the skin. The segmentation of the lesion region from the usual skin is a significant phase in the treatment of melanoma. In any field, a correct choice of the learning algorithm and its statistician validation are necessary for the use of learning techniques to improve performance. Computer-aided diagnosis[8] of skin cancer is difficult given the relative lack of labeling data for lesions, and therefore, the low quality of available training data.

Second, there are some areas that do not take into consideration the research domain for automated diagnosis of skin cancer in the balance between model complexity and generalization capability of the final classification model.

15.2 METHODS AND MATERIALS

15.2.1 NSCT ALGORITHM FOR FEATURES EXTRACTION

Extraction function is the first element from which dermoscopic input images remove the basic characteristic values. A Laplacian pyramid and directional characteristics can be used for detecting dot discontinuities using a directional filter bank for the curtain transition with a double-iterated filter bank. The comparison of the provided picture to a smooth contour at different decomposition rates is feasible.[4] This displays multi-scale and spatial characteristics while at specific sizes, this decomposes the dermoscopic pictures. As a consequence of the down-and-up sampler in both the directional filter bank and the Laplacian pyramid, the contourlet transform will have a redundant 33%, NSCT is being used. The sample architecture diagram is shown in Figure 15.1.

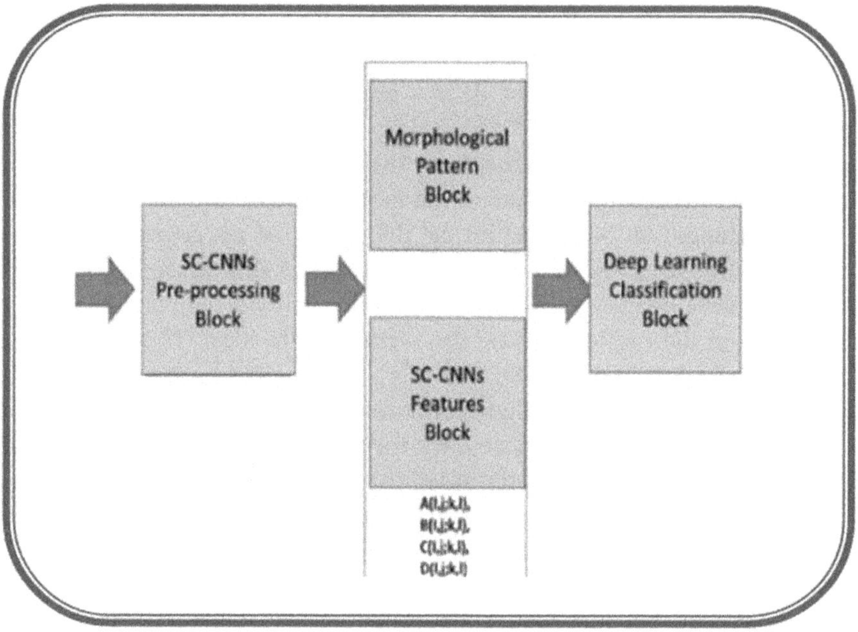

FIGURE 15.1 Architectural diagram for melanoma classification.

15.2.2 CLASSIFIER EXPLANATION

Gradient boosting is a machine learning technique that generates a predictive model in the form of a community of low predictor models usually decision trees for regression and classification. This constructs the model as other boosting approaches do and generalizes it by maximizing a differentiable loss function randomly.

15.3 RESULT AND DISCUSSION

Dermoscopic images are an easy way to capture melanoma images. In this study, we use 60 images normal and abnormal. From the procedure of pre-processing and feature extracting normal and abnormal images has been shown in Figures 15.2 and 15.3.

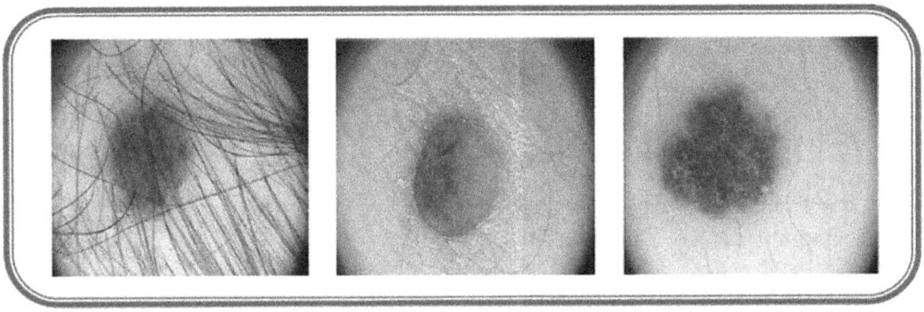

FIGURE 15.2 Normal images.

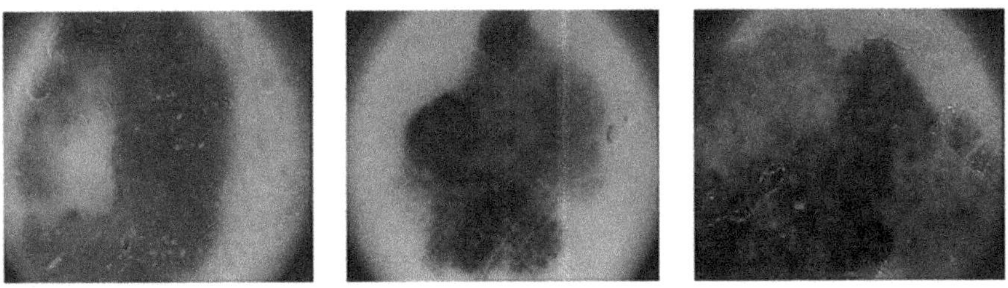

FIGURE 15.3 Abnormal images.

Classification of melanoma disease is done by deep learning classifier of gradient boost as shown in Table 15.1

TABLE 15.1 Classification of Melanoma.

Classifier	Accuracy	Specificity
Gradient Boost	92.4%	88.7%

15.4 CONCLUSIONS

Melanoma identification process is the one of the important things in biomedical field. In this study, we use deep learning libraries to classify the disease whether it is or not by analysis report of classifiers. We implement dermoscopic images as a dataset directory. In this directory, it is split up in to normal and abnormal images. Total of 60 dermoscopic images used in this directory. It is preprocessing in to 40 images as training and 20 images as testing. Then use NSCT algorithm to build a feature extraction and gradient boosting classifier to predict a result of melanoma disease. From the classifier, we got accuracy of 92.4% and specificity rate of 88.7% as a prediction result for melanoma disease classification.

KEYWORDS

- **dermoscopic**
- **NSCT**
- **deep learning**
- **gradient boost**
- **accuracy**
- **specificity**

REFERENCES

1. Codella, N. C.; Nguyen, Q. B.; Pankanti, S.; Gutman, D. A.; Helba, B.; Halpern, A. C.; Smith, J. R. Deep Learning Ensembles for Melanoma Recognition in Dermoscopy Images. *IBM J. Res. Dev.* **2017,** *61* (4/5), 5–1.
2. Liao, H. *A Deep Learning Approach to Universal Skin Disease Classification*; University of Rochester Department of Computer Science, CSC.
3. Esteva, A.; Kuprel, B.; Novoa, R. A.; Ko, J.; Swetter, S. M.; Blau, H. M.; Thrun, S. Dermatologist-Level Classification of Skin Cancer with Deep Neural Networks. *Nature* **2017,** *542* (7639), 115–118.
4. Sonia, R. Melanoma Image Classification System by NSCT Features and Bayes Classification. *Int. J. Adv. Signal Image Sci.* **2016,** *2* (2), 27–33.
5. Gal, Y.; Islam, R.; Ghahramani, Z. Deep Bayesian Active Learning with Image Data. *Proc. 34th Int. Conf. Mach. Learn.* **2017,** *70*, 1183–1192.
6. Premaladha, J.; Ravichandran, K. S. Novel Approaches for Diagnosing Melanoma Skin Lesions Through Supervised and Deep Learning Algorithms. *J. Med. Syst.* **2016,** *40* (4), 96.
7. Jafari, M. H.; Karimi, N.; Nasr-Esfahani, E.; Samavi, S.; Soroushmehr, S. M. R.; Ward, K.; Najarian, K. Skin Lesion Segmentation in Clinical Images Using Deep Learning. In *2016 23rd International conference on pattern recognition (ICPR)*; IEEE, 2016; pp 337–342.
8. Masood, A.; Al-Jumaily, A.; Anam, K. Self-Supervised Learning Model for Skin Cancer Diagnosis. In *2015 7th International IEEE/EMBS Conference on Neural Engineering (NER)*; IEEE, 2015; pp 1012–1015.

CHAPTER 16

IMAGE PROCESSING AND DEEP LEARNING TECHNIQUES FOR LUNG DISEASE SEGMENTATION USING KNN CLASSIFIER

G. SONIYA PRIYATHARSINI[1], N. NAVEENKUMAR[2], BALACHANDRA PATTANAIK[3], and M. KUMARASAMY[4]

[1]*Department of ECE, DR. M. G. R. Educational and Research Institute, Maduravoyal, Chennai, India*

[2]*Department of CSE, Muthayammal Engineering College, Rasipuram, India*

[3]*Department of Electrical and Computer Engineering, College of Engineering and Technology, Wollege University, Ethiopia*

[4]*Department of Computer Science, College of Engineering and Technology, Wollege University, Ethiopia*

ABSTRACT

Lung cancer refers to an uncontrollable breakdown of the cells in lungs. This causes tumor development that reduces a person's breathing capacity. Throughout 2015, 218,527 individuals in the United States were infected with lung cancer, according to the Centers for Disease Control. Early warning will enable a person care in the course of the illness as quickly as possible. However, it may be challenging to recognize the early lung cancer, because the indications may be identical with no indicators at all to those of

Computational Imaging and Analytics in Biomedical Engineering: Algorithms and Applications.
T. R. Ganesh Babu, PhD, U. Saravanakumar, PhD & Balachandra Pattanaik, PhD (Eds.)

a coughing disorder. In this study, we implemented a segmented lung disease prediction using deep learning techniques such as convolutional Neural Network (CNN) and KNN classifier.

16.1 INTRODUCTION

Lung cancer is the most prevalent disease not to be missed which induces late healthcare mortality. Now, CT is available support clinicians diagnose early stage lung cancer.[2] The test to diagnose lung cancer is also calculated by the Doctors' expertise, which certain patients can neglect and trigger problems. Broad awareness has proven to be common and solid process in many testing fields of medical imaging. Three forms of deep neural networks are included in this paper (e.g., CNN, DNN, and SAE) are the calcification of lung cancer.

The CT images recognition function includes these networks. One of the main methods used by pathologists for evaluating the stage, the types and subtypes of lung cancers are the visual analysis of histopathological slides of lung cell tissues. In this research, we learned a profound learning neural convolutionary network (CNN) model (first v3).

This model[1] was used to identify whole-slide pathological photos in correct adenocarcinoma, squamous cell carcinomas, and typical lung tissue from the Cancer Genome Atlas (TCGA). Deep learning in pattern recognition and classification is considered to be a popular and powerful method. However, in the area of medical diagnostics imagery, there are not many highly organized implementations because broad databases are not often visible in medical images. In this[3] research, we checked if deep learning algorithms are feasible for the diagnosis of lung cancer in cases.

16.2 RELATED WORK

Lung cancer is one of the world's main causes of death. At least, an appropriate distinction of the clinical treatment of lung cancer types of cancer (adenocarcinoma, squamous carcinoma, and small cell carcinoma) is common. Nonetheless, the quality increases yet diagnosis reliability is complicated.[4] We also established an automated classification method in this analysis for lung cancers. In microscopic photos, a big deeper learning technique is the deep CNN (DCNN).

Late diagnosis of lung cancer will make a big difference decreased death rate for lung cancer, and is blamed for more than 17% of overall deaths diagnosed with disease. Radiologists report a considerable number of incidents. Diagnosis with machine support (computer-aided diagnostic [CAD]), radiologists can be helped by offering a second opinion and speeding up the whole operation.[5] We suggest a CAD system with profound functionality extracted from auto encoder of be malignant or malignant lung nodules friendly. A never-ending number of chest CT scans will be carried out in the near future by the launch of lung cancer screening programs, which will require radiologists to read to decide on a patient monitoring strategy.

The function of screen-detected nodules is highly dependent on the nodule size and form according to the existing guidelines. We present in this[6] paper, a method of deep learning, focused on multi-stream convolutional network that classifies all nodule forms important to the function of nodules automatically. In order to obtain quantitative tumor[7] separation outcomes, a traditional CAD scheme includes many image processing and pattern recognition processes. Lung CT image segmentation is an essential first phase in the study of the plum image, a preconditions for effective plum CT[8] examination such as the detection of lung cancer. In the society, image segmentation plays a vital role in identification of objects, persons, etc. The image representation[9] is changed by the method of image segmentation. This will quickly alter restrictions and borders.

16.3 METHODS AND MATERIALS

16.3.1 PRE-PROCESSING

The first step is pre-processing the images for the intensity measurement of particles. The use of standard segmentation technology is to divide the processed image. The image is, therefore, filled with cancer nodules. Other characteristics such as area, perimeter, and eccentricities were extracted during extraction of features such as centroid, diameter, and median intensity of the pixels. This is accompanied by a classification module that distinguishes between benign and malignant tumors based on the CT scan images. Extracted functions are used as preparation and the corresponding training model is developed for the classification followed by model evaluation with increasing precision, accuracy, specificity, and sensitivity for detection and classification.

16.4 FEATURES EXTRACTION BY CONVOLUTIONAL NEURAL NETWORK (CNN)

A convolutionary neural network (ConvNet/CNN) is a deep learning algorithm that can take an input image and assign importance to various aspects/objects in the image (learnable weights and bias) and distinguish them between them. The requirements for pre-processing are much lower than other classification algorithms in a ConvNet, whereas filters are handcrafted in primitive processes, ConvNets can learn these filters/characteristics with adequate training.

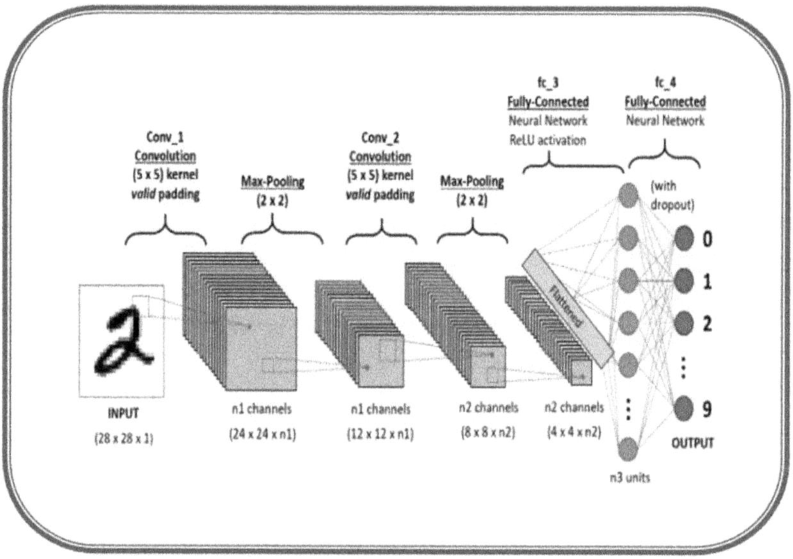

FIGURE 16.1 Architecture diagram for CNN.

Source: Adapted from https://www.cgupta.tech/quickdraw.html

The ConvNet architecture Figure 16.1 is analogous to that of the neuron's communication system in human brain which was focused on the visual cortex organization. In a small area of the visual field known as the receptive zone, each human neuron reacts to stimulus only. The entire visual area covers a collection of these fields. ConvNet measures the spatial and temporal dependencies of a picture efficiently by utilizing the necessary filters. The architecture fits in better with the image dataset because the number of parameters involved is reduced and weights are reusable. This means that the network can better understand the image's sophistication.

16.5 CLASSIFICATION TECHNIQUES USING RANDOM FOREST CLASSIFIER

The KNN algorithm is a simple machine learning algorithm, easily managed, which can be used to solve issues of classification as well as regression. The KNN algorithm believes that there are related items nearby. In other words, the KNN algorithm is based on that assumption that the algorithm is true enough to help. KNN incorporates some of the algebra we might have studied during our childhood—calculating the distance between points, which is the same as size, closeness, and similarity (or similarity) in a line. The implementation process has been shown in Figure 16.2.

16.6 RESULT AND DISCUSSION

Lung cancer is an early form of lung cancer. The lungs are two oxygen-treating spongy bodies in the chest that inhale and release carbon dioxide as you exhale. In this study, we implemented deep learning techniques for lung disease prediction using lung CT images as a dataset. It is used to segment images as shown in Figure 16.3. Also predict sensitivity, specificity, and accuracy values using KNN classifier as shown in Table 16.1.

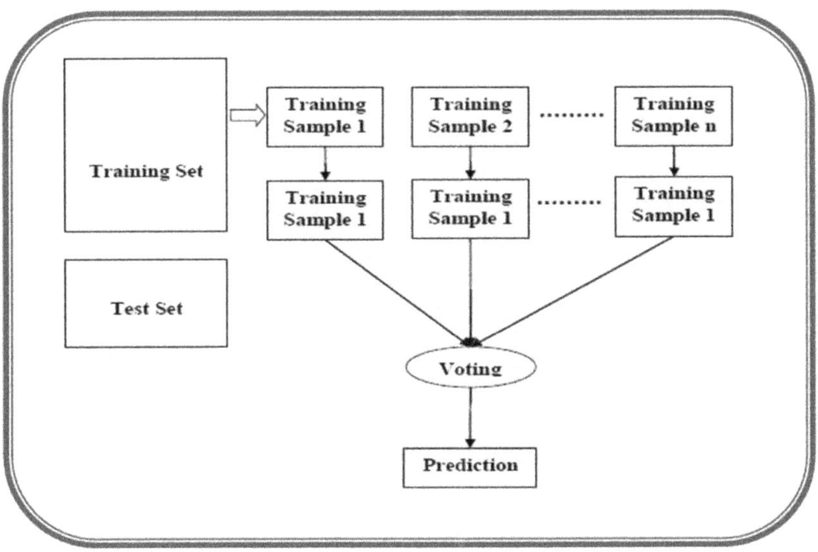

FIGURE 16.2 Prediction analysis of lung cancer using KNN classifier.

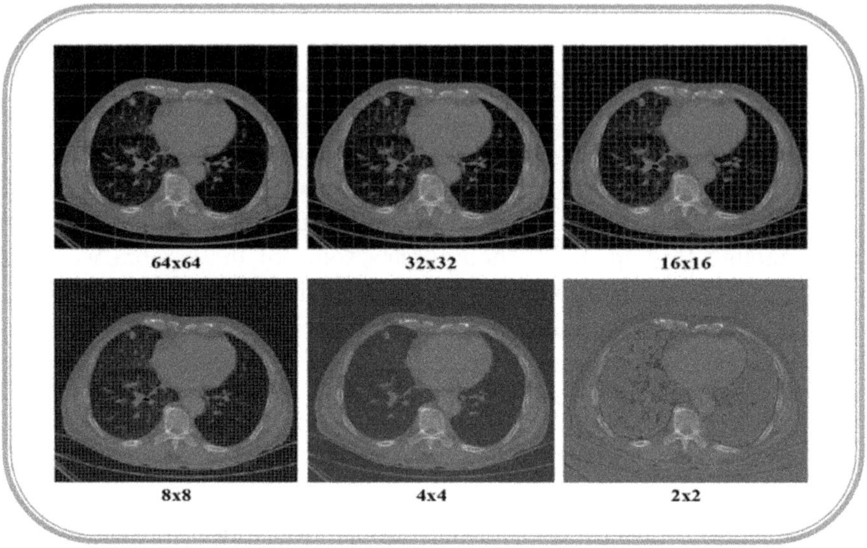

FIGURE 16.3 Lung Segmented images using CT scan images dataset.

TABLE 16.1 Analysis Report of Random Forest Classifier

Epoch	Sensitivity (%)	Specificity (%)	Accuracy (%)
0–25	67	72	88
25–50	76	78	89.6
50–100	78.5	82.5	94.3

16.7 CONCLUSIONS

Deep learning is an algorithmic field of machine science focused on brain structure and function, defined as artificial neural networks. In this study, we propose a deep learning structure to predict a segmented lung disease using CT scan images. It contains 100 images in dataset directory. It is pre-processed to 60 images as training and 40 images as testing. Then the features are extracted by both CNN and random boost classifier. Then the classifier also predicts the value of sensitivity, specificity, and accuracy value. The maximum accuracy was achieved through epoch process 50–100 for 94.3%. For the better prediction result, we may use multiple classifier to analyze and classification of lung disease.

KEYWORDS

- **deep learning**
- **CNN**
- **random boost classifier**
- **breathing**
- **lung cancer**

REFERENCES

1. Coudray, N.; Ocampo, P. S.; Sakellaropoulos, T.; Narula, N.; Snuderl, M.; Fenyö, D.; Moreira, A. L.; Razavian, N.; Tsirigos, A. Classification and Mutation Prediction from Non–Small Cell Lung Cancer Histopathology Images Using Deep Learning. *Nat. Med.* **2018,** *24* (10), 1559–1567.
2. Song, Q.; Zhao, L.; Luo, X.; Dou, X. Using Deep Learning for Classification of Lung Nodules on Computed Tomography Images. *J. Healthcare Eng.* **2017,** *2017.*
3. Sun, W.; Zheng, B.; Qian, W. Computer Aided Lung Cancer Diagnosis with Deep Learning Algorithms. In *Medical Imaging 2016: Computer-Aided Diagnosis*, Vol. 9785; International Society for Optics and Photonics, 2016; p 97850Z.
4. Teramoto, A.; Tsukamoto, T.; Kiriyama, Y.; Fujita, H. Automated Classification of Lung Cancer Types from Cytological Images Using Deep Convolutional Neural Networks. *BioMed Res. Int.* **2017,** *2017.*
5. Kumar, D.; Wong, A.; Clausi, D. A. Lung Nodule Classification Using Deep Features in CT images. In *2015 12th Conference on Computer and Robot Vision*; IEEE, 2015; pp 133–138.
6. Ciompi, F.; Chung, K.; Van Riel, S. J.; Setio, A. A. A.; Gerke, P. K.; Jacobs, C.; Scholten, E. T.; Schaefer-Prokop, C.; Wille, M. M.; Marchiano, A.; Pastorino, U. Towards Automatic Pulmonary Nodule Management in Lung Cancer Screening with Deep Learning. *Sci. Rep.* **2017,** *7*, 46479.
7. Hua, K. L.; Hsu, C. H.; Hidayati, S. C.; Cheng, W. H.; Chen, Y. J. Computer-Aided Classification of Lung Nodules on Computed Tomography Images via Deep Learning Technique. *OncoTargets Therap.* **2015,** 8.
8. Skourt, B. A.; El Hassani, A.; Majda, A. Lung CT Image Segmentation Using Deep Neural Networks. *Procedia Comput. Sci.* **2018,** *127*, 109–113.
9. Manikandan, M. Image Segmentation and Image Matting for Foreground Extraction using Active Contour Based Method. *Int. J. MC Square Sci. Res.* **2011,** *3* (1), 18–38.

CHAPTER 17

DESIGN DETECTING AND CLASSIFYING MELANOMA SKIN CANCER USING CNN WITH K MEANS CLUSTERING

R. SUMATHI[1] and SRIDHAR P. ARJUNAN[2]

[1]*Department of Computer Science and Engineering, Kalasalingam Academy of Research and Education, Krishnankoil, India*

[2]*Department of Electronics and Instrumentation Engineering, SRM Institute of Science and Technology, Kattankulathur, India*

ABSTRACT

Many people in the universe are highly affected with skin-related cancer diseases at present. Many researchers used many soft computing approaches for detecting the cancer part. To aid the dermatologist, we propose an automated approach with Cuckoo search optimization and K-means clustering for detecting the skin cancer and classifying the cancer as normal and abnormal by support vector machine. For preprocessing, median filters are used to reduce noise inference in the input image. For validating the accuracy we utilize the IISC-DSI dataset and our proposed approach yields 98.2% segmentation accuracy and 98.6% classification accuracy and compared with ABC with K-means, FCM for segmentation and for classification CNN and KNN. Our approaches take the average of 7s for processing the images using hybrid approach for cancer detection.

Computational Imaging and Analytics in Biomedical Engineering: Algorithms and Applications.
T. R. Ganesh Babu, PhD, U. Saravanakumar, PhD & Balachandra Pattanaik, PhD (Eds.)

17.1 INTRODUCTION

According to the world medical council, skin cancer is placed in the top place of cancer. Among various skin cancers, melanoma is the most dangerous skin cancer causing more than 80% peoples to lose their lives. It looks like a mole and appears anywhere in the skin where the sunrays do not fall, and it is brown in color which spreads fast and affects the blood vessels in the skin. With the features like asymmetry, border, color, dark, and elevation are the basic symptoms for melanoma skin cancer. Clustering is one of the fundamental methods for understanding various objects and categorizes it based on their similarity. Many clustering methods are used in various applications like image processing,[1–3] text mining, and so on. K-means clustering provides fast convergence rate compared with other clustering like hierarchical clustering. Cuckoo search optimization[4] is recently developed optimization algorithm which is based on brood parasitic behaviors of cuckoo. It is used in many medical applications for its global optimal solution.[5] Deep neural network solved many medical problems more globally and yields the expected outcome. To overcome the local optimal problem, this paper proposes a hybrid approach with the integration of Cuckoo Search with K-means for detecting the cancer part and classifies them using the DNN model. Many optimization techniques are applied to medical images to yield accurate output; one among them is cuckoo search optimization which imitates the swarm behavior and yields the global optimization. Integrating FCSO with K-means[6] for classifying the melanoma as malignant and benign with 10 datasets utilized from online dataset for validating the performance proves that this integration provides the result faster than K-means and its limitation is computational time and number of iterations. With the utilization of IDB2 and UCI dataset combination of firefly and K-means[7] is used for enhancing the exploitation and also reduces the cluster distance measures. To detect the cancerous part nearby unwanted noisy information are removed by applying morphological based fuzzy threshold[8] is applied for the enhancement and integration of the K-means with the firefly optimization technique for detecting the skin lesions with ISIC online dataset and proves its accuracy is superior than PSO[9] using K-means clustering for categorizing cancers and predicting the cancer using total dermoscopic value (TDO) for various levels of cancer and enhances the features using GLCM concept and proves its detection and prediction is superior than various state-of-the-art techniques and its limitation is that very few images are only used for validation. Feature selection[10] is done with the integration of BBA (Binary BAT Algorithm) with threshold based approach for segmenting the cancer part and toclassify them using Radial Basis Function Network (RBFN) and support vector classifier (SVM) with 300 images and proved its specificity rate is far better than CNN, KNN etc. DCNN methods[11] are for classifying the tumor part with three steps: in the first step, enhancing the color transformation is done, in the second step lesion boundary is detected by using CNN, and in the last step to extract the deep features and yield accurate output DNN is used. Based on skin color[12] lesions are extracted with K-means and Multi-class Support Vector Machine for classification and obtained 96% classification accuracy with ISIR dataset.[13] With the combination of genetic algorithm with PSO enhanced K-means is used for segmenting the skin lesions and back propagation neural network is used for classification and yields 87.5% accuracy.[14] Using hybrid techniques like dynamic particle swarm optimization and K-means is performed for skin segmentation with high quality and it ensured that K-means is superior than K-mean clustering.[15] With 8 colors and 200×150 pixels the image is used as the input image and with CNN skin cancers are segmented and analyzed its performance with various existing start-of-the-art techniques.[16] With the combination of grab cut and K-means skin lesions are segmented and obtained 0.8236 and 0.9139 dice and Jaccard coefficient.

17.2 MATERIALS AND METHODS

i. **Image Description**

For validating the efficiency and accuracy of skin cancer using the hybrid approach, we collect various skin-related disease images from online dataset like from IISC-DSI dataset. We utilized 320 cancer images and 100 normal images for this approach. Dermoscopy or Epiluminescence Light Microscopy is an imaging method which examines the skin lesion in detail; the working process of dermoscopy is to place an oil immersion between sin and the optical lens. Microscope is used to identify the pigmented surface in terms of color, shape, and structure in detail for analyzing the lesions intensively for prevention or for diagnosis. Flow diagram briefs our hybrid approach and it is shown in Figure 17.1.

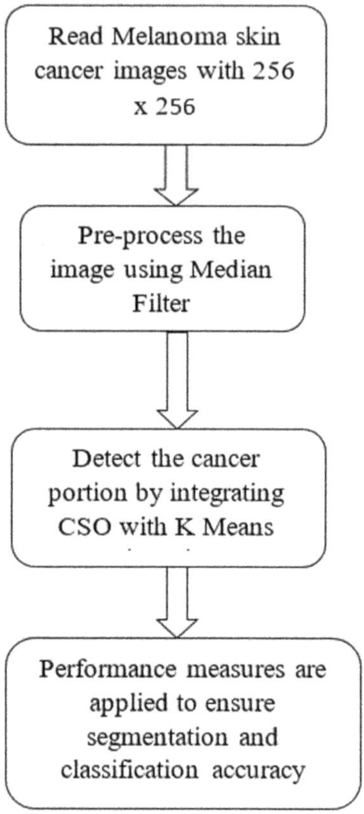

FIGURE 17.1 Flow diagram for CSO+K means.

ii. Median Filter for Preprocessing

Median filter is mainly used for enhancing the contrast based on median pixel value, which aids to detect the boundary and lesion parts in detail for further processing the image. The median filter selects a random pixel from the input image and sorts the pixels in an increasing order and then selects the middle value and replaces the existing pixel value with the obtained median value.

$$f(x,y) = median(s(x,y)) \tag{17.1}$$

where F(x,y) is the median filtered output and s(x,y) is the input image for preprocessing. To remove the hair, air bubbles median filtering is used and also aids in shaping and sharpening the edges of images for extracting the affected part in accuracy.

iii. Cuckoo Search Optimization

Cuckoo search optimization algorithm was developed by Yang and with three golden rules and levy distribution CS optimization is created and applied in many medical image processing applications[17]

Cuckoo search optimization pseudo code is as follows:

Step 1: To define the objective function f(x)

Step 2: To generate the initial solution with "n" nests

Step 3: While stop criteria are met do

3.1 Get a random cuckoo and compute levy flight

3.2 Compute the fitness function

3.3 Randomly choose a nest j among "n"

3.4 Check if fi > fj then

Replace the current solution

End if

3.5 Worst nests portion pa will be eliminated and new nests to be created and keep the best Solution

Step 4: End the while loop

iv. K-Means Clustering

K-Means is a clustering method which categorizes the classes into k clusters for classification based on their characteristics. K-means integrated with CS highly aids to detect the accurate part of skin cancer. The following steps are used to brief the K-means clustering method.

Step 1: Randomly choose number of cluster K

Step 2: Randomly choose K pixels from various intensities as centroids

Step 3: Find the distance between two points and assign clusters based on the minimum distance

Step 4: Compute the mean of each cluster and centroid by calculating k clusters using the following formula:

$$\frac{\sum_{i=2}^{m} 1(C_i = j)x^i}{\sum_{i=1}^{m} 1(C_i = j)} \quad (17.2)$$

Step 5: Perform the above step repeatedly until all the pixels in each cluster is converged

Cuckoo search optimization is integrated with K-means to yield global optimal solution, K-means aids in finding the initial cluster and to update the centroid CS is integrated for yielding best outcome of this hybrid approach. Features set (12) are extracted from the segmented result and fed to SVM for classification with 70% training set and 30% testing images. Various features like energy, autocorrelation, contrast, skewness, Kurtosis, Difference Entropy, Variance, Standard deviation, Correlation, Homogeneity, dissimilarity are computed for classification. The formulas for these features are mentioned in Table 17.1:

TABLE 17.1 Features Extraction Formula Table.

Name of the feature	Formula	
Cluster prominence	$\sum_{i,j=0}^{g-1} h_{ij}(i - f_i + j - f_j)^3$	(17.3)
Energy	$\sum_{ij=0}^{g-1} -log\left(h_{ij}\right)^2$	(17.4)
Autocorrelation	$\sum_{ij=0}^{g-1} (ij)\,h(ij)$	(17.5)
Contrast	$\sum_{ij=0}^{g-1} -hlj\,(i-j)^2$	(17.6)
Difference Entropy	$-\sum_{i=0}^{g-1} h_{i+j}(i)\log\left(h_{i+j}(i)\right)$	(17.7)
Skewness	$\sum_{h=0}^{g-1} (o_i - mean)^3\, h(oi)$	(17.8)
Kurtosis	$\sum_{h=0}^{g-1} (o_i - mean)^4\, h(oi)$	(17.9)
Variance	$\sum_{h=0}^{g-1} (o_i - mean)^2\, h(oi)$	(17.10)

TABLE 17.1 *(Continued)*

Name of the feature	Formula	
Standard deviation	$\sqrt{\sum_{h=0}^{g-1}(o_i - mean)^2 h(oi)}$	(17.11)
Correlation	$\sum_{i,j=0}^{g-1} h_{ij} \frac{(i-\mu)(j-\mu)}{1+(i-j)^2}$	(17.12)
Homogeneity	$\sum_{i,j=0}^{g-1} \frac{h_{ij}}{1+(i-j)^2}$	(17.13)
dissimilarity	$\sum_{ij}^{g} \lvert i-j \rvert h(ij)$	(17.14)

v. SVM Classifier

An SVM classifier classifies the skin cancer into benign and malignant based on their illumination shape and various features extracted from the segmented image. The following steps brief the working principle of the SVM classifier

Step 1: Collect the images from dataset and partition the data into training (70%) and testing (30%).

Step 2: All features extracted from segmented images are classified and label it.

Step 3: Compute support value and estimate it.

Step 4: Iterate the following steps until instance value is null.

Step 4.1: Check if support value is equal to similarity between each instance, if so means, find the total error value.

Step 5: Check the instance if it is less than zero and then calculate the FA where

$$FA = \frac{Support\_value}{Total\ Error} \tag{17.15}$$

Repeat step 5 until it becomes empty else stop the process.

In total 320 tumor and 100 normal images are collected from IISC-DSI dataset for validating the proposed CSO + K means and the SVM classifier.

The optimized features are taken from segmented output and partitioned as 70% for training and 30% for testing ratio. The performance measures ensured the efficiency of the proposed approach. Sensitivity, specificity, and accuracy measure ensure both segmentation and classification accuracy and proved the efficiency of the proposed approach.

$$Specivicity = 1 - FPR * 100 \tag{17.16}$$

$$Sensitivity = \frac{TP}{TP + FN} * 100 \tag{17.17}$$

$$Accuracy = \frac{TP + TN}{TP + FN + TN + FP} * 100 \tag{17.18}$$

where TP represents the correctly detected part, FN represents the incorrectly detected part, FPR represents the total number of wrongly detected parts, and the classification comparison diagram is shown in Figure 17.2.

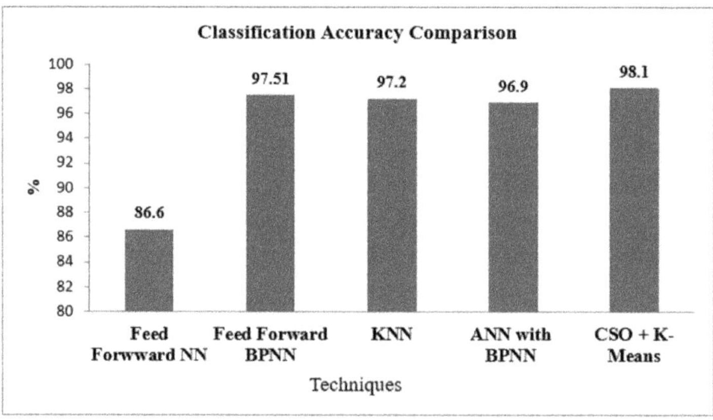

FIGURE 17.2 Comparison of classification accuracy.

From Figure 17.2 it was found that our SVM classifier yields good accuracy measures compared with existing approaches.[17–19] From the segmentation result it was proved that the CSO with K-means is highly suitable for skin cancer detection and its segmentation accuracy is compared with existing approaches like FCM, K-means, adaptive K-means, and its comparison chart is shown in Table 17.2.

TABLE 17.2 Segmentation Accuracy Comparison.

Techniques	Segmentation accuracy (%)	Computational time (S)
Fuzzy C-means	97	11
K Means	95.3	9
Adaptive K-means	967	8
Proposed CSO + K means	98.2	7

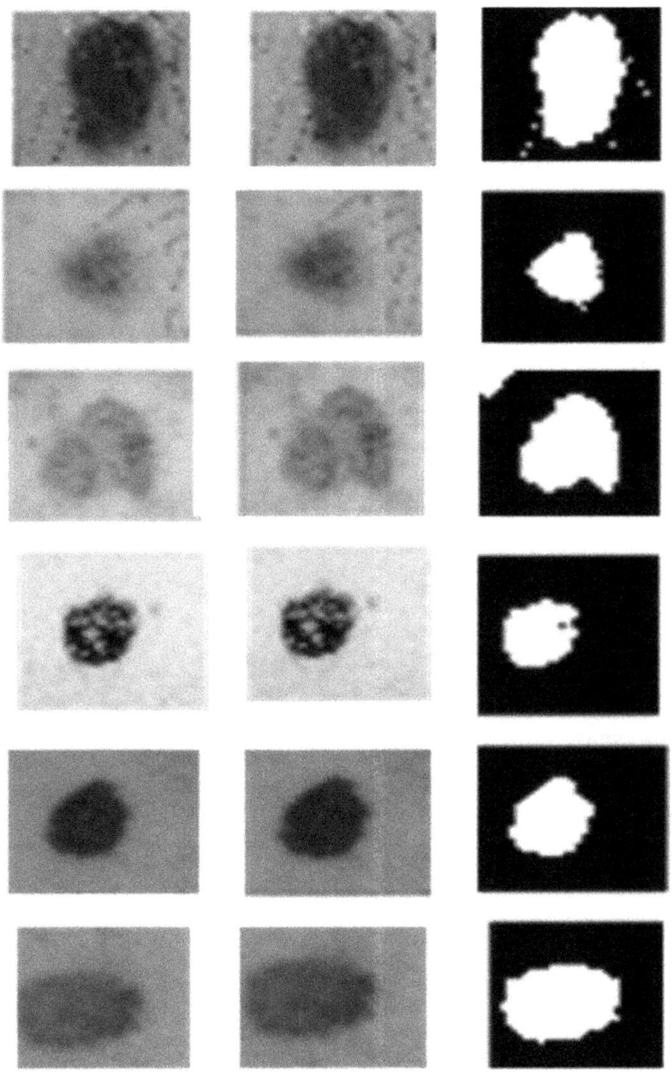

FIGURE 17.3 Proposed CSO + K means segmented results.

Figure 17.3 represents the segmented result using CSO + K means with median filters, where the first column represents the gray scale input images, the second column represents the median filtered images, and the third column represents the proposed output. From the integration of CSO + K –means it was observed that based on cluster and number of iterations the processing and accurate result will be modified. In our proposed approach number of clusters is chosen as 2 and maximum number of iterations for segmenting the tumor part is 90. It was observed that few hair and skin artifacts are detected in some benign images; to reduce that problem we need to include morphological reconstruction filter[20] for exact detection.

17.3 CONCLUSIONS

Integration of CSO with K-Means Clustering approach is suitable for segmenting the skin cancers with 98.2% segmentation accuracy and with SVM classifier it ensures 98.6% classification accuracy. Efficiency of segmentation is proved with existing techniques like PSO with K-means, Fuzzy C-Means, GA with K-Means, and ABC with K-Means. From the computational time comparison it was proved that classification accuracy is far superior to the existing techniques. In the future DNN models are used for skin cancer classification.

KEYWORDS

- **dermoscopy imaging**
- **median filter**
- **cuckoo search optimization**
- **K-means clustering**
- **performance measures**

REFERENCES

1. İlkin, S.; Gençtürk, T. H.; Gülağız, F. K.; Özcan, H.; Altuncu, M. A.; Şahin, S. SVM: Bacterial Colony Optimization Algorithm Based SVM for Malignant Melanoma Detection. *Eng. Sci. Technol.* **2021,** *24*, 1059–1071.

2. Chitra, A.; Rajkumar, A. Paraphrase Extraction Using Fuzzy Hierarchical Clustering. *Appl. Soft Comput.* **2015,** *34*, 426–437.
3. Lakshmi, S. P.; Karthikeyani, K.; Visalakshi, N.; Shanthi, S. Clustering Categorical Data Using K-Modes Based on Cuckoo Search Optimization Algorithm. *ICTACT J. Soft Comput.* **2015,** *8*, 1561–1566.
4. Esteva, A.; Kuprel, B.; Novoa, R. Dermatologist-Level Classification of Skin Cancer with Deep Neural Networks. *Nature* **2017,** *542*, 115–118.
5. Sumathi, R.; Venkatesulu, M.; Arjunan, S. P. Segmenting and Classifying MRI Multimodal Images Using Cuckoo Search Optimization and KNN Classifier. *IETE J. Res.* **2021,** 1–8.
6. Gaonkar, R.; Singh, K.; Prashanth, G. R.; Kuppili, V. Lesion Analysis Towards Melanoma Detection Using Soft Computing Techniques. *Clin. Epidemiol. Glob. Health* **2020,** *8*, 501–508.
7. Girsang, A. S.; Yunanto, A.; Aslamiah, A. H. A Hybrid Cuckoo Search and K-Means for Clustering Problem. *Int. Conf. Electr. Eng. Comput. Sci. (ICECOS)*, 2017; pp. 120–124.
8. Xie, H.; Zhang, L.; Lim, C. P.; Yu, Y.; Liu, C.; Liu, H.; Walters, J. Improving K-Means Clustering with Enhanced Firefly Algorithms. *Appl. Soft Comput.* **2019,** *84*, 105763.
9. Ali, Md. S.; Miah, Md. S.; Haque, J.; Rahman, Md. M.; Islam, Md. K. An Enhanced Technique of Skin Cancer Classification Using Deep Convolutional Neural Network with Transfer Learning Models. *Mach. Learn. App.* **2021,** *5*, 100036.
10. Dorj, U.-O.; Lee, K.-K.; Choi, J.-Y.; Lee, M. The Skin Cancer Classification Using Deep Convolutional Neural Network. *Multimedia Tools App.* **2018,** *77*, 9909–9924.
11. Krishna Monika, M.; Arun Vignesh, N.; Usha Kumari, Ch.; Kumar, M. N. V. S. S.; Laxmi Lydia, E. Skin Cancer Detection and Classification Using Machine Learning. *Mater. Today: Proc.* **2020,** *33*, 4266–4270.
12. Aljawawdeh, A.; Imraiziq, E.; Aljawawdeh, A. Enhanced K-Mean Using Evolutionary Algorithms for Melanoma Detection and Segmentation in Skin Images. *Int. J. Adv. Comput. Sci. App.* **2017,** *8* (12), 447–483.
13. Lam, Y. K.; Tsang, P. W. M.; Leung, C. S. PSO-Based K Means Clustering with Enhanced Cluster Matching for Gene Expression Data. *Neural Comput. App.* **2013,** *22*, 1349–1355.
14. Li, H.; He, H.; Wen, Y. Dynamic Particle Swarm Optimization and K-Means Clustering Algorithm for Image Segmentation. *Optik Int. J. Light Electron Opt.* **2015,** *126*, 4817–4822.
15. Setiawan, W.; Faisal, A.; Resfita, N. Effect of Image Downsizing and Color Reduction on Skin Cancer Pre-screening. In *2020 International Seminar on Intelligent Technology and Its Applications*, 2020; pp 148–151.
16. Jaisakthi, S. M.; Mirunalini, P.; Aravindan, C. Automated Skin Lesion Segmentation of Dermoscopic Images Using Grab Cut and K-Means Algorithms. *IET Comput. Vis.* **2018,** *12*, 1088–1095.
17. Hoshyar, A. N.; Al-Jumaily, A. Comparing the Performance of Various Filters on Skin Cancer Images. *Procedia Comput. Sci.* **2014,** *42*, 32–37.
18. Mehta, A.; Parihar, A. S.; Mehta, N. Supervised Classification of Dermoscopic Images Using Optimized Fuzzy Clustering Based Multi-Layer Feed-Forward Neural Network. In *2015 International Conference on Computer, Communication and Control (IC4)*, 2015; pp 1–6.

19. Sreelatha, T.; Subramanyam, M.; Prasad, M. G. Early Detection of Skin Cancer Using Melanoma Segmentation Technique. *J. Med. Syst.* **2019,** *43*, 190–205.
20. Sumathi, R.; Arjunan, S. Towards Better Segmenting Low Depth of Filed Images Using Morphological Reconstruction Filters. *Int. J. Signal Syst. Eng.* **2014,** *7*, 189–194.

CHAPTER 18

DETECTION OF LUNG CANCER USING FUSION METHODS FOR CT AND PET IMAGES

V. RAMESH BABU[1], G. GUNASEKARAN[1], and SREENITHI R.[2]

[1]*Department of Computer Science and Engineering, Dr. M. G. R. Educational and Research Institute, Maduravoyal, Chennai, India*

[2]*Department of Computer Technology, Madras Institute of Technology, Chennai, India*

ABSTRACT

Lung cancer is caused due to anomalous growth of cells that develop into a tumor. Various researches report that the death rate of lung cancer is the highest among all other types of cancer. In the first part of work segmentation of lungs, a tumor in CT image is used for Spatially Weighted Fuzzy C Means Clustering (SWFCM) techniques. The overall accuracy, sensitivity, and predictive values achieved are 86.082, 85.636, and 92.673%, respectively. In the second part of the work segmentation of lungs, a tumor in PET image is used for Spatially Weighted Fuzzy C Means Clustering (SWFCM) techniques. The overall accuracy, sensitivity, and predictive values achieved are 89.31, 87.27, and 95.88%, respectively. In the third part of the work, the diagnosis is strengthened for mass screening; the CT and the PET images are fused effectively. The four fusion methods namely Wavelet Transform, Curvelet Transform, Non Subsample Contourlet Transform (NSCT), and Multimodal image fusion are applied. The performance analysis Entropy, Peak Signal Noise Ratio (PSNR), Standard Deviation (SD), Structural

Computational Imaging and Analytics in Biomedical Engineering: Algorithms and Applications.
T. R. Ganesh Babu, PhD, U. Saravanakumar, PhD & Balachandra Pattanaik, PhD (Eds.)

Similarity Index Measure (SIM), and Root Mean Square Error (RMSE) are computed.

18.1 INTRODUCTION

The initial stage of cancer is when cells unusually grow out of control anywhere in the body. The growth of cancer cell and normal cell differs. The cancer cells do not die but instead they grow and form several abnormal cells whereas normal cells die. Cancer cells can also possibly invade into other organs, but normal cells cannot. The process of cells abnormally growing without control and affecting other tissues leads to the formation of a cancer cell. Utilization of anatomical priors in the process of segmentation of PET lung tumor images was proposed by Ref. [1]. This method combines both the anatomical and functional images and hence provides effective tumor staging and improved treatment planning. PET/CT detects tumor invasion into adjacent tissues successfully as well as leads to precise localization of lesions, even though there are no morphological changes found in CT[2] as presented in a PET and CT images based automated lung nodule detection. It was an automatic method to detect lung nodules in PET and CT. Here, the nodules that are in proximity region and similar are combined into one by a split-up postprocessing step, and the time of the localization can be minimized to a greater extent from more than one hour to maximum five minutes. This method when executed and authenticated on real clinical cases in Interview Fusion clinical evaluation software (Mediso) proved successful in detecting lung nodules and may be a valuable aid for physicians for the daily routine of oncology. Ref. [3] presented a method of Statistical Texture Features analysis for Automatic Lung Cancer Detection with the PET/CT images. The overall endurance rate of lung cancer patients is observed to be only 14%.[4] Early detections increase the chance of choosing the apt treatment for the cancer patients. Computational systems diagnoses are highly useful for radiologists in the elucidation of images. Image preprocessing methods namely Contrast Limited Adaptive Histogram Equalization (CLAHE) and Wiener filtering have been performed to remove the artifact due to contrast variations and noise. Haralick statistical texture features were chosen as they could extract more texture information from the cancer regions than the visual assessment. In order to classify the regions as normal or abnormal, Fuzzy C-Means (FCM) clustering was used.[5]

18.2 DETECTION OF LUNGS TUMOR USING CT IMAGES

Figure 18.1 shows the detection of lung tumor from CT image. The lungs' tumor from CT image is segmented by using Spatially Weighted Fuzzy–C Means Clustering.[6]

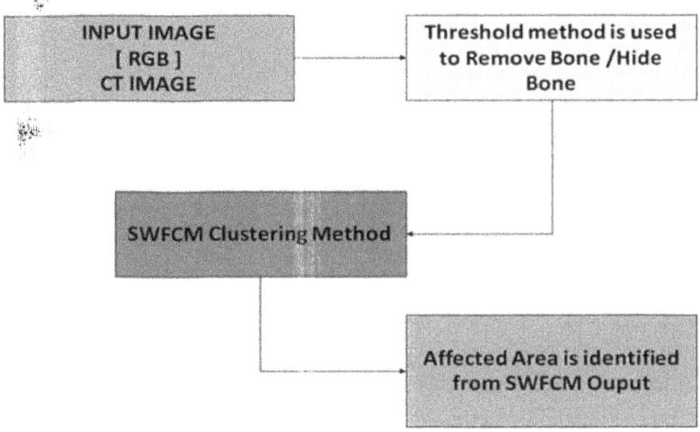

FIGURE 18.1 Flow chart: lung tumor detection from CT image.

18.2.1 PREPROCESSING REMOVAL OF BONE REGION

The first step is the removal of bone region from the lung CT image since the bone region affects the segmentation accuracy. Toward this, R-plane, G-plane, and B-plane are separated from the RGB image. In this entire plane, the bone region is detected. By subtracting all these images, the resultant image is obtained. Figure 18.2 shows R-plane, G-plane, and B-plane images. Figure 18.3 shows input CT image.[7]

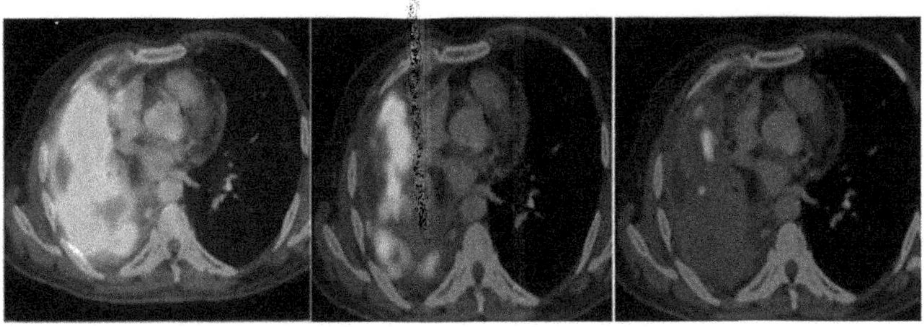

FIGURE 18.2 R-plane, G-plane, and B-plane.

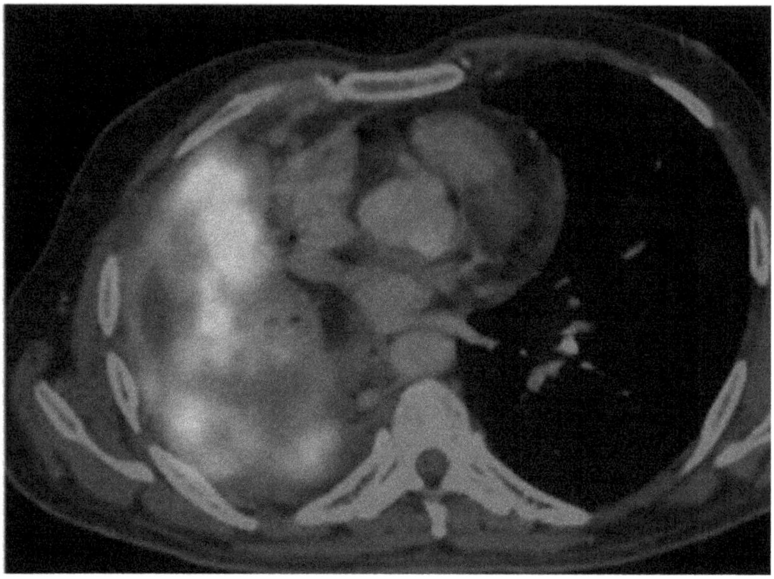

FIGURE 18.3 Input lung CT image.

Figure 18.4 shows the enhanced and bone removal image. T is added with difference between G-plane and B-plane as given in eq 3.1 to find the affected area of the disease. S=T+[G-B]

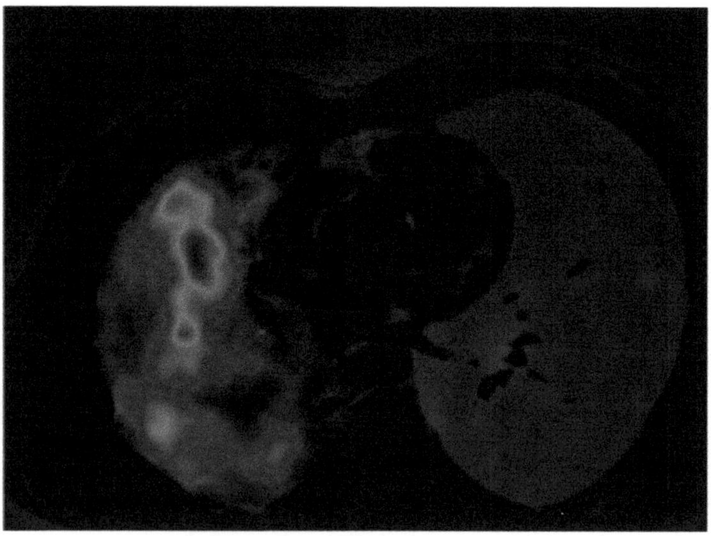

FIGURE 18.4 Enhanced and bone-removed image.

18.2.2 LUNG CANCER DETECTION USING SPATIALLY WEIGHTED FUZZY C-MEAN ALGORITHM

The quality of the CT image will not be very good, so the SWFCM method is adopted for the corresponding region.[8]

In SWFCM, to exploit the spatial information, a spatial function is defined as eq 18.1

$$h_{ij} = \sum\nolimits_{k \in NB(x_j)} u_{ik} \tag{18.1}$$

where $NB(x_j)$ represents a square window centered on pixel x_j in the spatial domain. The spatial function is incorporated into membership function as eq 18.2

$$u'_{ij} = \frac{u_{ij}^{p} h_{ij}^{q}}{\sum\nolimits_{k=1}^{c} u_{kj}^{p} h_{kj}^{q}} \tag{18.2}$$

where p and q are controlling parameters of both functions.

SWFCM—Spatially Weighted Fuzzy C-Means Clustering Algorithm

Step 1: Create the random number with the range from 0 to1 to be the initial memberships. Let us consider the number of clusters as N and then calculate V_i using (18.3)

$$V_i = \frac{\sum\nolimits_{j=1}^{N} u_{ij}^{m} x_j}{\sum\nolimits_{j=1}^{N} u_{ij}^{m}} \tag{18.3}$$

where

V_i = i^{th} cluster center, m = fuzziness parameter m = 2

where u_{ij} is by using eq 18.4

$$u_{ij} = \frac{1}{\sum\nolimits_{k=1}^{c} \left(\frac{\|x_{j-} v_i\|}{\|x_j - v_k\|} \right)^{2/(m-1)}} \tag{18.4}$$

Step 2: Map u_{ij} into the pixel position and calculate the modified membership u_{ij} using (3.19). Compute objective function J using eq 18.5

$$J = \sum\nolimits_{j=1}^{N} \sum\nolimits_{i=1}^{C} u_{ij}^{m} \left\| x_j - \upsilon_i \right\|^2 \tag{18.5}$$

Step 3: Update the cluster center using (18.6)

Step 4: Repeat step 2 to step 4 until the following termination criterion is satisfied:

$$\left\| J_{new} - J_{old} \right\| < \varepsilon \tag{18.6}$$

Figure 18.5 shows the SWFCM output image. The image with three clusters is shown in Figure 18.5.

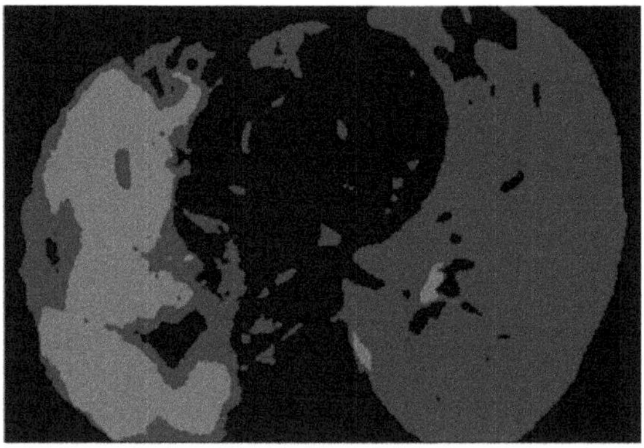

FIGURE 18.5 SWFCM image.

The segmented image has three clusters, namely the background region of interest and small regions. The background and small regions are eliminated and only large size of cluster is considered. Figure 18.6 shows tumor-affected area in the CT image. Figure 18.7 shows the tumor-affected area in identified cluster superimposed with the CT input image. Table 18.1 shows the data sets of cancer-affected area of CT image.

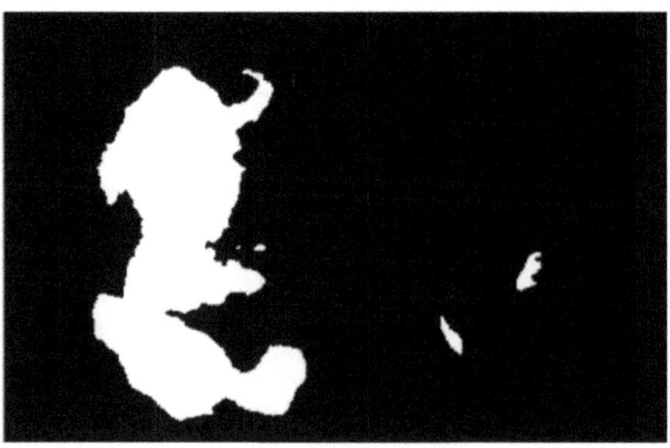

FIGURE 18.6 Identified clustered image.

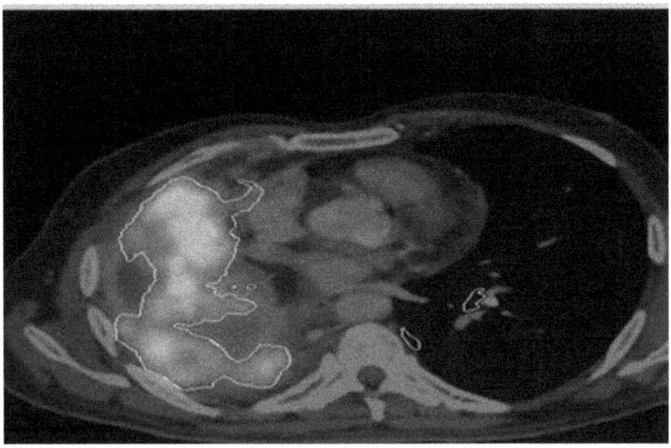

FIGURE 18.7 Super imposed image.

TABLE 18.1 Data Sets of Cancer-Affected Area of CT Image.

Image	T	R	M	TP	FP	FN	S (%)	PV (%)
1	1102	1067	87.31	1011	56	90	91	94.57
2	1151	1104	91.92	1080	24	71	93.83	94.75
3	1053	1071	91.01	1012	59	41	96.11	94.49
4	1104	1062	86.89	1007	55	97	91.21	94.82
5	1162	1082	86.89	1007	55	97	91.29	98.24
6	5324	4864	88.81	4792	72	532	90.01	98.52
7	1151	1052	81.92	992	60	159	86.39	94.30
8	517	520	89.18	486	37	34	93.42	92.89
9	1322	1288	70.55	1340	149	183	86.23	88.51
10	1673	1518	89.94	1511	8	163	90.31	99.55
11	1214	1183	93.62	1159	25	56	95.38	97.98
12	2131	2769	76.24	2117	647	15	99.35	76.62
13	1175	1156	84.56	1062	90	108	90..90	92.48
14	1565	1372	86.19	1359	19	207	86.93	99.05
15	1974	1378	86.20	1365	984	138	93.08	73.33

T, ground truth region; R, region is proposed method; M, accuracy; TP, true positive, FP, false positive; FN, false negative; S, sensitivity; PV, predictive value

Figure 18.8 shows the Bar Graph of the above-obtained result.

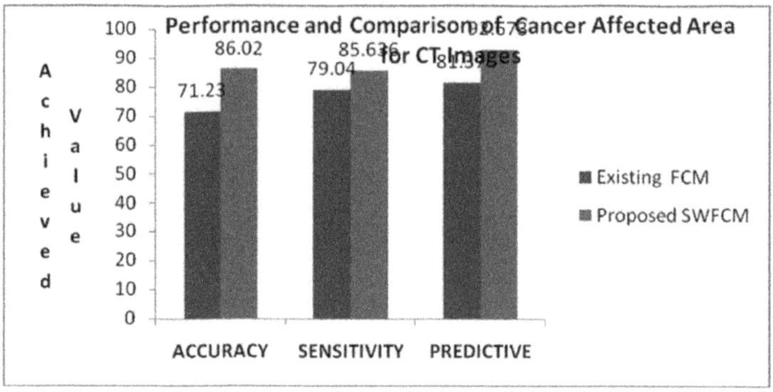

FIGURE 18.8 Bar graph-comparison of lung cancer-affected area for the CT image.

18.3 DETECTION OF LUNGS TUMOR FROM PET IMAGES

The proposed method is presented in a flowchart as shown in Figure 18.9. In this thesis Fuzzy Local Information C-means Clustering is used for segmentation of lungs tumor from PET images.[9,10]. The size of the input PET images is 512*512.

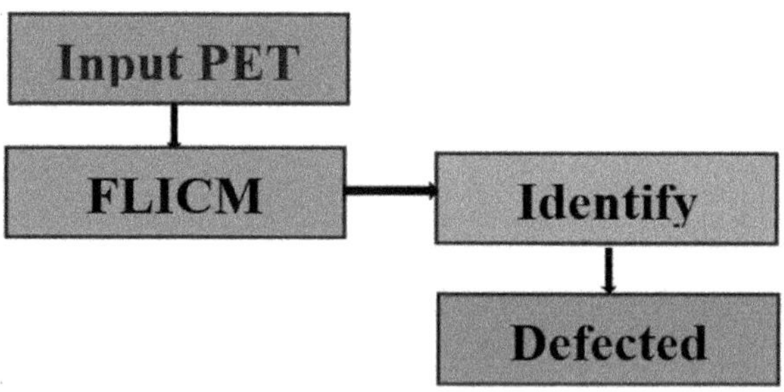

FIGURE 18.9 FLICM flowchart.

18.3.1 LUNG CANCER DETECTION USING FLICM METHOD

The FLICM is applied to the PET image. The major characteristic of FLICM is that it guarantees noise immunity and preserves image detail and it is free of any parameter selection.[11]

FLICM incorporates local spatial and gray level information into its objective function as defined in eqs 18.7 and 18.8

$$J_m = \sum_{i=1}^{N} \sum_{k=1}^{c} [U_{ki}^m \left\| x_i - v_k \right\|^2 + G_{ki}] \tag{18.7}$$

where the *i*th pixel is the center of the local window, k is the reference cluster, and the jth pixel belongs to the set of the neighbors falling into a window around the ith pixel (N_j). $d_{i,j}$ is the spatial Euclidean distance between pixels i and j, U_{kj} is the degree of membership of the jth pixel in the kth cluster, m is the weighting exponent on each fuzzy membership shown in eq 18.9, and Vk is the prototype of the center of cluster k shown in eq 18.10

Here G =fuzzy factor

$$G_{ki} = \sum_{\substack{j \in N_i \\ i \neq j}} \frac{1}{d_{ij} + 1} \left(1 - U_{kj}\right)^m \left\| X_j - V_k \right\|^2 \tag{18.8}$$

$$U_{ki} = \frac{1}{\sum_{j=1}^{C} \left(\frac{\left\| x_i - v_k \right\|^2 + G_{ki}}{\left\| x_i - v_j \right\|^2 + G_{ji}} \right)^{\frac{1}{m-1}}} \tag{18.9}$$

$$V_k = \frac{\sum_{i=1}^{N} U_{ki}^m x_i}{\sum_{i=1}^{N} U_{ki}^m} \tag{18.10}$$

The details of the algorithm are given in eq 18.7.

Algorithm: FLICM algorithm

Step 1. Set the cluster prototypes c, fuzzification parameter m, and the ending condition .

Step 2. Set randomly the fuzzy partition matrix.

Step 3. Set the loop counter b = 0.

Step 4. Determine the cluster prototypes using 20

Step 5. Calculate membership values using 19

Step 6. If max $\{U^{(b)} - U^{(b+1)}\} < \varepsilon$ then stop, otherwise set b = b + 1 and go to step 4.

When the algorithm has converged, a defuzzification process takes place to convert the fuzzy partition matrix U to a crisp partition FLICM.

Figure 18.10 shows the result of FLICM clustering output. From the clusters, the tumor is identified by selecting the cluster index at the location of the darkest point in the PET image because the tumor belongs to lower intensity regions. Figure 18.10 shows the segmented tumor region from the FLICM output. From the FLICM output corner cluster is removed, the zero pixels are tumor regions. Figure 18.11 shows the segmented tumor region from the FLICM output. Table 18.2 illustrate data sets of cancer-affected area of the PET image.

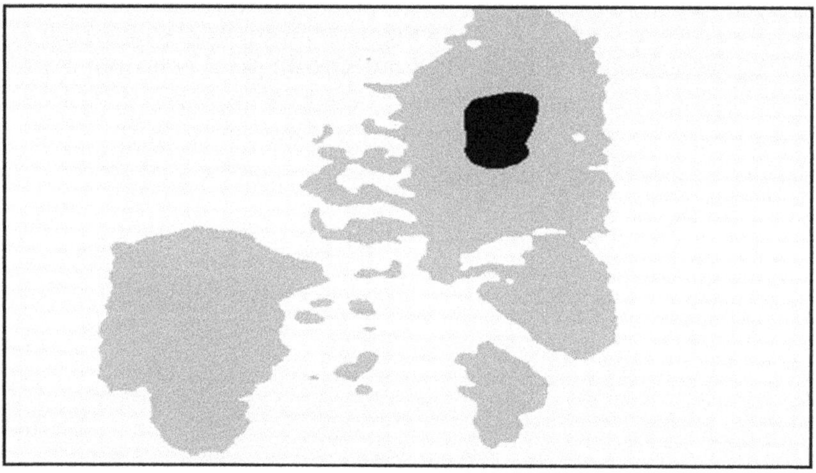

FIGURE 18.10 The result of FLICM with white background.

FIGURE 18.11 Detection of tumor in PET image.

TABLE 18.2 Data Sets of Cancer-Affected Area of PET Image.

Image	T	R	M %	TP	FP	FN	S (%)	Pv (%)
1	1152	1092	92.10	1101	59	95	91.52	95.56
2	1110	1108	94..32	1086	29	82	94.95	95.85
3	1192	1065	94.62	1020	59	41	96.01	94.32
4	1164	1083	95.52	1007	55	97	91.55	96.85
5	1180	1170	97.32	1069	20	103	92.48	98.24
6	5355	5325	98.52	4856	79	562	93.35	98.52
7	1168	1142	96.25	900	70	168	90.37	94.3
8	560	530	96.23	486	40	56	93..86	93.89
9	1352	1340	97.23	1131	150	192	92.42	93.56
10	1690	1562	95.23	1510	15	172	95.36	98.54
11	1252	1199	96.32	1170	34	65	95.37	97.21
12	2146	2956	97.01	2119	676	45	99.34	96.52
13	1196	1175	94.55	1098	102	108	91.23	93.58
14	1582	1499	92.16	1468	25	252	92.13	99.12
15	2001	1965	96.54	1954	978	152	93.01	92.18

T, ground truth region; R, region is proposed method; M, accuracy; TP, true positive; FP, false positive; FN, false negative; S, sensitivity; PV, predictive value

Figure 18.12 shows the Bar Graph of the above obtained result.

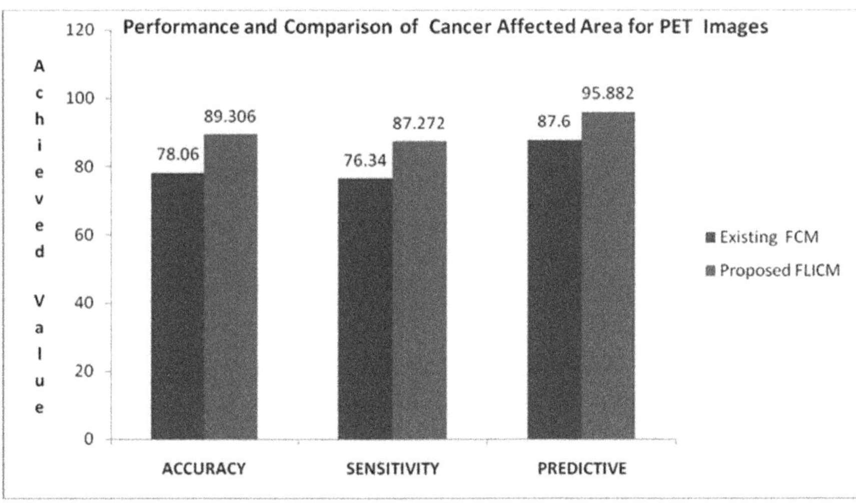

FIGURE 18.12 Bar graph: comparison of lung cancer-affected area for PET image.

18.4 IMAGE FUSION OF CT AND PET IMAGES

Image fusion is defined as the process of combining the relevant information from a set of images into a composite image, where the resultant fused image will be more informative and complete than any of the input images. Image fusion techniques can improve the quality and increase the application of these data. Image fusion finds applications in the area of navigation guidance, object detection and recognition, satellite imaging for remote sensing and civilian etc. In this paper four model works are proposed: the wavelet curvelet, NSCT and multi-model image method, and two fusion rules namely PCA and maximum absolute rule.[12–21]

18.5 RESULT AND DISCUSSION

The first column in Figure 18.13 shows CT images, the second column shows the PET images, the third column shows the fused image for wavelet transform based maximum absolute fusion rule, and the fourth column shows the PCA fusion rule

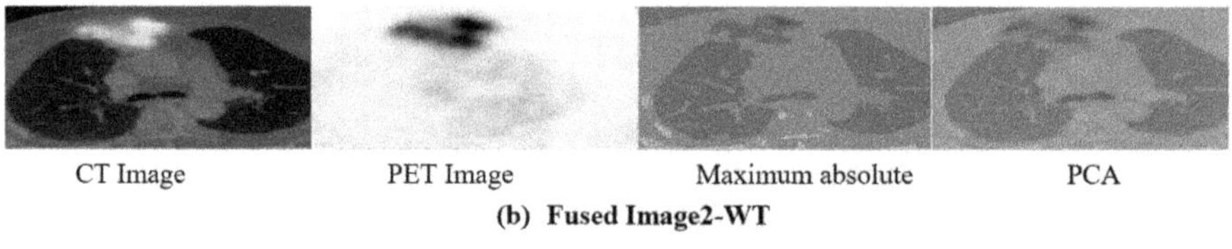

FIGURE 18.13 The result for fused image in wavelet transform.

The first column in Figure 18.14 shows CT images, the second column shows the PET images, the third column shows the fused image for Fast Discrete Curvelet Transform based maximum absolute fusion rule, and the fourth column shows the PCA fusion rule.

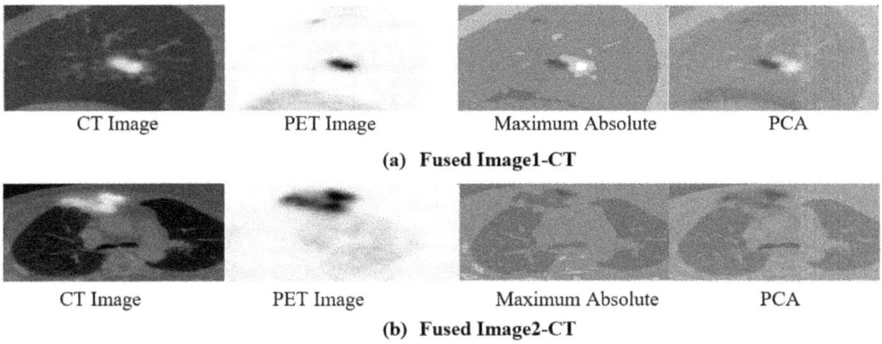

FIGURE 18.14 The result for fused image in curvelet transform.

The first column in Figure 18.15 shows CT images, the second column shows the PET images, the third column shows the fused image for Non Sub-sampling Contourlet Transform (NSCT) based maximum absolute fusion rule, and the fourth column shows the PCA fusion rule.

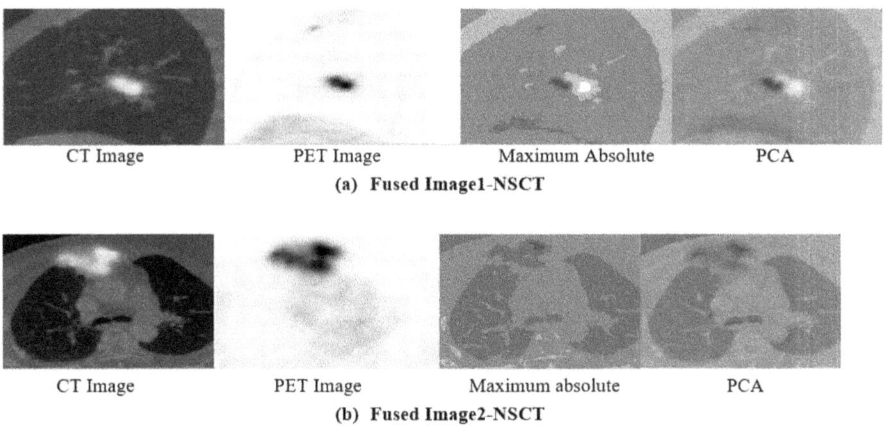

FIGURE 18.15 The result for fused image in NSCT.

The first column in Figure 18.16 shows CT images, the second column shows the PET images, the third column shows the fused image for multi-modal based maximum absolute fusion rule, and the fourth column shows the PCA fusion rule.

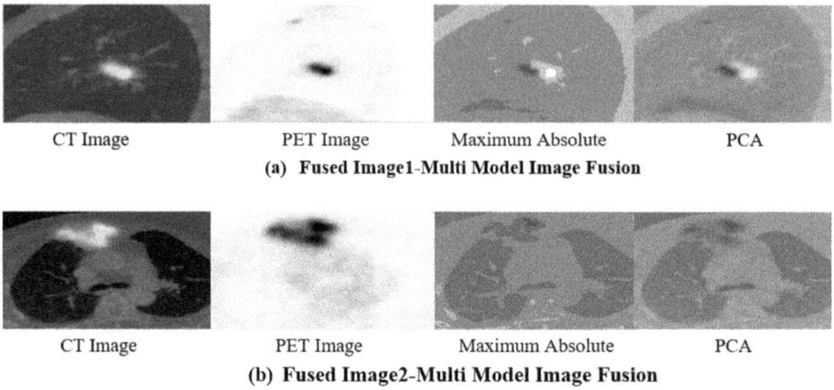

FIGURE 18.16 The result for fused image in multi-model image fusion based.

Evaluating the performance of the fusion algorithm can be ascertained effectively via Image Quality Assessment (IQA) of the fused image (Sudeb Das and Malay Kumar Kundu, 2013). Image fidelity metrics based on error estimation, that is, peak signal-to-noise ratio (PSNR) is commonly employed for objective evaluation of fused image quality. Other than this, fusion metrics based on entropy (E), standard deviation (SD), and root mean square error (RMSE) account for the restored information content in the fused image. Table 18.3 shows the performance analysis of different fusion methods. Figures 18.17, 18.18, 18.19, 18.20, and 18.21 show bar charts for various parameters. The graph indicates that the proposed method is better than other methods.

TABLE 18.3 Performance Analysis for Different Fusion Methods.

Transform	Fusion rule	Entropy	PSNR	SD	Similarity index measure	RMSE
Wavelet	Maximum absolute	5.832	25.96	34.7321	0.421	3.256
	PCA	5.956	26.72	36.888	0.432	3.168
Curvelet	Maximum absolute	6.328	28.47	54.1765	0.521	2.527
	PCA	6.786	29.56	56.192	0.556	2.496
NSCT	Maximum absolute	6.543	22.78	55.682	0.435	2.478
	PCA	7.498	23.16	56.182	0.479	1.986
Multi Model Image Fusion	Maximum absolute and PCA	7.568	49.58	60.282	0.623	0.978

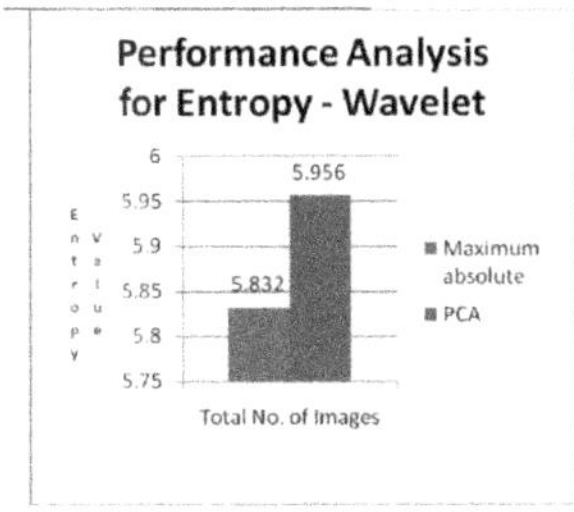

(a) Entropy - Wavelet

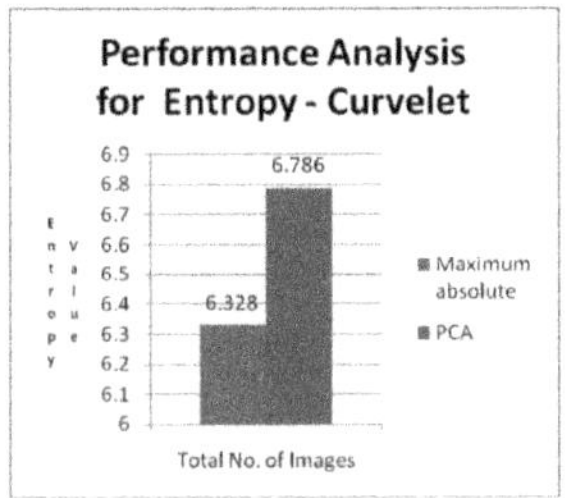

(b) Entropy - Curvelet

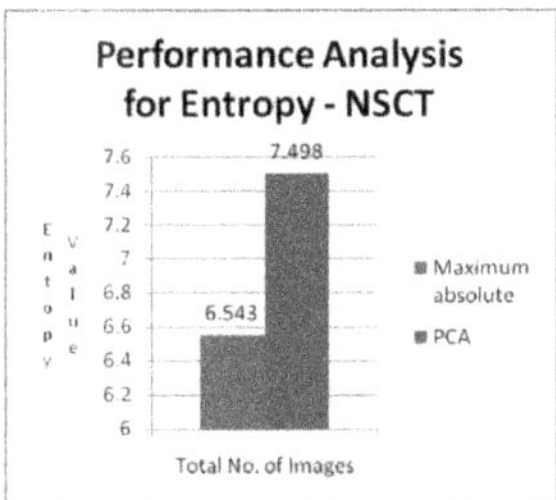

(c) Entropy - NSCT

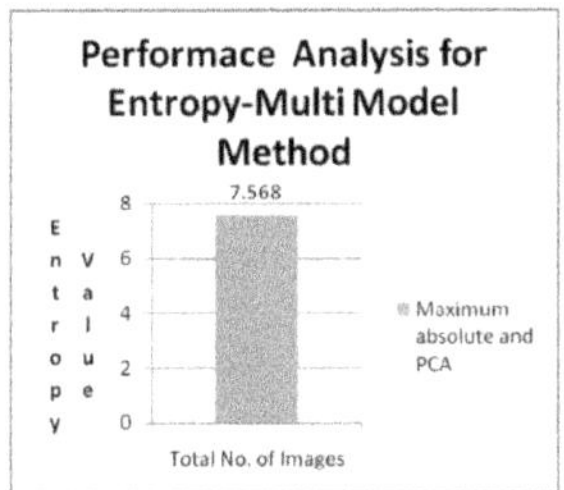

(b) Entropy - Multi Model Method

FIGURE 18.17 Performance analysis of entropy.

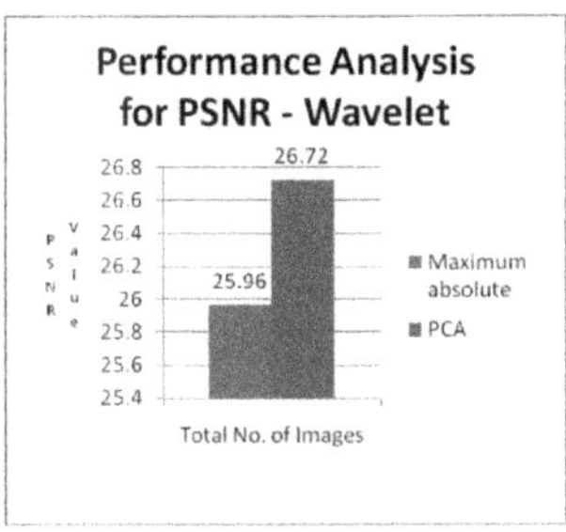

(a) PSNR - Wavelet

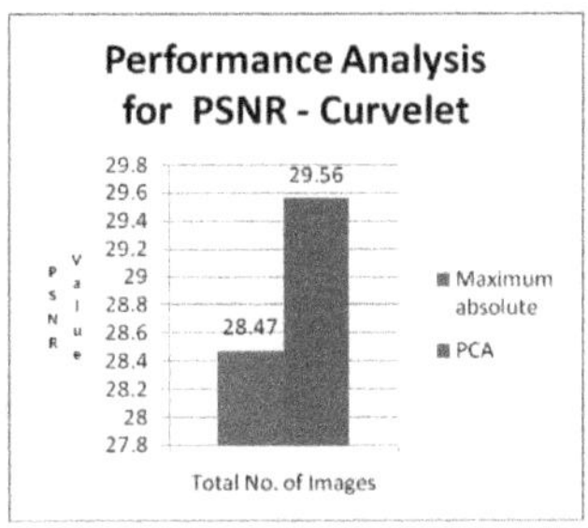

(b) PSNR - Curvelet

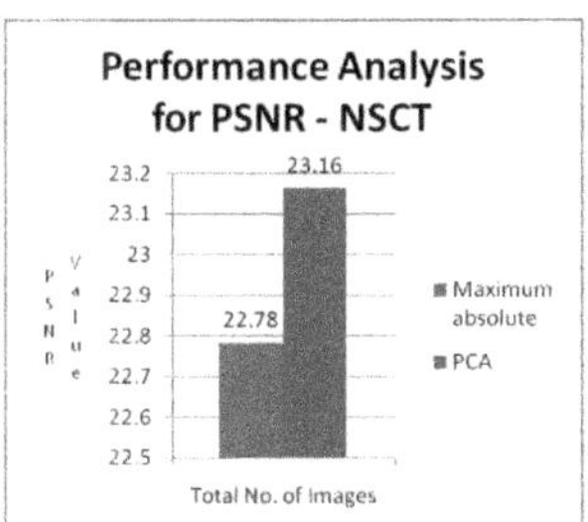

(c) PSNR - NSCT

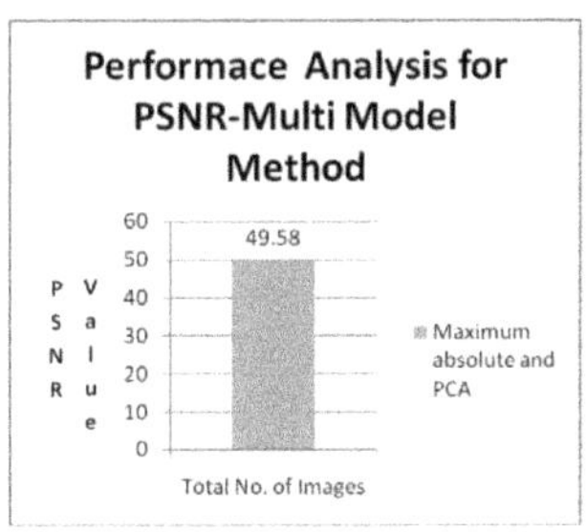

(b) PSNR - Multi Model Method

FIGURE 18.18 Performance analysis of PSNR.

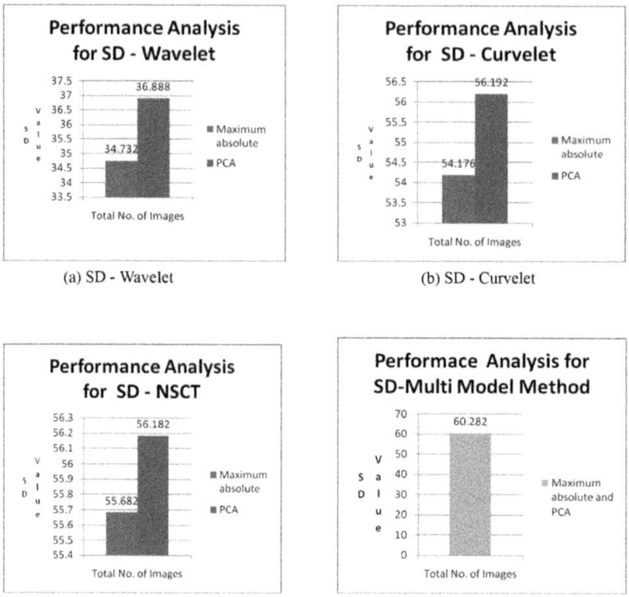

(a) SD - Wavelet (b) SD - Curvelet

(c) SD - Wavelet (d) SD - Curvelet

FIGURE 18.19 Performance analysis of SD.

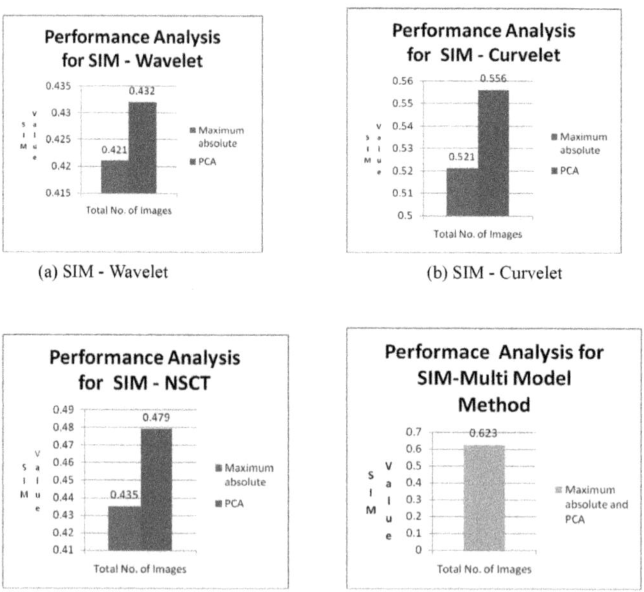

(a) SIM - Wavelet (b) SIM - Curvelet

(c) SIM - NSCT (d) SIM - Multi Model Method

FIGURE 18.20 Performance analysis of SIM.

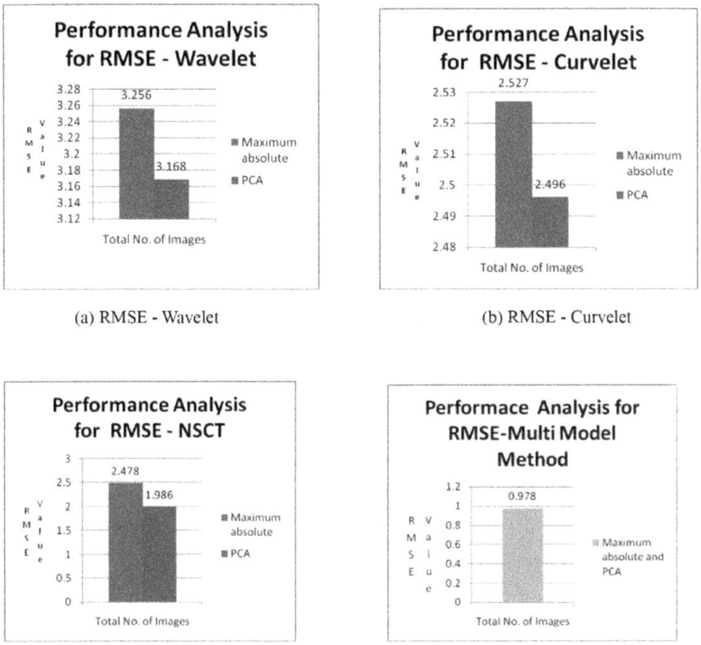

(a) RMSE - Wavelet (b) RMSE - Curvelet

(c) RMSE - NSCT (b) RMSE - Multi Model Method

FIGURE 18.21 Performance analysis of RMSE.

18.6 CONCLUSIONS

In this work, first phase lung cancer is detected by CT images by using SWFCM clustering. In second phase lung cancer is detected by PET image by using FLICM algorithm. In third part of work image fusion is done by CT and PET images. The values of entropy, PSNR, SD, SIM, and RMSE for wavelet transform under fusion rule of maximum absolutes are 5.832, 25.96, 34.7321,0.421, and 3.256 and principal component analyses are 5.956, 26.72, 36.888, 0.432, and 3.168, respectively. The values of entropy, PSNR, SD, SIM, and RMSE for curvelet transform under fusion rule of maximum absolutes are 6.328, 28.47, 54.1765, 0.521, and 2.527 and principal component analyses are 6.786, 29.56, 56.192, 0.556, and 2.496, respectively. The values of entropy, PSNR, SD, SIM, and RMSE for NSCT under fusion rule of maximum absolutes are 6.543, 22.78, 55.682, 0.435, and 2.478 and principal component analyses are 7.498, 23.16, 56.182, 0.479, and 1.986, respectively. The values of entropy, PSNR, SD, SIM, and RMSE for multi-model image fusion method under fusion rule of maximum absolute and PCA are 7.568, 49.58, 60.282, 0.623, and 0.978, respectively. Higher resolution of CT and PET image may be used for future work to increase the level of accuracy for the result.

KEYWORDS

- **CT**
- **PET**
- **SWFCM**
- **FLICM**
- **RMSE**
- **PSNR**

REFERENCES

1. Aggarwal, P. et al. Sardana, Semantic and Content-Based Medical Image Retrieval for Lung Cancer Diagnosis with the Inclusion of Expert Knowledge and Proven Pathology. In *Proceedings of the IEEE Second International Conference on Image Information Processing ICIIP'*2013, 2013; pp 346–351.
2. Zsoter, N.; Bandi, P.; Szabo, G.; Toth, Z.; Bundschuh, R. A.; Dinges, J.; Papp, L. PET-CT Based Automated Lung Nodule Detection. In *Annual International Conference of the IEEE Engineering in Medicine and Biology*, 2012; pp 4974–4977.
3. Punithavathy, K.; Ramya, M. M.; Poobal, S. Analysis of Statistical Texture Features for Automatic Lung Cancer Detection in PET/CT Images. In *International Conference on Robotics, Automation, Embedded & Control RACE* 2015, **2015**.
4. Akram, S.; Nicolini, S.; Carolia, P.; Nannia, C.; Massarob, A.; Marzolab, M.-C.; Rubellob, D.; Fantia, S. PET /CT Imaging in Different Types of Lung Cancer: An Overview. *Eur. J. Radiol.* **2013,** *81*, 988–1001.
5. Akram, S.; Javed, M.-Y.; Qamar, U.; Khanum, A.; Hassan, A. Artificial Neural Network based Classification of Lungs Nodule Using Hybrid Features from Computerized Topographic Images. *Appl. Math. Inf. Sci.* **2015,** *9* (1), 183–195.
6. Sharma, D.; Jindal, G. Identifying Lung Cancer Using Image Processing Techniques. *Int. Conf. Comput. Techniq. Artif. Intell. (ICCTAI'2011)* **2011,** *17*, 872–880.
7. Tong, J.; Da-Zhe, Z.; Ying, W.; Xin-Hua, Z. Xu, W. Computer-Aided Lung Nodule Detection Based on CT Images. In *IEEE/ICME International Conference on Complex Medical Engineering*, 2007; pp 816–819.
8. Ganeshbabu, T. R. Segmentation of Optic Nerve Head for Glaucoma Detection using Fundus images. *Biomed. Pharmacol. J.* **2014,** *7* (2), 1–9.

9. Kanakatte, A.; Mani, N.; Srinivasan, B.; Gubbi, J. Pulmonary Tumor Volume Detection from Positron Emission Tomography Images. *Int. Conf. Bio Med. Eng. Inf.* **2008,** *2* (8), 213-217.
10. Kumar, A.; Kim, J.; Wen, L.; Dagan, D.; Feng, F. A Graph-Based Approach to the Retrieval of Volumetric PET-CT Images. In *International Conference of the IEEE Engineering in Medicine and Biology*, 2012; pp 5408–5411.
11. Ganesh Nabu, T. R.; Tamil Thendral, M.; Vidhya, K. Detection of Lung Cancer Tumorusing Fuzzy Local Information C- Means Clustering. *Int. J. Pure Appl. Math.* **2018,** *118* (17), 389–400.
12. Ahmed, H.; Hassan, E. N.; Badr, A. A. Medical Image Fusion Algorithm Based on Local Average Energy-Motivated PCNN in NSCT Domain. *Int. J. Adv. Comput. Sci. App.* **2016,** *7* (10), 269–276.
13. Patel, K. Fusion Algorithms for Images Based on Principal Component Analysis and Discrete Wavelet Transform. *Int. J. Innov. Res. Sci. Technol.* **2015,** *1*, 180–182.
14. Nahvi, N.; Sharma, O. C. Comparative Analysis of Various Image Fusion Techniques for Biomedical Images: A Review. *Int. J. Eng. Res. App.* **2015,** *4* (5), 81–86.
15. Rajesh, K.; Ravichandran, C. G. Curvelet Transform Based Image Fusion With Noise Reduction Using Gaussian Filter. In *Australian J. Basic Appl. Sci.* **2015,** *9* (35), 161–166.
16. Rajalingam, B.; Priya, R. A Novel Approach for Multimodal Medical Image Fusion using Hybrid Fusion Algorithms for Disease Analysis. *Int. J. Pure Appl. Math.* **2017,** *117* (15), 599–619.
17. Wang, W.; Chang, F. A Multi-Focus Image Fusion Method Based on Laplacian Pyramid. *J. Comput.* **2011,** *6* (12), 2559–2566.
18. Sun, Y.; Zhao, C.; Jiang, L. A New Image Fusion Algorithm Based on Wavelet Transform and the Second Generation Curvelet Transform. In *IEEE International Conference on Image Analysis and Signal Processing*, 2010; pp. 438–441.
19. Fei, Y.; Wei, G.; Zongxi, S. Medical Image Fusion Based on Feature Extraction and Sparse Representation. *Int. J. Biomed. Imaging* **2017**.
20. Yang, Y.; Huang, S.; Gao, J.; Qian, Z. Multi-Focus Image Fusion Using an Effective Discrete Wavelet Transform Based Algorithm. *Measurement Sci. Rev.* **2014,** *14* (2), 102–108.
21. Liu, Z.; Chai, Y.; Yin, H.; Zhou, J.; Zhu, Z. A Novel Multi-Focus Image Fusion Approach Based on Image Decomposition. *Inf. Fusion* **2017,** *35*, 102–116.

CHAPTER 19

A FRAMEWORK PROMOTING POSITION TRUST EVALUATION SYSTEM IN CLOUD ENVIRONMENT

S. R. SRIDHAR, S. PRAGADEESWARAN, and M. GANTHIMATHI

Department of CSE, Muthayammal Engineering College, Namakkal, Tamil Nadu, India

ABSTRACT

Trust management is one of the biggest obstacles to the adoption and growth of cloud computing. Both consumers and corporations commonly use cloud computing. Protecting the customer's privacy would not be a simple task due to the sensitive data contained in the interaction between the customer and the trust evaluation service. It might be challenging to defend cloud services against malicious users. The design and development of CloudArmor, a legacy trust evaluation framework that offers a set of functions to provide Trust as a Service, will be discussed in this chapter (TaaS). This chapter also discusses the problems with calculating trust based on input from cloud users. This technology effectively protects cloud services by detecting hostile and inappropriate behavior by the use of trust algorithms, which may recognize on/off assaults and colluding attacks by using different security criteria. In conclusion, the findings demonstrate that the suggested trust model system may deliver high security by lowering security risk and enhancing the decision-making capabilities of cloud users and cloud operators.

Computational Imaging and Analytics in Biomedical Engineering: Algorithms and Applications.
T. R. Ganesh Babu, PhD, U. Saravanakumar, PhD & Balachandra Pattanaik, PhD (Eds.)

19.1 INTRODUCTION

Two major players have emerged in the new computing method known as cloud computing. cloud-end consumers and cloud service providers. There are multiple definitions given forth by various authors to precisely clarify what cloud computing is. In the realm of computers, cloud computing is a new business model. According to NIST's official definition, "public cloud is a platform which enables widespread, comfortable, on-demand access to a common cloud computing model" (i.e., connections, servers, storage, applications, and services) that can be quickly provisioned and released with little interaction from service providers.

Cloud computing adoption raises privacy problems. Customers and cloud service providers may engage in dynamic exchanges that include sensitive data. There have been a number of instances of privacy violations, including the disclosure of behavioral data or sensitive information (such a person's address and birth date). Services that use customer information (such as interaction history) should unquestionably protect that information.

Collusion attempts become a threat whenever an individual or group of evil elements tries to undermine the system. A reputational system's credibility is typically put at greater risk when several evil actors work together than when they act maliciously alone. Here are a few instances of particular cooperation attacks.

Collusive slandering attacks, also known as corrupt business badmouthing attacks, take place when dishonest users team up to disseminate negative testimonials about a reliable person in an effort to badly damage that user's reputation. They also hope to boost their own reputations by complimenting one another.

Several industries, including e-commerce, human sociology, wireless systems, and others employ trust management extensively. Finding a service provider you can trust in the cloud environment requires careful consideration of your level of trust. When evaluating a service provider's reliability, cloud service user evaluations play a big role. This study looks at a number of risks that may arise when trust is determined by user input from the cloud.

The link among cloud user's services is offered by a trust management system (TES) for efficient trust management. However, because to the unpredictability of user numbers and the extremely dynamic character of the cloud environment, ensuring TES uptime is a challenging task. It is unsuitable for cloud systems to evaluate user preferences and capabilities using indicators of success or functional durability measures. TES must be extremely scalable and adaptive in order to function in cloud settings.

19.2 TRUST EVALUATION

19.2.1 EXISTING SYSTEM

One of the top 10 related to the implementation of cloud computing, according to Berkeley researchers, is trust and security. Service-Level Agreements do exist (SLAs). Feedback from customers is a reliable source for determining how trustworthy cloud services are overall. Several researchers have acknowledged the need of managing trust and have put forth strategies to gauge and regulate effect on participant feedback.

Two distinct types of trust exist. (1) Indirect Trust and (2) Direct Trust. Indirect trust states that when a person lacks any direct experience, he must rely on the direct trust of others. Direct trust is approach on personal experience. Indirect trust is the name for this kind of trust. There are several service providers available in a cloud environment. Finding a reliable service provider is so crucial. Since trust is more irrational, dependent on context, asymmetric, and uncertain than other cloud entities, it is a complex interaction between them.

Given that cloud solutions are very dynamic, dispersed, and opaque, maintaining confidence in public clouds is a significant challenge. Feedback from customers is a reliable source for determining how trustworthy cloud services are overall. Several researchers have acknowledged the value of managing trust and have put forth strategies to gauge and maintain trust situated on participant feedback. In practice, it is not unusual for a cloud service to encounter malevolent user actions (such as collusion or Sybil assaults).

There are several methods for measuring trust. By reviewing user reviews posted to the cloud, we may assess the trustworthiness in the e-commerce sector. There is a risk for different evaluation base attacks when trust is determined via cloud user feedback ratings. The potential attacks on the feedback-based trust assessment are the focus of the next section. Users of cloud solutions may be handling remarkably confidential material during using trust maintenance services, so it is imperative that the highest level of security be offered to them; obtaining cloud services by efficiently identifying fraudulent and unsuitable behaviors by the use of trust methodologies that can recognize on/off and collusion threats.

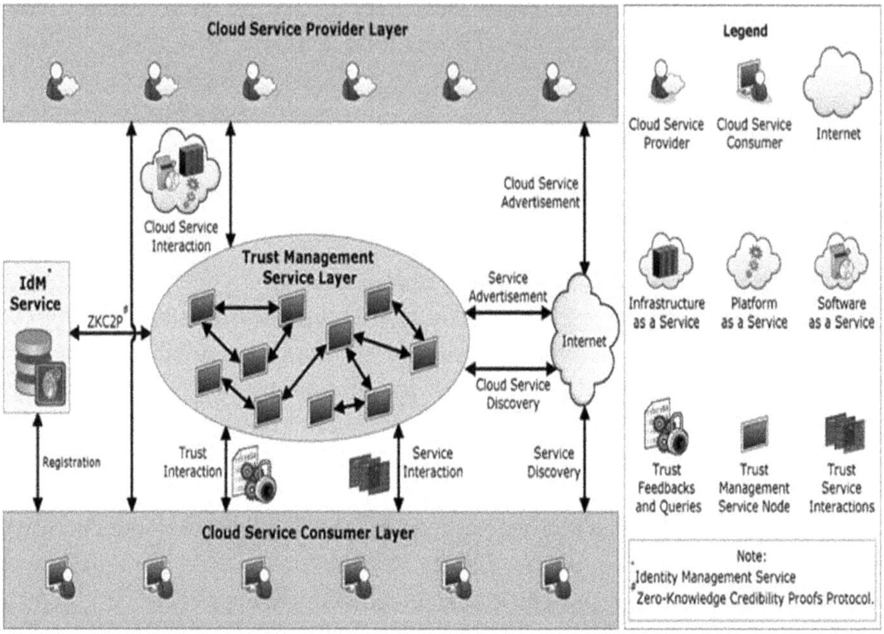

FIGURE 19.1 System architecture.

19.2.2 PROPOSED SYSTEM

The opinions of cloud service consumers are a reliable source for determining how trustworthy cloud services are overall. As demonstrated in Figure 19.1, unique methodologies are introduced in this paper that assist in identifying reputation-based assaults and enable users to quickly select reliable cloud services. It presents a version of success that not only recognizes deceptive trust feedbacks arising from differential attack but also recognizes Sybil attacks whether they occur over a lengthy or brief period of time, also provides a model of availability that keeps the trust comprehensive suite operating at the desired level; it also provides a model of availability that keeps the trust management solutions operating at the desired level.

A model of credibility. The effectiveness of the trust management is significantly influenced by the veracity of feedback. Therefore, suggest a number of metrics, such as Output Volume and Frequent Feedback Colluding, for the detection of feedback collusion. These measurements separate deceptive user feedback from malevolent user behavior.

Benefits of the Proposed System

- TES will validate user comments without knowing the user's credentials.
- Processes credentials without storing sensitive information.
- Uses anonymized data and consistently hashes it to find Sybil assaults, whether they last for a long time or just a short while.

19.3 DESIGN METHODOLOGIES

19.3.1 TRUST EVALUATION SYSTEM (TES)

The trust model has additional layers added to it to improve the system's overall efficacy. The following section provides descriptions of the various TES subsections.

The central repository serves as the place where interactions are stored. For later use by the deciding engine trust for determining the values of the tasks and roles, it maintains all types of trust data and interaction histories produced by interaction skills and duties. Elements not found in the TES are inaccessible to the central repository.

- Role Behavior Analyzer: This part examines how the simplest levels of trust laws apply to shared resources, including roles and functions. Based on the feedback provided by the service in the central repository, it assesses those regulations that have been recognized in the level of trust. The role behavior checker links the roles in order to learn more about them and find any leaks that might exist. In order to easily follow unauthorized users or attackers and provide proof of any type of data loss, it is crucial to identify the user and keep track of every actions they do.
- Task Behavior Analyzer: The task behavior analyzer is in charge of assessing tasks and functions in light of minimal trust level laws while gaining access to shared resources. By computing the trust value and storing it in the central repository, the tasks indicated only within confidence level are examined in terms of such feedback of owners. When determining the histories of users with relation to the stored data, it gathers data from the channels, though there are two in this case reports from development comes information leakage and data from the role behavior analyzer. The task behavior analyzer can help identify customers, and it's important to keep track of the

activities completed. This makes it simpler to locate attackers or unwelcome clients and to provide evidence of data leakage, should it have happened. Updated client accounts that have been registered are checked to see if a customer account was responsible for the occurrence.

- Feedback Collector: Before being automatically assigned, feedback from service owners to a depository headquarters must be managed by the feedback collector. The feedback provided on duties and tasks reveals the user's credibility. Its integrity is protected by the job and role feedback collector integrity to guarantee security. This element makes sure that the person submitting feedback further into system is authorized. It is able to identify false feedback and remove it off the system. Prior to uploading the data to the central repository, the role feedback collector also gathers information regarding the job and role assignments.
- Trust Decision Engine: In this section, the trustworthiness of the cloud users, their responsibilities, and the entity's values are examined and identified. Before deciding what sort of answer the system should provide, it gathers all types of data about the contact records that are contained in the repository center and the respect values of a given consumer.
- Detection of service: This layer is made up of many cloud service consumers. For instance, a young company with meager funding can use cloud services. This layer's interactions consist of: (1) service discovery, which allows customers to hunt up more cloud services and other services online. (2) Trust and counter, where users can provide feedback or access a cloud service's trust score; and (3) registration, where users confirm their identity by logging in to IdM before using TES.
- IDM Registration: The system suggests utilizing the Identity Management Service (IdM) to assist TES in determining the validity of a customer review. Processing the IdM data, however, has the potential to violate user privacy. Using cryptographic encryption techniques is one way to protect privacy. However, processing encrypted data cannot be done effectively. Another option to process IDM data without violating users' privacy is by using anonymization techniques. It is obvious that utility and great anonymity are trade-offs.
- Service Announcement and Communication: This layer comprises of several cloud service providers who openly provide IaaS

(Infrastructure services), PaaS (Platform as a Service), and SaaS (Software as a Service) cloud services, as well as combinations of these services, over the Internet Web portals provide access to these cloud services, which are also listed on search engines like Google, Yahoo, and Baidu. This layer's interactions with users and TES are regarded as cloud service interactions.

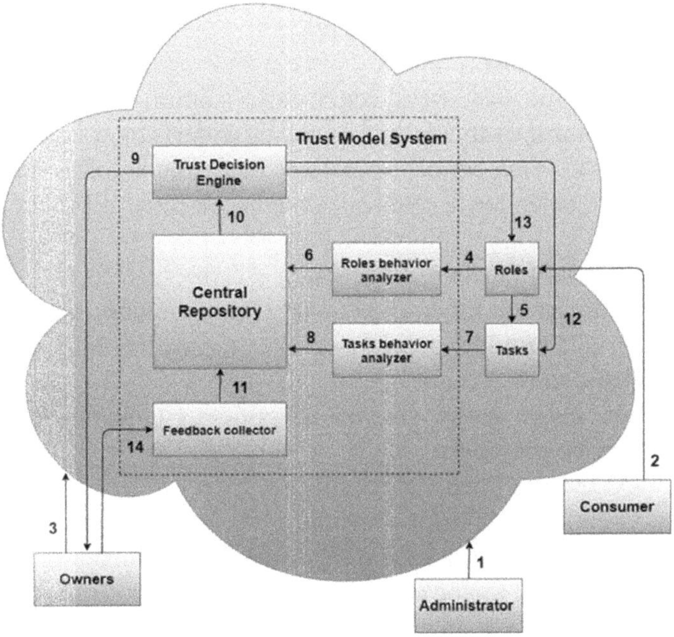

FIGURE 19.2 Trust model system architecture.

19.4 FEEDBACK BASE TRUST EVALUATION

A. *Feedback Storage and Evaluation*

In current methodologies, a service's trustworthiness is largely evaluated centrally, whereas feedback is provided by dispersed trust participants. The centralization of power may cause scalability and security problems. On a customized virtualization that protects the processor architecture, a suggested framework for safe application execution is based suggested that trust paradigm for private cloud environment data security proposed a framework for managing trust that uses credibility and feedback to determine the value of trust. The percentage of recognized consensus feedback can help determine the veracity of other feedback.

B. ***Accuracy of Trust Result***

Because the cloud environment is changing, selecting the reliable feedbacks from among the many supplied feedbacks is the most difficult problem. Accuracy of Trust and Trustworthiness are significantly correlated with others. By minimizing potential threats, we can obtain an accurate trust result.

C. Collusion Attacks

When several users band together to submit false feedback in an effort to boost or lower the service provider's trust score, the attack takes place. This conduct is referred to as tacit collusion malicious feedback behavior. Collusion assaults of three different types are feasible.

a) A promotion-focused attack
Entire group Enter all encouraging comments to help the cloud service provider grow.

b) Slanderous Assault
The entire group entered all critical comments for the cloud service provider.
Occasional Feedback Collusion.

c) Oblique Feedback Collusion
Cloud services occasionally experience collusion assaults. Time is a crucial factor in recognizing sporadic collusion attacks.
Irissappane suggested a clustering strategy that would differentiate submitted feedback from malicious feedback presented clustering approach to increase the performance of the trust system. This approach creates clusters based on the variances in all the ratings. The value of trust is increased by combining reliable weighted ratings. We can calculate the irregular change in submitting all feedback loops within the overall feedback behavior to identify Occasional Feedback Collusion Sybil Attacks.

d) Such an attack takes place when malicious people leave several false reviews while using different identities (i.e., producing numerous false ratings while making a limited number of different purchases in a brief period of time) in an effort to boost or lower the trust rating.

e) These remarks are made for many different reasons, including user ineptitude, deception, or carelessness, among others. Attackers might assume several different identities to cover up their previous bad reputations. Sybil can attack in one of three ways. Attack That Promotes Itself Users provide encouraging comments to promote cloud service providers. This assault is sometimes referred to as a ballot-stuffing assault.

f) A Slandering Attack
Users can promote a cloud service provider by leaving unfavorable reviews. This attack is also known as a verbal assault.

g) Occasional Sybil Attacks
By making numerous accounts and submitting false comments in a short amount of time, malicious users can influence the outcome in terms of trust. By employing previously saved identity records for strong identity authentication, we might be able to prevent the Sybil attack. By calculating the sporadic variations in all submitted IDs and identity behavior, it can be detected.

D. Intoxication Attacks

In this strategy, a member first acts appropriately for a set amount of time to build up a high reputation before beginning to act inappropriately. Due to their high reputation, intoxication makes it challenging for the system to locate such misbehaving users.

On-off attack is another name for this assault. A member first exhibits positive behavior, but as time goes on, the user starts to behave badly. Such members are challenging to spot since they have historically maintained a positive reputation. This attack is referred to as the hold a positive attitude of peers in P2P network systems.

The forgetting factor is one of the most common tactics used to stop this onslaught. However, malevolent people frequently took advantage of forgetting. Forgetting factor, put out by Sun and Liu states that when trust results go below.

E. Discrimination Attacks

This attack occurs when a service provider offers customers better services in some geographical regions while offering subpar services in others. These guaranteed conflicting feedbacks between these regions, which may have resulted in a coordinated attack. No defense against such an attack has been discovered.

F. ***Newcomer Attacks***

If a person can readily register under a new identity even when the member already has registered with the service provider and displayed some unsavory behavior in the past, the member can launch a newbie assault. Re-entry attack is another name for this assault. It succeeded in Sybil's attack in the end. By matching the credential recodes with various parameters, such as location and unique id, we can lessen newbie attacks.

The system is launched by the administrator, and defines the hierarchy of roles and responsibilities for it. Channel 1 facilitates uploading to the cloud of the roles and task parameters established by the system;

Customers must first make an access permission via Port 2 depending on their responsibilities and roles if they want access privileges to cloud data. The role entity transfers the demand to the task's entity via Channel 5 if the customer request is approved. In reply, the cloud offers users an assignment access control (T-RBAC) scheme that is encrypted. The owner can only authorize encryption and data uploading through Channel 3 if they believe the job or task can be respected. The proprietor also discloses Owners should contact the feedback collectors through Channel 14 if they discover that one of their customers has leaked their data. To preserve each trust assessment and interaction record created whenever the roles interact, the input collector sends new feedback from an approved owner to the central database using Channel 11.

Then, utilizing Channel 10, the confidence decision engine uses the interaction logs to calculate the trust level of roles and tasks; the database is then centrally located.

The roles' entity may at any time approve the TMS's confidence assessments for the roles, at which point it responds to the TMS through Channel 13 to get feedback from the TMS. Until the input from the trusted decision engine is received, the role entity evaluates the role parameters that influence a consumer's cloud role membership, and any harmful consumer membership is removed; When a data breach results in negative feedback from an owner regarding a role, the role entity sends information about the leak to the role behavior the responsibilities entity may at any moment accept the TMS's confidence evaluations for the tasks through Channel 4, at which point it responds to the TMS through Channel 12 to get input from of the trusted decision engine.

Until input from the confidence decision engine is received, the task entity analyses the task's parameters that determine a consumer's cloud task membership, and any harmful consumer membership is removed.

When a task's owner complains about it because of information leaks, the task entity sends Channel 7 information about the leak to the task behavior analyzer. Then, the analyzers utilize Channels 6 and 8 to continuously update the trust information again for roles and responsibilities in the centralized database;

The TMS does a trust analysis whenever an owner desires that his data be transferred and protected in the cloud. Following receipt of the request, the TMS contacts the proprietors through Channel 9; the trusted decision engine informs the owners of the results of the trust management for their respective responsibilities. Based on the findings, the data owners decide whether or not to grant customers access to their services.

19.5 SIMULATION RESULTS

In C#.net, we created a Windows Forms application.net to test the outcomes of our system, and we used big data to verify all the precautions taken to prevent these attacks. To ascertain the trust model's resilience to reputational threats, experiments were used. The TMS predicated the penalty for on attack criteria: whether the connection significance was greater than or equivalent to the risk rate (DR), or whether the recommender's feedback (F) was less acceptable than interface importance (II). The trust model computed the confidence decline penalty PTD when the recommender's feedback (F) was less significant than the importance of the interaction (II). The contact trust values are affected by the cost of the on attack and trust deterioration.

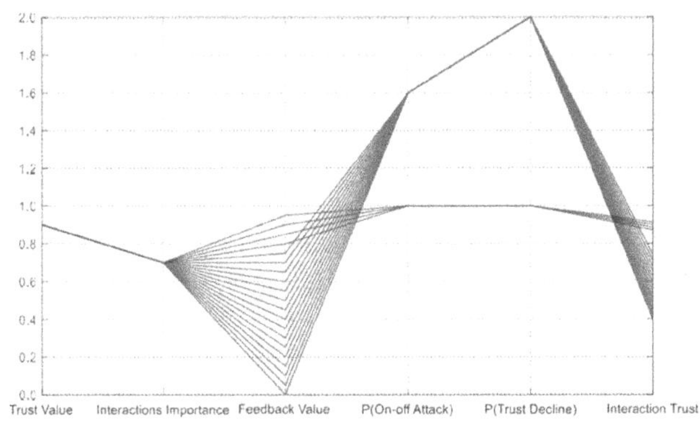

FIGURE 19.3 Penalties for on/off attack and declining confidence.

The trust model uses the penalties for bad behavior to establish the worth of contact trust for malicious users. Figures 19.4 and 19.5 show how fresh feedback affects interactions between trustworthy users.

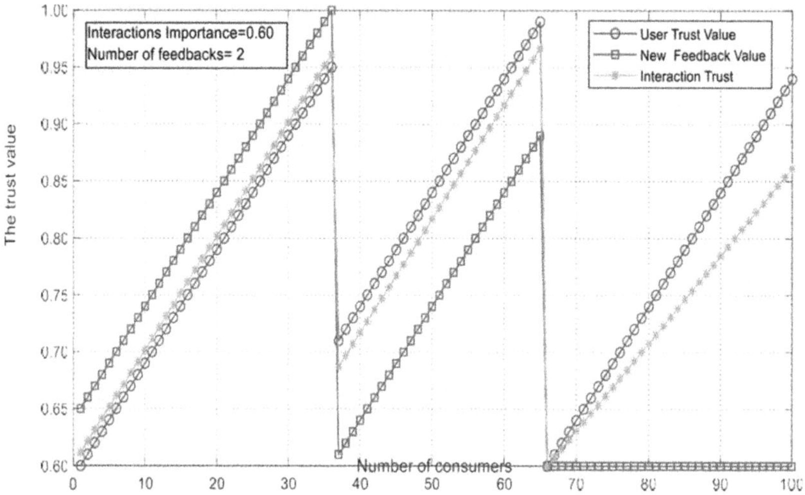

FIGURE 19.4 Values of interaction trust for 100 consumers.

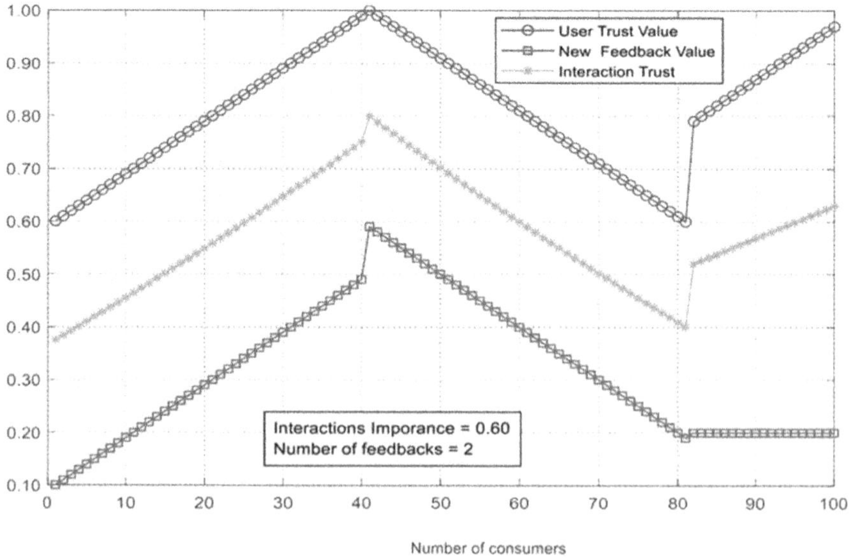

FIGURE 19.5 Trust scores for interactions with 100 malevolent customers.

The resilience of the trust model against reputational attacks is investigated in this section through experimentation. To stop collusion attacks, the TMS determined the cooperation attack frequency (CAF). Here, feedback frequency had an inverse connection with credibility and a direct link with collusion in the feedback. The amount of recommendation systems feedback items and the overall number of input pieces within the suspect set were used to calculate the feedback frequency (SS).

Seven questionable recommenders' feedback frequency is shown Figure 19.6 shows how, as we presume, in Table 19.1. The fact that five suspected recommenders' feedback frequency exceeded the feedback limit (FL) suggests that the TMS will move the feedback they provided to the collusion set (CS) or that the trust. The suspected feedback will be moved by the model. provided by a specific recommender for a specific user to the feedback set (FS).

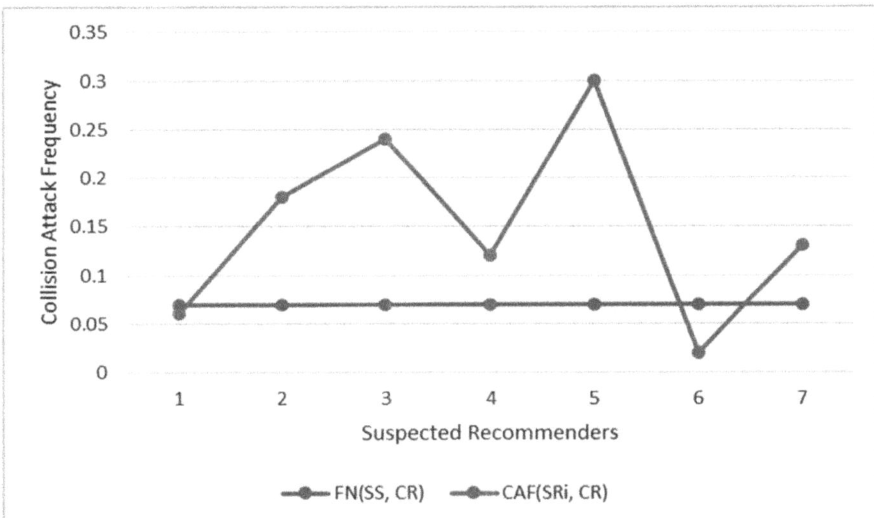

FIGURE 19.6 Collusion attack frequency.

TABLE 19.1 Collusion Attack Frequency.

FN(SRi, CR)	7	20	32	13	36	1	17
FN(SS, CR)	126	126	126	126	126	126	126
CAF(SRi, CR)	0.06	0.18	0.24	0.12	0.3	0.02	0.13

Figure 19.7 demonstrates how the attacker scale (AS), which is the size of the attack scale for distinct collusion sets, is determined using the trust model. To assault and undermine the trust model, the harmful recommenders in a recommender's community must make up a sizeable portion of all recommenders. This is known as the collusion set (CS).

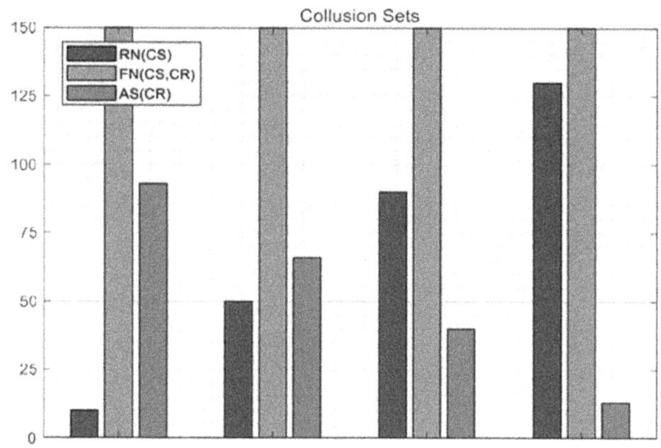

FIGURE 19.7 Attack scale (AS).

Figure 19.8 gives the result of the attack target scale (ATS), which provides the destructive feedback rate from one collusion set (CS) for a certain user.

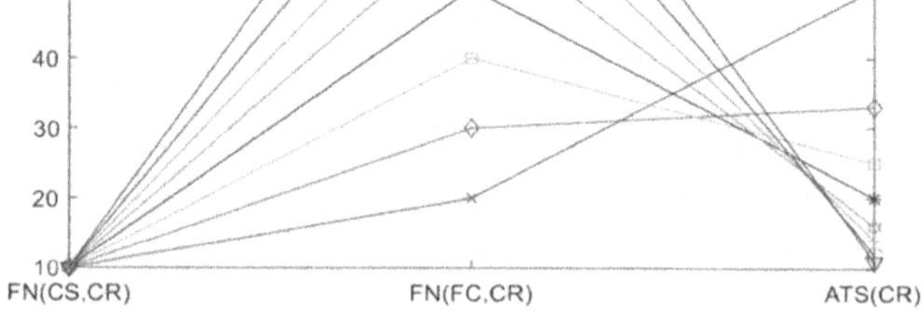

FIGURE 19.8 Attack target scale (ATS).

Figure 19.9 gives the result of a collusion attack strength (CAS), which is calculated by calculating the rate of all harmful feedback from various collusion sets (CS) for a certain user.

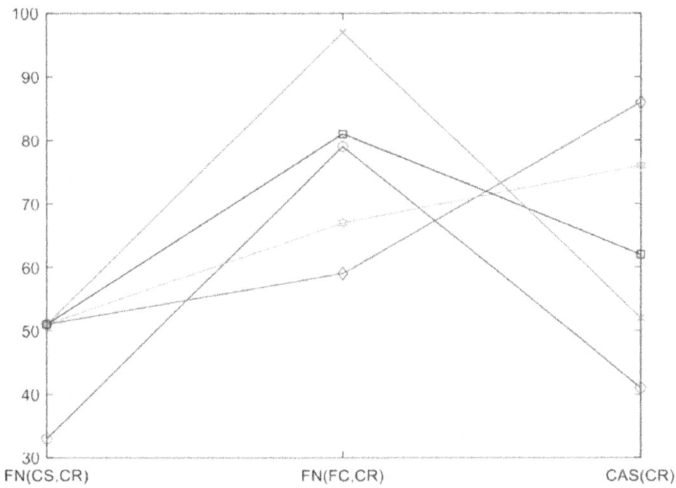

FIGURE 19.9 Collusion attack strength (CAS).

19.5.1 COMPARISON OF SECURITY AND ACCURACY

A number of assaults can be launched against any kind of reputation management service. These assaults have the power to either improve or harm a particular entity's reputation. We concentrate on applying different measures to stop these attacks in order to develop a stable, dependable, and correct trust model framework, which helps us to develop an efficient and reliable trust model system. Table 19.2 compares our suggested TMS to those mentioned in similar studies.

TABLE 19.2 Difference of Accuracy and Security.

Addressed metrics	[1]	[2]	[3]	[4]	[5]	Ours
Collaboration importance		✓				✓
On/off the attack		✓	✓		✓	✓
Trust failure	✓	✓				✓
Collusion outbreak	✓	✓		✓	✓	✓
Conspiracy attack frequency				✓		✓

Our extensive research project, Cloud Armor, which gives a framework for prestige trust administration of cloud services, includes the deployment of the trust management service. Users of the platform can submit suggestions and ask for a trust assessment of a specific cloud service in a safe environment provided by the platform. Specifically, the management of trust The Trusted Data Provision and the Trust Evaluation Function are the two main parts of the Trust Management Service (TMS).

The Trusted Data Provisioning is in charge of gathering trust data and cloud service information. To enable the system to automatically find clouds services on the Internet, we created the Cloud Computing Crawl which ensures the right on the Open-sourced Crawler for Java (crawler4j). To make the crawling process easier and the data more thorough, we provided a number of capabilities (such as add Seeds (), pick Crawling Domain (), and add Crawlers Time ()). Additionally, we created the Confidence Feedbacks Collection module to gather user feedback directly and store it in the Trust Legit reviews Database as historical records: In fact, users often need to register in order to verify their identities the very first time they want to use submitting their credentials for storage in the Trust at the Access Control Service (IdM).

Identity Registry: Furthermore, we established the Identification Info Recycler subsystem to gather the overall number of founded personalities among the whole identification behavior (i.e., all founded personalities for customers who tried to give feedback loops to a specific cloud service).

The Trust Assessment Function is in charge of processing user requests for trust assessments, which compare the reliability of cloud services and compute the trust feedback factors (i.e., the credibility factors). We established the Variables Calculations for proactive actions based on the set of variables (more specifics on how well the believability variables are measured can be found). In addition, we developed the Confidence Assessment to compare the trust in cloud services by acquiring the aggregated weighted factors from the Components Estimator to weigh feedbacks, and then figuring out the mean of all the feedbacks given to each cloud service. The trust outcomes for each cloud service are stored in the trust conclusions and component weights storage, together with the component weights for trust feedback.

a) Experimental Evaluation:

We paid special attention to validating and researching the suggested credibility model's resistance to various harmful behaviors, such as collusion and Sybil assaults under various behaviors, as well as the functionality of our available model.

b) Credibility Model Experiments:

We put our credibility model to the test using actual customer reviews of cloud services. We specifically crawled a number of review websites, including CloudHostingReviewer.com, fog. findthebest.com, cloud storage supplier sreviews.com, and cloud computing. findthebest.com, where consumers submit feedback on cloud services they have utilized. The gathered information is shown as a field H, where the feedback corresponds to the various QoS criteria stated previously and is supplemented with a credential for each associated customer. We were able to compile 10,076 comments submitted by 6982 people about 113 actual cloud services. The project website has made the gathered dataset available to the scientific community. The gathered information was split into six categories of cloud computing, three of which were used for experimental reasons.

Where each group, consisting of 100 users, was used to validate the criteria needed against differential attack, as well as the other three categories have been used to model validation against Sybil assaults. Each cloud storage group served as a representation of a different type of assaulting behavior, including Waves, Uniform, and Peaks. The behavior models show how many harmful feedbacks were introduced overall during a specific time instance, for example, $|V(s)| = 50$ malicious feedbacks.

When testing against collusion attacks $Ti = 50$. When testing against Sybil assaults, the behavior models also show the total number of names created by attackers over a period of time (for instance, $|I(s)| = 78$ malevolent individuals when $Tj = 30$. We modeled malicious feedback to improve the trust results for cloud services in collusion attacks (i.e., a self-promotional attack), while we modeled hostile feedback to decrease the trust results for Sybil attacks (i.e., slandering attack). To assess how resilient our credibility model is to malevolent conduct (such as conspiracy and Sybil assaults),

We carried out two sets of experiments: (1) testing the robustness of the credibility model using a standard model Con(s, t0, t) (i.e., setting Cr(c, s, t0, t) to 1 for all confidence feedbacks); and (2) assessing the performance of the model using accuracy and recall (i.e. how well TMS detected attackers) (i.e., how many detected attacks are actual attacks). In our trials, TMS started paying cloud services that engaged in harmful behavior once the attack percentage reached 35% (i.e., et(s) = 35%), so the awarding procedure would only occur when there was a significant drop in the trust result.

Six of the 12 experiments we ran were to test how well our credibility model stood up to collusion assaults, while the other six were to test how well our model stood up to Sybil attacks. According to Table 19.3, each study is identified by a letter ranging from A to F.

TABLE 19.3 Behavior Experimental Design.

Malicious behaviors	Experimental setting	Waves	Uniform	Peaks
Collusion	I	A	B	C
outbreaks	II	A′	B′	C′
Sybil	I	D	E	F
outbreaks	II	D′	E′	F ′

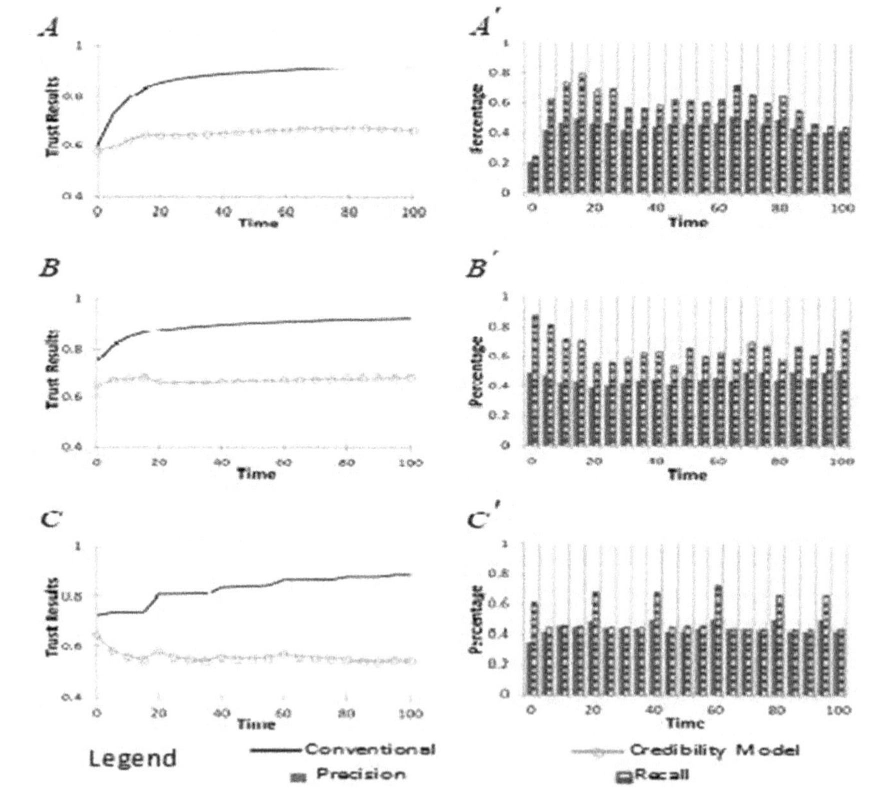

FIGURE 19.10 Robustness against collusion attacks.

High resistance against Collusion Attacks: To raise the results of cloud service trust (i.e., a self-promotional attack), we simulated hostile users who gave input in the [0.8, 1.0] range. The assessment of six tests, which was done to determine how resilient our model was against collusion attacks, is shown in Figure 19.10. The trust results for experimental environment I are shown in Figure 19.10 as A, B, and C, whereas the findings for experiment setting II are shown as A′, B′, and C′. We observe that the stronger the trust, the closer the temporal instance is to 100. which was determined using the standard technique and the results are produced. This happens due to dishonest customer feedback that is offered in an effort to increase the trust rating of the cloud service. However, there is essentially no change when the proposed credibility model is applied to estimate the outcomes of trust Figures 19.10A, 19.10B, and 19.10C. These facts prove that our reputational model is responsive to differential assault and capable of recognizing such malicious behaviors. Additionally, we may make the exciting discovery that the Normal behavior model works best in terms of recall while our feature of sustainability perform great in terms of accuracy when the Waves approach is utilized

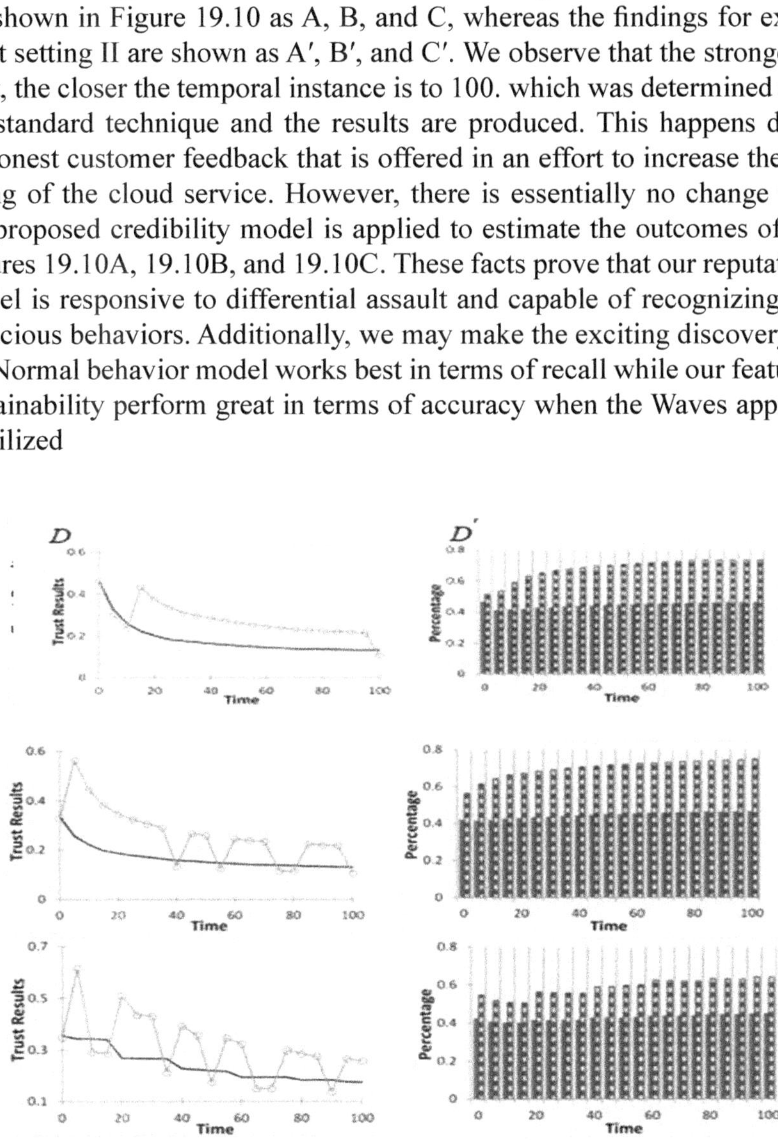

FIGURE 19.11 Robustness against Sybil attacks.

Structural rigidity against Attacks: For the experiments involving Sybil attacks, we designed to simulate malicious users who sought to lower the trust ratings of cloud services (i.e., a slandering attack) by creating multiple identities and disseminating one malevolent feedback with a level of [0, 0.3] per individuality. The assessment of six tests that were run to test our model's resistance against Sybil assaults is shown in Figure 19.11, which illustrates the trust results for the experimental setting I, whereas Figure 19.11 illustrates the results for the experimental setting II.

As the time instance approaches 100, as seen in Figure 19.11, the trust results obtained using the standard model decline. This is due to malevolent users that provide false feedback in an effort to lower the cloud service's trust score.

On the other hand, the findings of our suggested credibility model in terms of trust are better than those of the traditional approach Figures 19.11D, 19.11E, and 19.11F. Here this is so because when the attacks took place, the cloud service received a payment.

Actual Availability Compared to (a) Estimated Trust result accessibility (b) Rate of caching errors.

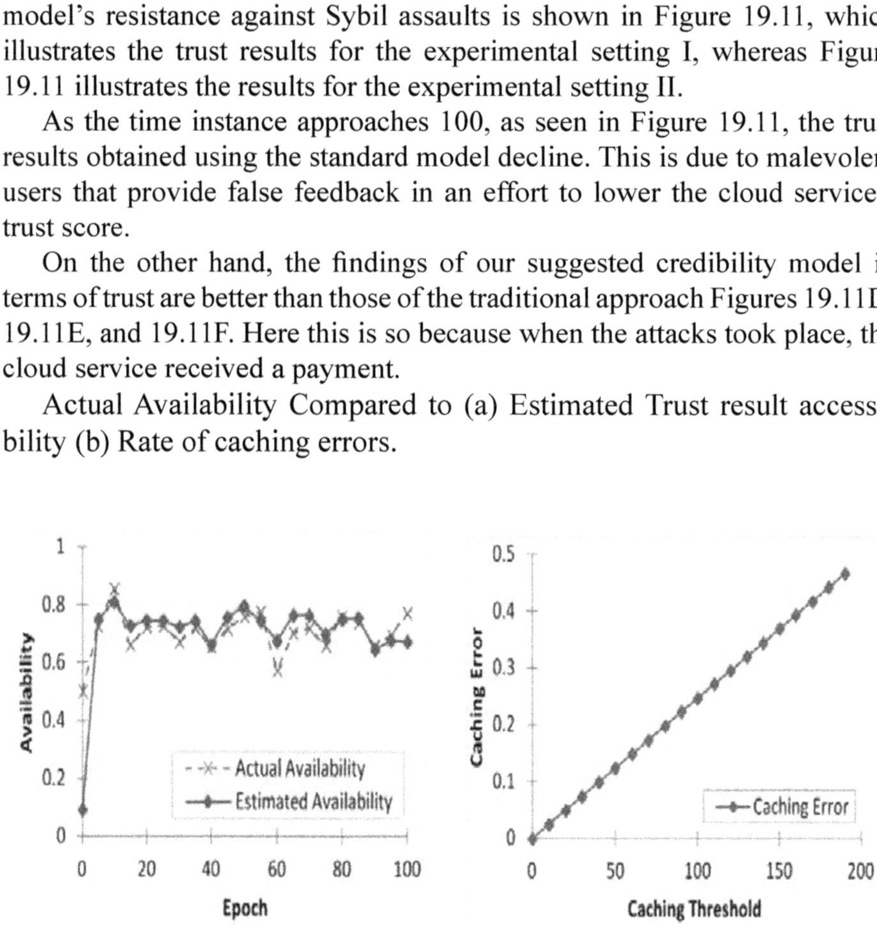

FIGURE 19.12 Availability prediction and caching accuracy.

The greatest amount of liquid is documented when the Peaks behavior model is applied (i.e., we can see five Figure 19.11F, which shows drops which matches exactly the drops in the Peak position behavior as shown in Figure 19.11. We can also see that some dramatic drop in the confidence results obtained by taking into account our criteria are needed. This occurs as a result of TMS's requirement that the percentage of assaults over the same period of time exceeds a certain threshold before rewarding the impacted cloud services (i.e., which is set to 25% in this case). This indicates that TMS has given the impacted cloud service a reward based on the factor for the change rate of trust outcomes. Additionally, Figure 19.11D′, 19.11E′19.11D′, and 19.11F′ show that our credibility model performs best in terms of accuracy when the Waves behavior model is applied (i.e., 0.48; see Fig. 19.11D′), whereas the maximum recall score is obtained when the Uniform behavior model is applied (i.e., 0.75; see Fig. 19.11A′). The ability of TMS to commend the impacted cloud service that uses the rate of confidence results factor shows that our model can identify Attacks (i.e., either corporate strategy attacks like those in the Waves and Uniform behavior models or infrequent attacks like those in the Peaks behavior model) successfully.

c) Availability Model Experiments:

Using the same dataset, we used to verify the credibility model, we evaluated our availability model. However, we concentrated on verifying the availability forecast accuracy, trust outcomes cache accuracy, and re-allocation efficiency of the available model for the availability tests. Accuracy of Availability Prediction: We simulated 500 nodes housing TMS instance and set the probability of failure for the nodes at 3.5%, which is consistent with the results to test the predictive performance of the availability model.

This experiment was designed to investigate the applicability of our method for estimate. For 100 time steps, we tracked the availability fluctuations of TMS nodes that we had simulated. We gathered and compared the anticipated accessibility of TMS nodes produced by our particle filter technique with their real availability. The outcome of one specific TMS node is depicted in Figure 19.12. This figure shows that the projected availability and the TMS node's actual availability are extremely similar. This indicates that our method is effective at tracking and forecasting TMS node availability.

Results of Trust Accurate Caching: To assess the accuracy of the availability model's prefetching, we varied the cache management threshold to identify the optimum number of new trust feedback mechanisms that TMS actually received to recalibrate the trust result for a particular cloud storage service without having experienced a significant error in the trust results. By assessing the difference between the predicted trust values and actual trust result of a particular cloud service, or the underlying cause error (RMSE), also known as the cache management error, the accuracy of the trust result caching is determined. The trustworthiness result caching is more accurate than the smaller RMSE value. The trustworthiness outcome caching reliability of one specific cloud service is displayed in Figure 19.13. The chart shows that when the cache threshold rises, the caching error grows approximately linearly.

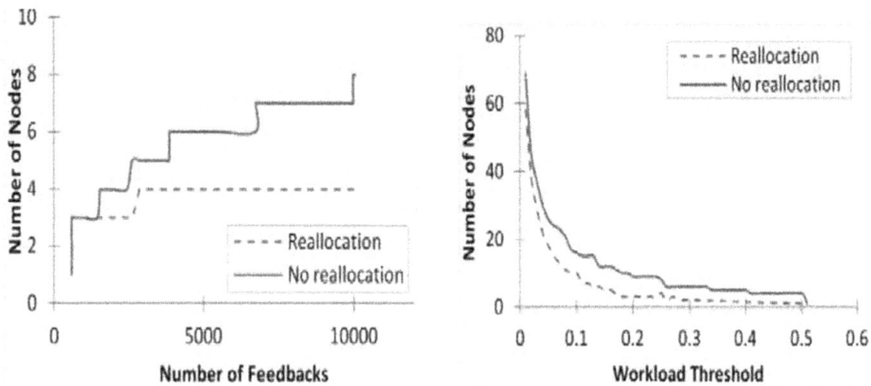

FIGURE 19.13 Re-allocation performance.

Based on a tolerable caching error rate, the results enable us to select the best caching threshold. If a 10% error margin, for instance, is acceptable, the cache criterion can be set at 50 feedbacks. It is important to note that the cache error was calculated using real users' reviews of actual cloud services.

The outcomes let us choose the ideal caching threshold based on an acceptable caching error rate. The cache criteria can be set at 50 feedbacks, for example, if a 10% error margin is acceptable. It is significant to highlight that the cache fault was determined using assessments made by actual customers of cloud services.

Reallocation Effectiveness: When the activity flow threshold ew (stms) = 25%, we calculated the number of TMS nodes using distribution of wealth of competence positive feedbacks but without re-allocation while increasing the quantity of feedbacks, and when using reorientation of competence feedback loops and without re-allocation while varying ew (stms). The more TMS nodes there are, the more cost-effective TMS becomes. Figure 19.13 shows the outcomes of

I. The experimental settings show that the total number of TMS nodes when utilizing the redistribution of trusted feedback systems method is fairly low and much more reliable than the overall number of TMS nodes when reallocation is not employed. The outcomes of experiments II are displayed in Figure 19.13. The graphic shows that the number of TMS nodes decreases as the workload threshold increases. The number of TMS nodes, however, is lower when the reshuffling of trust feedback systems technique is used than when reshuffling is not taken into account. This means that by lowering the number of TMS nodes, our solution provides advantages in cutting the bandwidth cost.

19.6 CONCLUSIONS

Accessing cloud storage for users is severely hampered by authorization problems, especially when it comes to significant data, which really is frequently highly sensitive. The trust architecture presented in this study enables reliable defences against on/off and collaboration assaults, which are key security issues that users of cloud services must contend with. Specifically, this paper introduces a criteria needed that recognizes Sybil attacks in addition to false trust feedback loops from collusion attempts. Whether they take place over a long or short period of time make an availability model that maintains the appropriate degree of performance for the TES. The experiment's results demonstrate the viability of this tactic and demonstrate the capacity to spot such malevolent activities. Controlling models should be combined with trust evaluation for decentralized systems through the suggested trust algorithms, which can spot on/off and collusion assaults and guarantee the maximum degree of anonymity for cloud service customers, in order to appropriately address these concerns.

KEYWORDS

- **trust evaluation**
- **CloudArmor**
- **colluding attacks**
- **trust management**
- **cloud computing**

REFERENCES

1. Noor, T. H.; Sheng, M.; Alfazi, A. , In Proceedings of the 12th IEEE International Conference on Trust; Security and Privacy in Computing and Communications, Melbourne, Australia, July 16–18, 2013; pp 469–476.
2. Chang, E. , In Proceedings of the IEEE International Conference on Computer Communications, Valencia, Spain, July 29–August 1, 2019.
3. Mahajan, S.; Mahajan, S.; Jadhav, S.; Kolate, S. Trust Management in E-commerce Websites. **2017,** , 2934–2936.
4. Habib, S. M.; Hauke, S.; Ries, S.; Muhlhauser, M. Trust as a Facilitator in Cloud Computing: A Survey. **2012,** , 19.
5. Khan, K.; Malluhi, Q. Establishing Trust in Cloud Computing, Qatar University, IEEE IT Professional, 2010; vol (5).
6. Manuel, P.; Somasundaram, T. S. A Novel Trust Management System for Cloud Computing – IaaS Providers. **2011,** , 3–22.
7. Chong, S. K.; Abawajy, J.; Hamid, I. R. A.; Ahmad, M. A Multilevel Trust Management Framework for Service Oriented Environment. **2013,** , 396–405.
8. Wang, D.; Mullerr, T.; Liu, Y.; Zhang, J. Towards Robust and Effective Trust Management for Security: A Survey. **2014.**
9. Kotikela, S. N. S.; Gomathisankaran, M. In , International Conference on Cyber Security, 2012.
10. Muchahari, M. K.; Sinha, S. K. In , IEEE International Symposium on Cloud and Services Computing (ISCOS0), 2012.
11. Canedo, E. D.; de Sousa, R. T.; de Carvalho, R. R.; de Oliveira, A. R. In , IEEE International Conference on Cyber Security; Cyber Warfare and Digital Forensic(CyberSec), 2012.
12. Noor, T. H.; Sheng, Q. Z.; Yao, L.; Dustdar, S.; Ngu, A. H. H. CloudArmor: Supporting Reputation-Based Trust Management for Cloud Services. **2014**.
13. Xiong, L.; Liu, L. Peertrust: Supporting Reputation-Based Trust for Peer-to-Peer Electronic Communications. **2004,** (7), 843–857.
14. Irissappane, A. A.; Jiang, S.; Zhang, J. In , Proceedings of the 2014 International Conference on Autonomous Agents and Multi-Agent Systems, 2014; pp 1385–1386.
15. Liu, S.; Zhang, J.; Miao, C.; Theng, Y. L.; Kot, A. C. In , Proceedings of the 10th International Conference on Autonomous Agents and Multi agent System, 2011; vol 3, pp 1151–1152.

CHAPTER 20

EFFICIENT MACHINE LEARNING TECHNIQUES FOR MEDICAL IMAGES

GURURAMA SENTHILVEL P., K. SAMPATH KUMAR, and T. GANESH KUMAR

Department of Computing Science and Engineering, Galgotias University, Greater Noida, Uttar Pradesh, India

ABSTRACT

Medical image analysis plays a pivotal role in modern healthcare, aiding in diagnosis, treatment planning, and disease monitoring. With the exponential growth in medical data, there is an increasing demand for efficient machine learning techniques to extract valuable insights from medical images while minimizing computational resources. This abstract provides an overview of recent advancements in the realm of efficient machine learning techniques for medical image analysis. Efficient machine learning techniques for medical images are not only crucial for improving patient care but also for overcoming challenges posed by limited computational resources, data privacy concerns, and the ever-increasing volume of medical data. Future research in this field will likely continue to focus on enhancing the efficiency, interpretability, and generalizability of machine learning models for medical image analysis. Deep Learning Architectures: Convolutional Neural Networks (CNNs) have revolutionized medical image analysis by providing state-of-the-art results. Efficient architectures like Mobile Nets, Efficient Nets, and Squeeze-and-Excitation networks have been adapted to reduce model size and computational demands while maintaining high accuracy.

Computational Imaging and Analytics in Biomedical Engineering: Algorithms and Applications.
T. R. Ganesh Babu, PhD, U. Saravanakumar, PhD & Balachandra Pattanaik, PhD (Eds.)

Multi-modal fusion: Integrating information from various imaging modalities (e.g., MRI, CT, X-ray) through fusion techniques enhances diagnostic accuracy and reduces reliance on single-modal data. Explainable AI: Interpretable machine learning models provide insights into why a particular decision was made, fostering trust and acceptance among medical professionals

20.1 INTRODUCTION

AI may be a system for seeing plans that may be applied to medical images. The same manner that associate necessary resource will facilitate in transfer clinical findings, it will normally be twisted. AI often begins with the machine learning calculation system schemingthe image remembers that are conceded to be of significance for creating the supposition or examination of interest. The Machine Learning calculation system additionally, at that time, perceives the fashionable mixture of these image options for requesting the image or enrolling some estimation for the given image space. There are a handful of designs that may be used, every with totally different characteristics and scarcities. There are open-supply metamorphoses of utmost of those machine learning systems that alter them to endeavor to use to film land. A handful of estimations for assessing the donation of a calculation live; anyhow, one ought to be apprehensive of the doable connected snares that may succeed in deceiving estimations. So a lot of recently, important accomplishments have begun to be used; this fashion enjoys the profit that it does not bear image purpose ID-associated computation as an underpinning advance; rather, options are worthy as a region of the accomplishment frame. AI has been employed in clinical imaging and can have an extra clear impact from then on out. Those operating in Medicalpictures ought to be apprehensive of however machine accomplishment capacities.

20.2 IMAGE PROCESS

Taking care of image may be a methodology to play out specific strategy on a picture, to urge a superior image or to get rid of several accommodating data from it.[1] It is a quite sign taking care of whereby input is and result could also be picture or characteristics options connected thereupon image. Currently, image addressing is among swiftly making marches.[2] It structures concentrate assessment locus within coming up with and programing disciplines too.

- Image addressing in an exceedingly general sense consolidates the going with three phases;
- Acquiring the image through image obtaining contrivances;
- Taking piecemeal and dominant image.

Yield, during which result will be altered image or report that depends upon image examination. There are two sorts of procedures used for picture dealing with specifically; straightforward and progressed picture taking care of. Basic picture dealing can be used for the printed duplicates like printouts and photographs.[3–6] Picture agents use various rudiments of interpretation while using these visual techniques. Progressed picture taking care of methodologies help in charge of the automated pictures by using PCs.[8] The three general stages that a wide scope of data need to go through while using progressed strategy are pre- taking care of, redesign, and show information extraction (Fig. 20.1).

20.3 KEY PHASES OF DIGITAL IMAGE PROCESSING

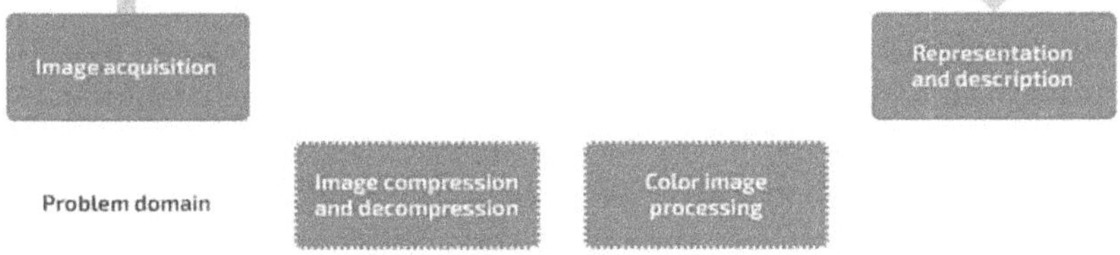

FIGURE 20.1 Key phases of digital image processing.

20.4 WHAT IS DIGITAL IMAGE?

The electronic pictures expect a fundamental part reliably. The clinical imaging taking care of suggests managing pictures by using the PC. This dealing with fuses numerous sorts of strategies and undertakings, for instance, picture getting, limit, show, and correspondence.[9] The image is a limit that implies an extent of properties like edification or concealing a saw sight. The mechanized pictures enjoy a couple of benefits, for instance, speedier and unassuming taking care of cost, basic taking care of and correspondence, brief quality examination, different imitating with saving the quality, fast and unobtrusive age, and flexible control. The bothers of cutting edge pictures are misuse copyright, weakness to resize with saving the quality, the need of huge breaking point memory, and the need of speedier processor for control.[10–15]

An image taking care of technique is the utilization of PC to control the modernized picture. This method enjoys many benefits, for instance, adaptability, adaptability, data taking care of, and correspondence.[16] With the advancement of different picture resizing strategies, the photos can be kept beneficially.[17] This method has numerous plans of rules to perform into the photos at the same time. The 2D and 3D pictures can be dealt with in various viewpoints. The image dealing with techniques were laid out during the 1960s.[18,19] Those systems were used for different fields like space, clinical purposes, articulations, and TV picture improvement. During the 1970s with the improvement of PC system, the cost of picture taking care of ended up being less and speedier. During the 2000s, the image taking care of ended up being speedier, humble, and less troublesome.[20–23]

20.5 WHAT IS DIGITAL IMAGE PROCESSING?

Progressed photography care of is that the operation of associate electronic laptop to manage progressive footage through associate algorithmic program.[24] As a subcategory or field of machine-driven signal taking care of, innovative image coping with partakes in colorful high grounds over introductory photography care of. It permits considerably more broad extentof calculations to be applied to the information and might avoid problems, for case, the advancement of bouleversement and wringing throughout coping with. Since footage are delineate north of 2 angles (perhaps more developed) image coping with is also shown as advanced systems. The age and advance of electronic photography care of chiefly plagued by 3 factors initial, the improvement of PCs; second, the progression of calculation (particularly the creation and improvement of separate numerical proposition); third, the interest for a large extentof functions in tract, cultivating, military, diligence, and clinical knowledge has extended.[25]

20.6 WHAT ARE MEDICAL IMAGES?

Medical imaging, in any case called radiology, is the field of medicine wherein clinical specialists recreate various pictures of parts of the body for characteristic or treatment purposes. Medical imaging technique consolidates innocuous tests that license experts to dissect wounds and diseases without being nosy.[27]

Medical imaging is a central piece of the better consequences of present- day drug.

Different sorts of clinical imaging strategies include:

- X-beams
- Attractive reverberation imaging (MRI)
- Ultrasounds
- Endoscopy
- Material imaging
- Electronic tomography (CT examine)

Other profitable clinical imaging systems consolidate nuclear prescription helpful imaging strategies, similar to positron surge tomography (PET) analyses.[28–30] Various reasonsfor clinical imaging fuse compasses to see how well your body is noting a treatment for a breakor infection (Fig. 20.2).

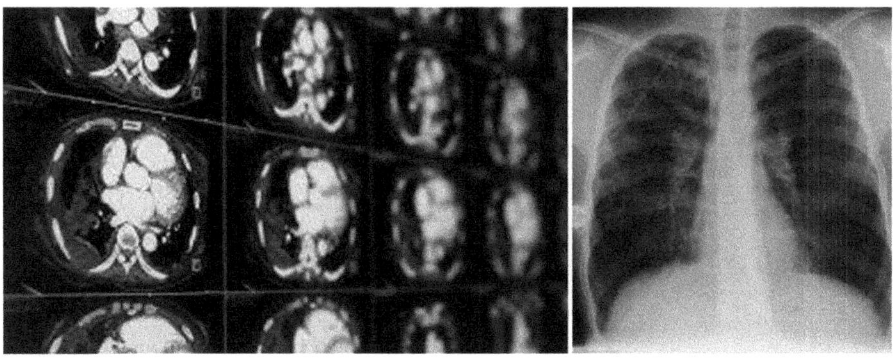

FIGURE 20.2 Medical image.

20.7 WHAT IS MEDICAL IMAGE PROCESSING?

Medical image handling includes the applying and essay of 3D image datasets of the mortalbody, got most typically from a reckoned Tomography

(CT) or glamourous Resonance Imaging (MRI) scanner to dissect pathologies or companion clinical negotiations like careful medication, or for exploration purposes (Fig. 20.3).[31] Clinical image running is completed by radiologists, specialists, and clinicians to promptly comprehend the life structures of either individual cases or crowd gatherings.[32]

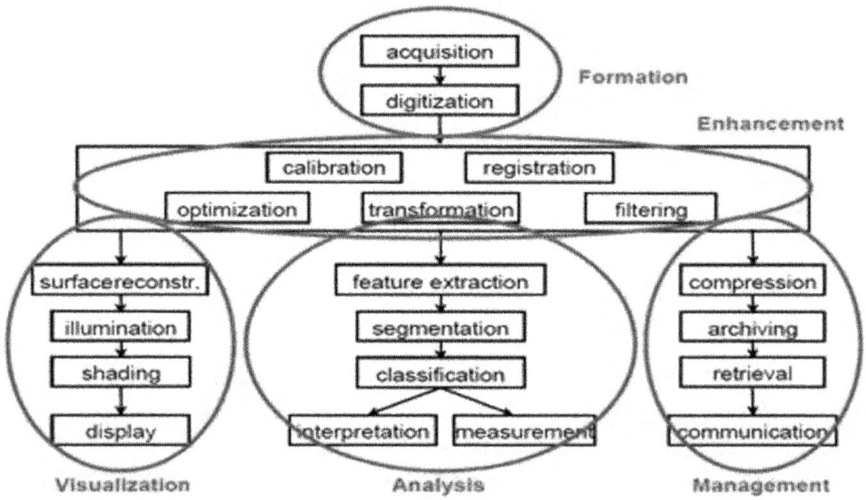

FIGURE 20.3 Medical image processing model.

Source: Reprinted with permission from Ref. [48]. Copyright © 2012, Society for Imaging Informatics in Medicine

20.8 MEDICAL IMAGING TECHNOLOGIES

Progresses used in clinical imaging join those having a spot with the area of radiography. X-bar and CT checks are fundamental resources, however because of ionizing radiation, they ought to be used sparingly.[33,34] Ionizing radiation conveys with it possibility of growths, cascades, cell change, and odd progression in incipient organisms. X-beams, including nuclear appealing resonation (NMR), offer diminished bets and no ionizing radiation. Ultrasound, using ultrasonic vibrations to make pictures, likely addresses the most reliable sort of clinical imaging.[35]

The use of surface-mounted sensors to evaluate electrical development is another safeguarded kind of clinical imaging and is used in electroencephalography (EEG) and electrocardiography (ECG), but these advances produce a change for a really long time chart rather than a graphical picture.[36]

In different clinical imaging progresses modernized thinking (AI) is working on the ability to interpret and analyze results. PC vision is being used to ostensibly break down conditions not yet recognizable to the regular eye.[37]

20.9 WHO USES MEDICAL IMAGING?

A radiographer in any case called a clinical imaging technologist or radiology technologist is obligated for controlling clinical imaging frameworks.[38] Radiographers are school ready with thorough data on the body's plan and how it is affected by different contaminations and wounds. They can have pragmatic involvement with the systems referred toabove— including MRIs and CT checks—as well as in locales, for instance,[39]

- Angiography—which incorporates imaging a patient's veins and heart.
- Adaptable radiography—which is the use of extraordinary machines to perform imaging frameworks on patients who are too crippled to even think about considering branching out to a crisis center.
- Fluoroscopy—which is a x-pillar that investigates the patient's internal body and shows moving pictures on a screen, like a film.
- Injury radiography—which frequently remembers work for emergency workplaces.

Radiographers perform clinical imaging techniques in accordance with a radiologist. Radiologists are experts arranged to examine and treat diseases and wounds using clinical imaging propels. Radiologists are moreover responsible for treating disorders—like harmful development and coronary ailment—using radiation or insignificantly meddlesome, picture drove an operation.[40]

At the point when the methods are done, the radiographer presents the photos to the radiologist. The radiologist then separates the results, presents an investigation of the disorder orinjury, and chooses the best treatment decisions for the patient.

20.10 MEANING OF MEDICAL IMAGING

Clinical imaging helps experts to all the more promptly review patients' bones, organs, tissue, and veins through effortless means. The methodologies help with choosing if operation would be a strong treatment decision; track down tumors for treatment and removal; noticeblood clusters or various blockages; direct experts overseeing joint replacements or the treatmentof breaks; and help various frameworks including the game plan of contraptions—like stents or catheters—inside the body.[41]

For the most part, clinical imaging has additionally evolved discoveries and prescriptions by phenomenally reducing how much secret done by subject matter experts, allowing them to even more truly deal with patients' injuries and afflictions.

20.11 ADVANTAGES OF MEDICAL IMAGE PROCESSING

The guideline benefit of clinical picture dealing with is that it considers each around, but inoffensive examination of internal life fabrics. 3D models of the actuality designs of interestcan be made and audited to also foster treatment results for the case, encourage better clinical contrivances and medicine movement structures, or achieve further tutored anatomize. It has come one of the crucial instruments used for clinical progress of late.[42]

The constantly dicing down at nature of imaging joined with state-of-the-art programming instruments works with exact progressed duplication of factual plans at colorful scales, too comparatively likewise with for the utmost part changing parcels including bone and fragile apkins. Assessment and development of re-sanctioning models which intertwine pukka factual calculations allow the important occasion to more complete the process of understanding, for illustration of relationship between determined life fabrics and clinical contrivances.[43]

20.12 HOW DOES MEDICAL IMAGE PROCESSING WORK?

The course of medical image includes taking care of thresholds by getting rough data from CT or MRI pictures and reproducing them into a plan proper for use in applicable programming.[44] A 3D bitmap of grey scale powers containing a voxel (3D pixels) network makes the normal commitment for picture taking care of. CT check grey scale power depends upon X-shaft ingestion, while in MRI still hanging out there by the strength of signs from proton patches during loosening up and after use of particularly strong charming fields.

For clinical guests, the duplicated picture volume is constantly dealt with to part out and modify different regions of factual interest, analogous to towel and bone.[45] In Synopsys Simpleware programming, for illustration, guests can do different picture taking care of undertakings at the 2D and 3D positions, including:

- Lessening and wiping out unfortunate clatter or antiquated aberrations with picturechannels.
- Managing and retesting input data to make it more direct and briskly to manage pictures.
- Using division instruments to perceive different actual regions, including motorizedtechniques using AI-based AI computations.
- Applying assessment and estimations gadgets to assess different bits of the image data,for example, centerlines.
- Acquiring CAD models, similar to embeds or clinical devices, to focus on how they communicate with individual life structures.
- Trading managed models for real science based diversion, further game plan work, or for 3D printing certifiable pantomimes of the presence structures being suggested to an incredible continuous representation of how clinical picture dealing with included patient-unequivocal hemodynamic entertainments of awesome aortic investigations, a piece of work finished at University College London into better getting hazardous vascular conditions.
- Experts used Simpleware programming to deal with CT yields and create models sensible for CFD assessment, with the going with propels taken.
- CT looks are gained from patient-unequivocal occurrences of aortic investigations.
- Information is imported to Simple wareScanIP to imitate patient estimation, including the treatment of noise, and division of areas of interest like the dismantled aorta and branches Scripting is used to normally finish smoothing computations to kill pixelation trinkets. Surface models are made from the dismantled aorta and imported to ANSYS® programming to set up CFD reenactments, including intraluminal strain and divider shear-stress-based records.
- Simulation results create hemodynamic pieces of information that can be used to help future clinical understanding.

20.13 HOW IS ARTIFICIAL INTELLIGENCE USED IN MEDICINE?

Man-made consciousness in medication is the utilization of AI models to look through clinical information and uncover bits of knowledge to assist with further developing wellbeing results and patient encounters. Because of ongoing advances in software engineering and informatics, man-made reasoning (AI) is rapidly turning into a vital piece of present-day medical care.[45] Artificial intelligence calculations and different applications fueled by AI are being utilized to help clinical experts in clinical settings and in continuous exploration.

At present, the most well-known jobs for AI in clinical settings are clinical choice help and imaging examination. Clinical choice help devices assist suppliers with settling on choices about medicines, drugs, psychological wellness, and other patient necessities by furnishing them with speedy admittance to data or examination that is pertinent to their patient. In clinical imaging, AI instruments are being utilized to investigate CT filters, x-beams,[46] MRIs, and different pictures for sores or different discoveries that a human radiologist could miss.

The difficulties that the COVID-19 pandemic made for some wellbeing frameworks likewise drove numerous medical services associations all over the planet to begin field-testing new AI-upheld advancements, for example, calculations intended to assist with checking patients and AI-controlled devices to screen COVID-19 patients.

The exploration and aftereffects of these tests are as yet being accumulated, and the general guidelines for the utilization AI in medication are as yet being characterized. However open doors for AI to help clinicians, scientists, and the patients they serve are consistently expanding. Now, there is little uncertainty that AI will turn into a center piece of the advanced wellbeing frameworks that shape and back current medication.[47]

20.14 MAN-MADE INTELLIGENCE APPLICATIONS IN MEDICAL FIELD

There are different ways AI can strongly influence the demonstration of medicine, whether it is through speeding up the speed of investigation or helping clinicians with making better decisions. The following are a couple of examples of how AI could be used:

20.14.1 WHAT IS MACHINE LEARNING?

Man-created intelligence may be an operation of man-created mindfulness (AI) that empowers systems to therefore acquire and ameliorate for a reality while not being expressly changed. Simulated intelligence revolves the sweetening of computer programs that may get to knowledge and use it to seek out every alone. Computer grounded intelligence is that the risk that a computer program will learn and accommodates new knowledge without mortal intervention. Simulated intelligence may be a field of man-created attentiveness (AI) that keeps a PC's empirical estimations current paying very little brain to changes within the generalprudence.

20.14.2 USES OF MACHINE LEARNING

- Artificial insight is employed in varied locales for various reasons. Exchanging constructions may be conversant in understanding new pursuit open doorways. Progressing and online business stages may be tuned to offer precise and re-tried plans to their shoppers considering the clients' net search history or past exchanges. Attributing affiliations will be part of AI to expect dreadful advances and assemble a credit risk model. Server ranch centers will utilize AI to hide tremendous extents of reports from all edges of the globe. Banks will create compulsion clear confirmation contraptions from AI techniques some machine learning ways. PC grounded know-how assessments are often mentioned as composed or freelance. Administered AI calculations will apply what has been conceded within the history to new knowledge victimization named counsels for anticipated unhatched events, ranging from the examination of a proverbial coming up with dataset, the skill estimation conveys a reasoned capability to create hypotheticals relating to the outgrowth values. The basic structure will provide centers to any new commitment once decent coming up with. The skill calculation will in like manner discrepancy its outgrowth and therefore the right, organized outgrowth and notice botches to vary the model meetly.

On the other hand, freelance AI calculations are used once the data won't to make preparations is neither gathered nor named. Freelance skill focuses on how structures will accumulate associate capability to depict a hid development from unlabeled knowledge. The circumstance does not understand the proper outgrowth; however, it researches and might attract derivatives from datasets to depict coated structures from unlabeled data. Semimanaged AI estimations fall several spots in coordinated and freelance skill, since they use each named and unlabeled knowledge for obtaining ready-naturally a confined quantum of named knowledge and a lot of unlabeled knowledge. The structures that use this technique will astoundingly additionally foster skill perfection picked once the noninheritable named knowledge needs blessed and applicable coffers for set it up/gain from it. Else, obtaining unlabeled knowledge overall does not bear recent coffers.

Support AI calculations may be a skill system that helps out its gift state of affairs by transferring exercises and tracks down bumbles or remunerates. Trial and error hunt and conceded value are the foremost applicable characteristics of facilitate skill. The procedure subventions machines and programming specialists to usually select the foremost ideal approach of acting within a particular setting to grow its show. Essential award analysis is reckoned upon for the knowledgeable to admit that movement is great; this can be called the assistance signal (Fig. 20.4).

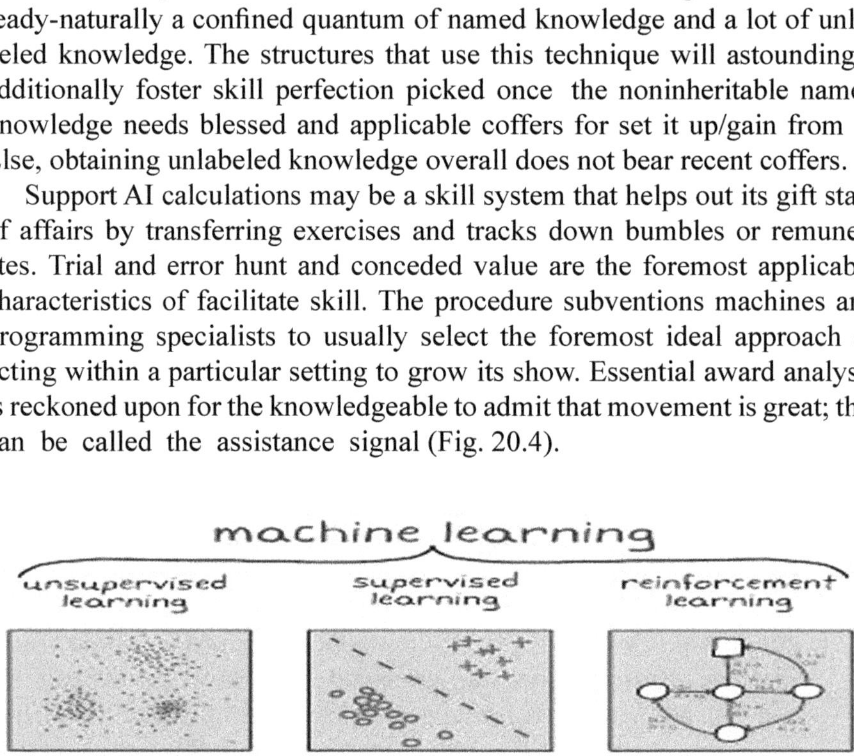

FIGURE 20.4 Machine learning types.

20.15 ARTIFICIAL INTELLIGENCE SURGERY

This is possibly the main locale for Machine Learning, and it will end up being impressively more notable in the near future. You can isolate robotized operation into the going with classes:

- Programmed stitching.
- Careful work process displaying.
- Improvement of automated careful materials.
- Careful expertise assessment.

Sewing basically suggests shutting everything down excruciating injury. Making this cycle robotized makes the whole methodology more restricted while eliminating pressure on the subject matter expert.

The best hardships for AI in healthcare:

Area	Challenges
Data governance	Clinical information is as yet private and taboo for access. In any case, as per a Wellcome Foundation study in the UK, just 17% of public respondents are against offering their clinical data to outsiders.
Transparent algorithms	The need for straightforward calculations is not simply expected to meet severe medication improvement guidelines, yet additionally as a general rule, individuals need to see how precisely calculations create ends.
Optimizing electronic records	There is still a great deal of split information between various informational indexes that need genuinely sorting out. At the point when the current situation improves, it will provoke propels in private treatment courses of action.
Embracing the power of data silos	The medical care industry should change its view on the worth of information and the manner in which it could bring esteem from the drawn out viewpoint. Drug organizations, for instance, are ordinarily hesitant to change their item methodologies and exploration without prompt monetary advantages.
Data science experts	Attracting more Machine Learning trained professionals and data science specialists is truly huge for both the clinical consideration and medication adventures.

20.15.1 TOP TEN OPERATIONS OF MACHINE LEARNING IN DRUG COMPANY AND MEDICINE

The persistent making range of functions of AI in clinical advantages grants United States to require a goose at a future wherever information, assessment, andadvance work indivisibly to assist bottomless cases while not them genuinely feting it. Once a brief time, it is going to be exceptionally standard to search out ML-based operations introduced with steady patient information open from completely different clinical advantages systems in colorful countries, consequently growing the ampleness of recent treatment opinions that were distant at one time. Then area unit the trendy ten functions of AI in clinical benefits.

1. Characteristic conditions and opinion

One in every of the focal cubic centimeter operations in clinical advantages is that the distinctive proof and assurance of diseases and affections that area unit for the utmost half flashed back to be arduous to anatomize. This could fuse something from cancers that area unit exhausting to trace down throughout the retired stages, to alternative ingrain afflictions. IBM Watson genetic science is a perfect depiction of however designing internal reckoning with genome-grounded nasty growth sequencing will facilitate in creating a speedy finding. Berg, the biopharma beast is victimization AI to cultivate important specifics in areas like medical specialty. P1vital's PReDicT (Predicting Response to Depression Treatment) hopes to cultivate AN economically possible strategy for designation and provides treatment in routine clinical conditions.

2. Drug discovery and producing

One in every of the abecedarian clinical uses of AI lies in planning stage drugs speech act method. This conjointly fuses R&D propels, for case, frontal line sequencing and perfection drug which may facilitate in chancing ex gratia ways for treatment of complex affections. As of now, the AI methods incorporate freelance admitting which may fete plans in information while not giving any hypotheticals. Project Hanover created by Microsoft is together with ML-grounded developments for quite whereas together with creating AI-grounded advancement for dangerous development treatment and tweaking drugs mix for AML (Acute Myeloid Leukemia).

3. Medical imaging designation

Man-created intelligence and vital accomplishment area unit each to blame for the high position advancement referred to as pc Vision. This has detected protestation within the InnerEye drive created by Microsoft that manages image demonstrative instruments for image assessment. As AI finally ends up being a lot of accessible and as they fill in their educational limit, want to check more information sources from varied clinical imagination come back a bit of this AI-driven scientific cycle.

4. Personalized drugs

The conventions will still the method that a lot of be possible by coordinative individual substance with even handed examination is additionally ready area unit for redundant essay and higher grievance assessment. This moment, specialists area unit restricted to poring a selected course of action of ends or check the bet to the case considering his intriguing history and open heritable info. Anyhow, AI in drug is creating implausible strides, and IBM Watson medical specialty is at the frontal line of this advancement by victimization patient clinical history to assist with making completely different treatment opinions. Sooner instead of late, we will see more contrivances and biosensors with current substance assessment capacities hit the request, permitting more information to open up for similar fashionable in school ML-grounded clinical advantages marches.

5. **AI-grounded behavioral revision**

Social revision could be a vital piece of preventative tradition, and since the time the addition of AI in clinical advantages, bottomless new associations area unit jumping up within the fields of infection balance and clear proof, patient treatment, etc. Somatix could be a B2B2C-grounded information examination association that has sent a ML- grounded operation to check movements that we tend to create in our everyday schedules, permitting United States to induce our careless approach to acting and do polar upgrades.

6. **Sensible health records**

Study the apprehensive of current substance records is an fierce cycle, and flashing back that advancement has had its impact in operating with the info member method, very so as of currently, an even bigger piece of the cycles place coffers into an opportunity to bring to a close. The introductory occupation of AI in clinical advantages is to ease cycles to avoid wasting time, trouble, and rich person. Record request systems victimization vector machines and ML-grounded OCR protestation ways that area unit sluggishly grouping brume, for case, Google's Cloud Vision API and MATLAB's AI grounded handwriting protestation development. MIT is moment at the terribly front of encouraging the exceptional amount of clever, sensible substance records, which is able to be part of ML-grounded sacrifices starting from the foremost stage to assist with finish, clinical treatment studies, etc.

7. **Clinical test and exploration**

PC-grounded intelligence includes a number of implicit operations within the field of clinical starters and assessment. As anybody within the drug company business would tell you, clinical fundamentals bring an enormous loading of your time and rich person and may invest in some occasion to complete, once in mistrustfulness. Applying ML- grounded perceptive assessment to fete implicit clinical starter contenders will facilitate consultants with drawing a pool from a good assortment of instructional effects, for case, past skilled visits, on-line recreation, etc. AI has conjointly detected use in icing nonstop checking and information access of the abecedarian individualities, seeing the trendy model size as tried, and victimization the ability of electronic records to lessen information grounded botches.

8. Crowdsourced information collection

Overtly supporting is veritably disreputable within the clinical field presently, permitting investigators and consultants to induce to a colossal proportion of knowledge enraptured by individuals considering their own concurrence. This live substance information has uncommon ramifications within the method drugs are going to be seen down the road. Apple's analysis Kit licenses guests to induce natural operations that apply ML-grounded facial protestation to endeavor to treat Asperger's and degenerative disorder impurity. IBM really helped out Medtronic to decompress, total, and create open polygenic disorder and hormone information persistently established on the freely upheld info. The movements being created in IoT, the clinical thought assiduousness is at now following down new habits by that to use this information and attack serious to- examine cases and facilitate within the overall improvement of finish and drug.

9. Higher radiotherapy

One in every of the foremost pursued uses of AI in clinical thought is within the area of Radiology. Clinical image assessment has colorful separate rudiments which may arise at a selected definition of your time. There are unit colorful injuries, illness foci, etc., which may not be primarily displayed victimization advanced circumstances. Since ML-grounded estimations gain from the massive range of various models open exhausting, it becomes less advanced to anatomize and notice the rudiments. One in every of the foremost notable functions of AI in clinical image assessment is that the depiction of papers, for case, injuries into orders like traditional or uncommon, injury or non-sore, etc. Google's DeepMind Health is de facto serving to experimenters in UCLH with creating calculations which may fete the isolation among sound and dangerous towel and farther foster radiation treatment for identical.

10. Irruption prediction

System should know how the grounded advancements and AI area unit moment conjointly being place to use in checking and prognosticating pestilences every round the earth. moment, scientists approach a good deal of information assembled from satellites, harmonious net primarily based diversion revives, web site info, etc., faux mind networks facilitate to appear at this info and anticipate everything from nature fever occurrences to outrageous nonstop overwhelming affections. Awaiting these eruptions is very precious in immature countries as they have introductory clinical structure and academic systems. A abecedarian illustration of this can be the ProMED-correspondence, AN Internet- grounded specifying stage that screens propellant impurities and arising bones and handovers flare reports unceasingly.

KEYWORDS

- **machine learning**
- **image processing**
- **artificial intelligence**
- **electronic tomography**
- **convolutional neural networks**
- **MRI Images**
- **CT Images**

REFERENCES

1. Saravanakumar, C.; Gururama Senthilvel, P.; Thirupurasundari, D.; Periyasamy, P.; Vijayakumar, K. In *Plant Syndrome Recognition by Gigapixel Image Using Convolutional Neural Network*, Proceedings of the Fist International Conference on Advanced Scientific Innovation in Science, Engineering and Technology, Chennai, India, ICASISET 2020, May 16–17, 2020; EAI, 2021. https://doi.org/10.4108/eai.16-5- 2020.2304207
2. Nagarajan, G.; Sampath Kumar, K. In *Security Threats and Challenges in Public Cloud Storage*, 2021 International Conference on Advance Computing and Innovative Technologies in Engineering (ICACITE); IEEE, 2021. https://doi.org/10.1109/icacite51222.2021.9404641
3. Kirubakaran, J.; Venkatesan, G. K. D. P.; Sampath Kumar, K.; et al. Echo State Learned Compositional Pattern Neural Networks for the Early Diagnosis of Cancer on the

Internet of Medical Things Platform. *J. Ambient Intell. Human Comput.* 2021, *12*, 3303–3316. https://doi.org/10.1007/s12652-020-02218-1
4. Vijayaraj, A.; Vasanth Raj, P. T.; Jebakumar, R.; Gururama Senthilvel, P.; Kumar, N.; Suresh Kumar, R.; Dhanagopal, R. Deep Learning Image Classification for Fashion Design. In*Wireless Communications and Mobile Computing*; Hashmi, M. F., Ed.; Hindawi Limited, 2022; vol. 2022, pp 1–13. https://doi.org/10.1155/2022/7549397
5. Saravanakumar, P.; Sundararajan, T. V. P.; Dhanaraj, R. K.; Nisar, K. Memon, F. H.; et al. Lamport Certificateless Signcryption Deep Neural Networks for Data Aggregation Security in WSN. *Intell. Autom. Soft Comput.* **2022,** *33* (3), 1835–1847.
6. Jeyaselvi, M.; Dhanaraj, R. K.; Sathya, M.; et al. A Highly Secured Intrusion Detection System for IoT Using EXPSO-STFA Feature Selection for LAANN to Detect Attacks. *Cluster Comput.* 2022. https://doi.org/10.1007/s10586-022-03607-1
7. Das, B.; Mushtaque, A.; Memon, F.; Dhanaraj, R. K.; Thirumalaisamy, M.; Shaikh, M. Z.; Nighat, A.; Gismalla, M. S. M. Real-Time Design and Implementation of Soft Error Mitigation Using Embedded System. *J. Circuits Syst. Comput.* World Scientific Pub Co Pte Ltd., 2022. https://doi.org/10.1142/s0218126622502802
8. Pichumani, S.; Sundararajan, T. V. P.; Dhanaraj, R. K.; Nam, Y.; Kadry, S. Ruzicka Indexed Regressive Homomorphic Ephemeral Key Benaloh Cryptography for Secure Data Aggregation in WSN. *J. Int. Technol.* **2021,** *22* (6), 1287–1297.
9. Dhanaraj, R. K.; Krishnasamy, L.; et al. Black-Hole Attack Mitigation in Medical Sensor Networks using the Enhanced Gravitational Search Algorithm. *Int. J. Uncertain. Fuzziness Knowledge-Based Syst.* **2021,** *29*, 397–315. https://doi.org/10.1142/S021848852140016X
10. Dhanaraj, R. K.; Ramakrishnan, V.; Poongodi, M.; Krishnasamy, L.; Hamdi, M.; Kotecha, K.; Vijayakumar, V. Random Forest Bagging and X-Means Clustered Antipattern Detection from SQL Query Log for Accessing Secure Mobile Data. In *Wireless Communications and Mobile Computing*; Jain, D. K., Ed.; Hindawi Limited., 2021; vol 2021, pp 1–9. https://doi.org/10.1155/2021/2730246
11. Dhanaraj, R. K.; Lalitha, K.; Anitha, S.; Khaitan, S.; Gupta, P.; Goyal, M. K. Hybrid and Dynamic Clustering Based Data Aggregation and Routing for Wireless Sensor Networks. *J. Intell. Fuzzy Syst.* **2021,** *40* (6), 10751–10765). https://doi.org/10.3233/jifs-201756
12. Krishnamoorthi, S.; Jayapaul, P.; Dhanaraj, R. K.; et al. Design of Pseudo-random Number Generator from Turbulence Padded Chaotic Map. *Nonlinear Dyn.* 2021. https://doi.org/10.1007/s11071-021-06346-x
13. Dhanaraj, R. K.; Krishnasamy, L.; Geman, O.; Izdrui, D. R. Black Hole and Sink Hole Attack Detection in Wireless Body Area Networks. *Comput. Mater. Continua* **2021,** *68* (2), 1949–1965. DOI: 10.32604/cmc.2021.015363
14. Ramasamy, M. D.; Periasamy, K.; Krishnasamy, L.; Dhanaraj, R. K.; Kadry, S.; Nam, Y. Multi-Disease Classification Model Using Strassen's Half of Threshold (SHoT) Training Algorithm in Healthcare Sector. *IEEE Access* DOI: 10.1109/ACCESS.2021.3103746.
15. Ramakrishnan, V.; Chenniappan, P.; Dhanaraj, R. K.; Hsu, C. H.; Xiao, Y.; Al- Turjman, F. Bootstrap Aggregative Mean Shift Clustering for Big Data Anti-pattern Detection Analytics in 5G/6G Communication Networks. *Comput. Electr. Eng.* **2021,** *95*, 107380. https://doi.org/10.1016/j.compeleceng.2021.107380
16. Krishnasamy, L.; Ramasamy, T.; Dhanaraj, R.; Chinnasamy, P. A Geodesic Deployment and Radial Shaped Clustering (RSC) Algorithm with Statistical Aggregation in Sensor Networks. *Turkish J. Electr. Eng. Comput. Sci.* **2021,** *29* (3). DOI: 10.3906/elk-2006-124

17. Kumar, D. R.; Krishna, T. A.; Wahi, A. Health Monitoring Framework for in Time Recognition of Pulmonary Embolism Using Internet of Things. *J. Computat. Theor. Nanosci.* 2018, *15* (5), 1598–1602. https://doi.org/10.1166/jctn.2018.7347
18. Krishnasamy, L.; Dhanaraj, R. K.; Ganesh Gopal, D.; Reddy Gadekallu, T.; Aboudaif, M. K.; Abouel Nasr, E. A Heuristic Angular Clustering Framework for Secured Statistical Data Aggregation in Sensor Networks. *Sensors* **2020,** *20* (17), 4937. https://doi.org/10.3390/s20174937
19. Sathyamoorthy, M.; Kuppusamy, S.; Dhanaraj, R. K.; et al. Improved K-Means Based Q Learning Algorithm for Optimal Clustering and Node Balancing in WSN. *Wirel. Pers. Commun.* 2022, *122*, 2745–2766. https://doi.org/10.1007/s11277-021-09028-4
20. Dhiviya, S.; Malathy, S.; Kumar, D. R. Internet of Things (IoT) Elements, Trends and Applications. *J. Computat. Theor. Nanosci.* **2018,** *15* (5), 1639–1643. https://doi.org/10.1166/jctn.2018.7354
21. Jena, S. R.; Shanmugam, R.; Dhanaraj, R. K.; Saini, K. Recent Advances and Future Research Directions in Edge Cloud Framework. *Int. J. Eng. Adv. Technol.* 2019, *9* (2), 439–444. https://doi.org/10.35940/ijeat.b3090.129219
22. Kumar, R. N.; Chandran, V.; Valarmathi, R. S.; Kumar, D. R. Bitstream Compression for High Speed Embedded Systems Using Separated Split Look Up Tables (LUTs). *J. Computat. Theor. Nanosci.* **2018,** *15* (5), 1719–1727. https://doi.org/10.1166/jctn.2018.7367
23. Irfan, S.; Dhanaraj, R. K. BeeRank: A Heuristic Ranking Model to Optimize the Retrieval Process. *Int. J. Swarm Intell. Res.* **2021,** *12* (2), 39–56. http://doi.org/10.4018/IJSIR.2021040103
24. Rajesh Kumar, D.; Shanmugam, A. A Hyper Heuristic Localization Based Cloned Node Detection Technique Using GSA Based Simulated Annealing in Sensor Networks. In *Cognitive Computing for Big Data Systems Over IoT*; Springer International Publishing, 2017; pp. 307–335. https://doi.org/10.1007/978-3-319-70688-7_13
25. Prasanth, T.; Gunasekaran, M.; Kumar, D. R. In *Big data Applications on Health Care*, 2018 4th International Conference on Computing Communication and Automation (ICCCA), Dec 2018. https://doi.org/10.1109/ccaa.2018.8777586
26. Sathish, R.; Kumar, D. R. In *Dynamic Detection of Clone Attack in Wireless Sensor Networks*, 2013 International Conference on Communication Systems and Network Technologies, Apr 2013 . https://doi.org/10.1109/csnt.2013.110
27. Rajesh Kumar, D.; ManjupPriya, S. In *Cloud based M-Healthcare Emergency Using SPOC*, 2013 Fifth International Conference on Advanced Computing (ICoAC), Dec 2013. https://doi.org/10.1109/icoac.2013.6921965
28. Sathish, R.; Kumar, D. R. In *Proficient Algorithms for Replication Attack Detection in Wireless Sensor Networks &#x2014: A Survey*, 2013 IEEE International Conference ON Emerging Trends in Computing, Communication and Nanotechnology (ICECCN), Mar 2013. https://doi.org/10.1109/ice- ccn.2013.6528465
29. Lalitha, K.; Kumar, D. R.; Poongodi, C.; Arumugam, J. Healthcare Internet of Things – The Role of Communication Tools and Technologies. In *Blockchain, Internet of Things, and Artificial Intelligence*; Chapman and Hall/CRC, 2021; pp 331–348. https://doi.org/10.1201/9780429352898-17
30. Rajesh Kumar, D.; Rajkumar, K.; Lalitha, K.; Dhanakoti, V. Bigdata in the Management of Diabetes Mellitus Treatment. In *Studies in Big Data*; Springer: Singapore, 2020; pp 293–324. https://doi.org/10.1007/978-981-15-4112-4_14

31. Chandraprabha, M.; Dhanaraj, R. K. In *Machine Learning Based Pedantic Analysis of Predictive Algorithms in Crop Yield Management*, 2020 4th International Conference on Electronics, Communication and Aerospace Technology (ICECA), Nov 5, 2020. https://doi.org/10.1109/iceca49313.2020.9297544
32. Dhanaraj, R. K.; Rajkumar, K.; Hariharan, U. Enterprise IoT Modeling: Supervised, Unsupervised, and Reinforcement Learning. In *Business Intelligence for Enterprise Internet of Things*; Springer International Publishing, 2020; pp 55–79. https://doi.org/10.1007/978-3-030-44407-5_3
33. Cynthia, J.; Sankari, M.; Suguna, M.; Kumar, D. R. In *Survey on Disaster Management Using VANET*, 2018 4th International Conference on Computing Communication and Automation (ICCCA), Dec 2018. https://doi.org/10.1109/ccaa.2018.8777331
34. Dhanaraj, R. K.; Shanmugam, A.; Palanisamy, C.; Natarajan, A. Optimal Clone Attack Detection Model using an Energy-Efficient GSA based Simulated Annealing in Wireless Sensor Networks. *Asian J. Res. Soc. Sci. Humanit.* 2016, *6* (11), 201. https://doi.org/10.5958/2249-7315.2016.01186.2
35. Sathya, K.; Kumar, D. R. In *Energy Efficient Clustering in Sensor Networks Using Cluster Manager*, 2012 International Conference on Computing, Communication and Applications, Feb 2012. https://doi.org/10.1109/iccca.2012.6179177
36. Lalitha, K.; Varadhaganapathy, S.; Santhoshi, S.; Kumar, D. R. In *A Review on Possibilities of Hearing Loss and Implantable Hearing Devices for Teenagers*, 2018 4th International Conference on Computing Communication and Automation (ICCCA), Dec 2018. https://doi.org/10.1109/ccaa.2018.8777336
37. Kumar, R. N.; Karthick, S.; Valarmathi, R. S.; Kumar, D. R. Design and Analysis of Multiply and Accumulation Units Using Low Power Adders. *J. Computat. Theor. Nanosci.* **2018,** *15* (5), 1712–1718.
38. Poongodi, C.; Lalitha, K.; Dhanaraj, R. K. The Role of Blockchains for Medical Electronics Security. In *Essential Enterprise Blockchain Concepts and Applications*; Auerbach Publications, 2021; pp 231–262.
39. Lalitha, K.; Kumar, D. R.; Poongodi, C.; Arumugam, J. Healthcare Internet of Things–The Role of Communication Tools and Technologies. In *Blockchain, Internet of Things, and Artificial Intelligence*; Chapman and Hall/CRC, 2021; pp 331–348.
40. Sathish, R.; Kumar, D. R. In *Proficient Algorithms for Replication Attack Detection in Wireless Sensor Networks—A Survey*, 2013 IEEE International Conference ON Emerging Trends in Computing, Communication and Nanotechnology (ICECCN); IEEE, Mar 2013; pp 1–7.
41. Kumar, D. R.; Sathish, R. Mitigation of Replication Attack Detection in Clusters Through a Mobile Agent in Wireless Sensor Networks. *Int. J. Eng. Res. Appl.* 2013, ISSN: 2248–9622.
42. Arvindhan, M.; Rajeshkumar, D.; Pal, A. L. A Review of Challenges and Opportunities in Machine Learning for Healthcare. In *Exploratory Data Analytics for Healthcare*, 2021; pp 67–84.
43. Arvindhan, M.; Rajesh Kumar, D. Analysis of Load Balancing Detection Methods Using Hidden Markov Model for Secured Cloud Computing Environment. In *Applications of Computational Methods in Manufacturing and Product Design. Lecture Notes in Mechanical Engineering*; Deepak, B. B. V. L., Parhi, D., Biswal, B., Jena, P. C., Eds.; Springer: Singapore, 2022. https://doi.org/10.1007/978-981-19-0296-3_53

44. Arvindhan, M.; Dhanaraj, R. K. The Firefly Technique with Courtship Training Optimized for Load Balancing Independent Parallel Computer Task Scheduling in Cloud Computing. *Int. J. Health Sci.* 2022, *6* (S1), 8740–8751. https://doi.org/10.53730/ijhs.v6nS1.6999
45. Juyal, V.; Pandey, N.; Saggar, R. In *Impact of Varying Buffer Space for Routing Protocols in Delay Tolerant Networks*, 2016 International Conference on Communication and Signal Processing (ICCSP); IEEE, 2016. https://doi.org/10.1109/iccsp.2016.7754562
46. Juyal, V.; Singh, A. V.; Saggar, R. In *Message Multicasting in Near-Real Time Routing for Delay/Disruption Tolerant Network*, 2015 IEEE International Conference on Computational Intelligence & Communication Technology (CICT); IEEE, 2015. https://doi.org/10.1109/cict.2015.79
47. Juyal, V.; Saggar, R.; Pandey, N. On Exploiting Dynamic Trusted Routing Scheme in Delay Tolerant Networks. *Wirel. Pers. Commun.* **2020,** *112*, 1705–1718. https://doi.org/10.1007/s11277-020-07123-6
48. Deserno T. Medical image processing. Optipedia, SPIE Press, Bellingham, WA. 2009.

INDEX

B

C

D

E

F

G

H

I

K

L

N

O

P

Q

R

S

T

For Product Safety Concerns and Information please contact our EU
representative GPSR@taylorandfrancis.com
Taylor & Francis Verlag GmbH, Kaufingerstraße 24, 80331 München, Germany

www.ingramcontent.com/pod-product-compliance
Ingram Content Group UK Ltd.
Pitfield, Milton Keynes, MK11 3LW, UK
UKHW022209190926
475702UK00003B/274

* 9 7 8 1 7 7 4 9 1 4 7 1 7 *